Recent Advances in the Growth and Characterizations of Thin Films

Recent Advances in the Growth and Characterizations of Thin Films

Editors

**Sheng-Rui Jian
Phuoc Huu Le**

Basel • Beijing • Wuhan • Barcelona • Belgrade • Novi Sad • Cluj • Manchester

Editors
Sheng-Rui Jian
I-Shou University
Kaohsiung
Taiwan

Phuoc Huu Le
Ming Chi University of Technology
New Taipei
Taiwan

Editorial Office
MDPI
St. Alban-Anlage 66
4052 Basel, Switzerland

This is a reprint of articles from the Special Issue published online in the open access journal *Coatings* (ISSN 2079-6412) (available at: https://www.mdpi.com/journal/coatings/special_issues/recent_advances_thin_films).

For citation purposes, cite each article independently as indicated on the article page online and as indicated below:

Lastname, A.A.; Lastname, B.B. Article Title. *Journal Name* **Year**, *Volume Number*, Page Range.

ISBN 978-3-0365-8582-6 (Hbk)
ISBN 978-3-0365-8583-3 (PDF)
doi.org/10.3390/books978-3-0365-8583-3

© 2023 by the authors. Articles in this book are Open Access and distributed under the Creative Commons Attribution (CC BY) license. The book as a whole is distributed by MDPI under the terms and conditions of the Creative Commons Attribution-NonCommercial-NoDerivs (CC BY-NC-ND) license.

Contents

About the Editors vii

Preface ix

Ting-Kai Lin, Huang-Wei Chang, Wan-Chi Chou, Chang-Ren Wang, Da-Hua Wei, Chi-Shun Tu and Pin-Yi Chen
Multiferroic and Nanomechanical Properties of $Bi_{1-x}Gd_xFeO_3$ Polycrystalline Films (x = 0.00–0.15)
Reprinted from: Coatings 2021, 11, 900, doi:10.3390/coatings11080900 1

Hui-Ping Cheng, Phuoc Huu Le, Le Thi Cam Tuyen, Sheng-Rui Jian, Yu-Chen Chung, I-Ju Teng, et al.
Effects of Substrate Temperature on Nanomechanical Properties of Pulsed Laser Deposited Bi_2Te_3 Films
Reprinted from: Coatings 2022, 12, 871, doi:10.3390/coatings12060871 15

Yeong-Maw Hwang, Cheng-Tang Pan, Bo-Syun Chen, Phuoc Huu Le, Ngo Ngoc Uyen, Le Thi Cam Tuyen, et al.
Effects of Stoichiometry on Structural, Morphological and Nanomechanical Properties of Bi_2Se_3 Thin Films Deposited on InP(111) Substrates by Pulsed Laser Deposition
Reprinted from: Coatings 2020, 10, 958, doi:10.3390/coatings10100958 27

Shu-Wei Cheng, Bo-Syun Chen, Sheng-Rui Jian, Yu-Min Hu, Phuoc Huu Le, Le Thi Cam Tuyen, et al.
Finite Element Analysis of Nanoindentation Responses in Bi_2Se_3 Thin Films
Reprinted from: Coatings 2022, 12, 1554, doi:10.3390/coatings12101554 39

Alexandr Stupakov, Tomas Kocourek, Natalia Nepomniashchaia, Marina Tyunina and Alexandr Dejneka
Large Negative Photoresistivity in Amorphous $NdNiO_3$ Film
Reprinted from: Coatings 2021, 11, 1411, doi:10.3390/coatings11111411 51

Charalampos Sakkas, Jean-Marc Cote, Joseph Gavoille, Jean-Yves Rauch, Pierre-Henri Cornuault, Anna Krystianiak, et al.
Tunable Electrical Properties of Ti-B-N Thin Films Sputter-Deposited by the Reactive Gas Pulsing Process
Reprinted from: Coatings 2022, 12, 1711, doi:10.3390/coatings12111711 65

Natalia Kovaleva, Dagmar Chvostova, Ladislav Fekete and Alexandr Dejneka
Bi Layer Properties in the Bi–FeNi GMR-Type Structures Probed by Spectroscopic Ellipsometry
Reprinted from: Coatings 2022, 12, 872, doi:10.3390/coatings12060872 79

Giji Skaria, Avra Kundu and Kalpathy B. Sundaram
Effect of Substrate Temperature on the Properties of RF Magnetron Sputtered p-$CuInO_x$ Thin Films for Transparent Heterojunction Devices
Reprinted from: Coatings 2022, 12, 500, doi:10.3390/coatings12040500 93

Cao Phuong Thao, Dong-Hau Kuo and Thi Tran Anh Tuan
The Effect of RF Sputtering Temperature Conditions on the Structural and Physical Properties of Grown SbGaN Thin Film
Reprinted from: Coatings 2021, 11, 752, doi:10.3390/coatings11070752 103

Pei-Hua Tsai, Chung-I Lee, Sin-Mao Song, Yu-Chin Liao, Tsung-Hsiung Li, Jason Shian-Ching Jang and Jinn P. Chu
Improved Mechanical Properties and Corrosion Resistance of Mg-Based Bulk Metallic Glass Composite by Coating with Zr-Based Metallic Glass Thin Fil
Reprinted from: *Coatings* **2020**, *10*, 1212, doi:10.3390/coatings10121212 117

Fariza Kalyk, Artūras Žalga, Andrius Vasiliauskas, Tomas Tamulevičius, Sigitas Tamulevičius and Brigita Abakevičienė
Synthesis and Electron-Beam Evaporation of Gadolinium-Doped Ceria Thin Films
Reprinted from: *Coatings* **2022**, *12*, 747, doi:10.3390/coatings12060747 129

Jin-Ji Dai, Cheng-Wei Liu, Ssu-Kuan Wu, Sa-Hoang Huynh, Jhen-Gang Jiang, Sui-An Yen, et al.
Improving Transport Properties of GaN-Based HEMT on Si (111) by Controlling SiH_4 Flow Rate of the SiN_x Nano- Mask
Reprinted from: *Coatings* **2021**, *11*, 16, doi:10.3390/coatings11010016 145

Ngo Ngoc Uyen, Le Thi Cam Tuyen, Le Trung Hieu, Thi Thu Tram Nguyen, Huynh Phuong Thao, Tho Chau Minh Vinh Do, et al.
TiO_2 Nanowires on TiO_2 Nanotubes Arrays (TNWs/TNAs) Decorated with Au Nanoparticles and Au Nanorods for Efficient Photoelectrochemical Water Splitting and Photocatalytic Degradation of Methylene Blue
Reprinted from: *Coatings* **2022**, *12*, 1957, doi:10.3390/coatings12121957 155

Tomas Sabirovas, Simonas Ramanavicius, Arnas Naujokaitis, Gediminas Niaura and Arunas Jagminas
Design and Characterization of Nanostructured Titanium Monoxide Films Decorated with Polyaniline Species
Reprinted from: *Coatings* **2022**, *12*, 1615, doi:10.3390/coatings12111615 171

Hou-Guang Chen, Yung-Hui Shih, Huei-Sen Wang, Sheng-Rui Jian, Tzu-Yi Yang and Shu-Chien Chuang
Van der Waals Epitaxial Growth of ZnO Films on Mica Substrates in Low-Temperature Aqueous Solution
Reprinted from: *Coatings* **2022**, *12*, 706, doi:10.3390/coatings12050706 183

William Vallejo, Carlos Diaz-Uribe and Cesar Quiñones
Optical and Structural Characterization of Cd-Free Buffer Layers Fabricated by Chemical Bath Deposition
Reprinted from: *Coatings* **2021**, *11*, 897, doi:10.3390/ coatings11080897 199

Dang Long Quan and Phuoc Huu Le
Enhanced Methanol Oxidation Activity of $PtRu/C_{100-x}MWCNTs_x$ (x = 0–100 wt.%) by Controlling the Composition of C-MWCNTs Support
Reprinted from: *Coatings* **2021**, *11*, 571, doi:10.3390/coatings11050571 211

Kuo-Jung Lee, Ming-Husan Lee, Yung-Hui Shih, Chao-Ping Wang, Hsun-Yu Lin and Sheng-Rui Jian
Fabrication of Carboxylated Carbon Nanotube Buckypaper Composite Films for Bovine Serum Albumin Detection
Reprinted from: *Coatings* **2022**, *12*, 810, doi:10.3390/coatings12060810 225

About the Editors

Sheng-Rui Jian

Sheng-Rui Jian (Prof. Dr.) is a Distinguished Professor at the Department of Materials Science and Engineering, I-Shou University, Kaohsiung, Taiwan. He received his PhD degree in Electrophysics from National Chiao Tung University, Hsinchu, Taiwan, in 2006. His research focuses on the nanoscale mechanical deformation behaviors and mechanisms of semiconductor materials and thin films using microscopic techniques and molecular dynamics simulations. He has published more than 180 journal papers, and is awarded the Word's Top 2% Scientists (Career Impact) 2021 and 2022. Jian can be reached by email at srjian@isu.edu.tw and srjian@gmail.com.

Phuoc Huu Le

Phuoc Huu Le (Assoc. Prof. Dr.) is Associate Professor at the Center for Plasma and Thin Film Technologies, Ming Chi University of Technology, New Taipei City, Taiwan. He joined Department of Physics and Biophysics, Faculty of Basic Sciences, Can Tho University of Medicine and Pharmacy, Can Tho City, Vietnam as a lecturer (3/2006–12/2019) and an associate professor (1/2020–*present*). He received his Ph.D. degree in Materials Science & Engineering from National Chiao Tung University, Taiwan in 7/2014. His research interest has been concentrated on Thin films and characterizations (i.e., Thermoelectric, topological insulator, transparent conducting oxide, complex oxide, and high-entropy alloy thin films; Nanomechanical, magnetotransport, ultrafast dynamics, and other functional properties of thin-film materials), and nanomaterials and applications (i.e., Nanophotocatalysts for degradation of emerging pollutants/contaminants; Nanomaterials for photochemical water splitting, energy conversion and storage, and electrochemical sensors). Dr. Le has published more than 40 journal papers, 8 book chapters (Google Scholar: >828 citations, h-index = 17).

Preface

At present, many innovative technologies are strongly dependent on thin films. The development of thin film deposition technologies enables the fabrication of various thin-film materials with enhanced desired properties by controlling key processing conditions for a wide variety of applications in electronics (e.g., HEMT devices), energies (e.g., solar cells and fuel cells), optoelectronics (e.g., photodetectors, LCD/TFT), spintronics, biomedical sensors, and photocatalytic water treatment, etc.

This Special Issue contains 18 original papers, representing recent advances in thin film growth and characterizations toward enhancing the film's functional properties and applications. Various thin film growth methods such as pulsed laser deposition, sputtering, electron-beam evaporation, chemical vapor deposition, anodization, etc., are employed to develop a wide range of thin film materials (i.e., $Bi_{1-x}Gd_xFeO_3$, Bi_2Te_3, Bi_2Se_3, $NdNiO_3$, Ti-B-N, Bi–FeNi, p-CuInOx, $Sb_{0.14}GaN$, Zr-based metallic glass, GaN, TiO_2, ZnO, Zn(O;OH)S, carbon nanotube, etc.). The studied thin films were developed by controlling the key deposition conditions, doping or surface/interface engineering. The studied properties of the thin films included structures, morphologies, compositions, (nano)mechanical properties (e.g., hardness, Young's modulus, pop-in phenomena, surface energy), multiferroics, photoconductivity, electrical conductivity, optical giant magnetoresistance, photoelectrochemical water splitting, photocatalytic and electrocatalytic activities. This series of publications provides a fundamental understanding of thin film growths and the film's resulting properties for some applications.

We are pleased to introduce this Special Issue to all interested readers and hope that this reprint book will serve as valuable support for students, engineers and researchers in the field of thin film technology to obtain new ideas for further developments with great significance and impact from both scientific and applicative/industrial points of view. We wish you a pleasant read.

Finally, we would like to acknowledge all the contributing authors for submitting their papers to the Special Issue "Recent Advances in the Growth and Characterizations of Thin Films", to all the reviewers, supporting editors and the staff members of *Coatings*, in particular, Ms. Flora Ao (the Section Managing Editor), for their excellent work and great assistance.

Sheng-Rui Jian and Phuoc Huu Le
Editors

Article

Multiferroic and Nanomechanical Properties of $Bi_{1-x}Gd_xFeO_3$ Polycrystalline Films ($x = 0.00$–0.15)

Ting-Kai Lin [1], Huang-Wei Chang [2,*], Wan-Chi Chou [3], Chang-Ren Wang [3,*], Da-Hua Wei [1,*], Chi-Shun Tu [4,5] and Pin-Yi Chen [5]

[1] Institute of Manufacturing Technology and Department of Mechanical Engineering, National Taipei University of Technology, Taipei 10608, Taiwan; t106569005@ntut.edu.tw
[2] Department of Physics, National Chung Cheng University, Chia-Yi 62102, Taiwan
[3] Department of Applied Physics, Tunghai University, Taichung 40704, Taiwan; s07210049@go.thu.edu.tw
[4] Department of Physics, Fu Jen Catholic University, Taipei 24205, Taiwan; 039611@mail.fju.edu.tw
[5] Department of Mechanical Engineering, Ming Chi University of Technology, New Taipei City 24301, Taiwan; pinyi@mail.mcut.edu.tw
* Correspondence: hwchang@ccu.edu.tw (H.-W.C.); wangcr@thu.edu.tw (C.-R.W.); dhwei@ntut.edu.tw (D.-H.W.)

Abstract: In this work, we adopted pulsed laser deposition (PLD) with a Nd:YAG laser to develop $Bi_{1-x}Gd_xFeO_3$ (BGFO) films on glass substrates. The phase composition, microstructure, ferroelectric, magnetic, and nanomechanical properties of BGFO films are studied. BGFO films with $x = 0.00$–0.15 were confirmed to mainly consist of the perovskite phase. The structure is transformed from rhombohedral for $x = 0.00$ to pseudo-cubic for $x = 0.05$–0.10, and an additional phase, orthorhombic, is coexisted for $x = 0.15$. With increasing Gd content, the microstructure and surface morphology analysis shows a gradual decrease in crystallite size and surface roughness. The hardness of 5.9–8.3 GPa, measured by nanoindentor, is mainly dominated by crystallized structure and grain size. Good ferroelectric properties are found for BGFO films with $x = 0.00$–0.15, where the largest remanent polarization ($2P_r$) of 133.5 $\mu C/cm^2$ is achieved for $x = 0.10$, related to low leakage and high BGFO(110) texture. The improved magnetic properties with the significant enhancement of saturation magnetization from 4.9 emu/cm^3 for $x = 0$ to 23.9 emu/cm^3 for $x = 0.15$ by Gd substitution is found and related to large magnetic moment of Gd^{3+} and suppressed spiral spin structure of G-type antiferromagnetism. Furthermore, we also discuss the mechanisms of leakage behavior as well as nanomechanical characterizations as a function of the Gd content.

Keywords: multiferroics; Gd-doped BFO; nanomechanical properties; microstructure

Citation: Lin, T.-K.; Chang, H.-W.; Chou, W.-C.; Wang, C.-R.; Wei, D.-H.; Tu, C.-S.; Chen, P.-Y. Multiferroic and Nanomechanical Properties of $Bi_{1-x}Gd_xFeO_3$ Polycrystalline Films ($x = 0.00$–0.15). Coatings 2021, 11, 900. https://doi.org/10.3390/coatings11080900

Academic Editor: Michał Kulka

Received: 3 July 2021
Accepted: 23 July 2021
Published: 28 July 2021

Publisher's Note: MDPI stays neutral with regard to jurisdictional claims in published maps and institutional affiliations.

Copyright: © 2021 by the authors. Licensee MDPI, Basel, Switzerland. This article is an open access article distributed under the terms and conditions of the Creative Commons Attribution (CC BY) license (https://creativecommons.org/licenses/by/4.0/).

1. Introduction

$BiFeO_3$ (BFO), recognized as a room-temperature (RT) multiferroic (MF) material, has two main ferro-ordering parameters, including ferroelectricity (FE) at temperature (T) below 830 °C [1] and antiferromagnetism (AF) at T below 370 °C [2]. However, it exhibits weak ferromagnetism at RT due to antisymmetric exchange or the Dzyaloshinskii–Moriya interaction [3]. Because of the exceptional FE properties, multiferroic nature, potential magnetoelectric (ME) coupling, and lead-free character, the single-phase MF BFO has received considerable attention last decade. The above properties present the potential applications in non-volatile ferroelectric data storage, the emerging area of spintronics, and ME sensors.

High FE appears in BFO due to the presence of stereo-chemically active Bi^{3+} ions with 6s lone pair electrons. Li et al. [4] reported that the epitaxial pseudo-cubic BFO(001) thin films, grown on $SrRuO_3/SrTiO_3$ (100) substrates by a pulse laser deposition (PLD), exhibit a surprisingly large remnant polarization ($2P_r$) response of up to $2P_r \sim 120$ $\mu C/cm^2$, which is one order of magnitude higher than that of bulk BFO (~6.1 $\mu C/cm^2$) at that time [5]. Shvartsman et al. [6] showed that large FE polarization ~40 µC/cm^2 was attained in ceramic BFO, polycrystalline, and epitaxial films. The above results indicate that larger polarization comes from the intrinsic nature of BFO, which could also be supported by the first-principles calculations [7]. On the other hand, the magnetic nature of BFO is a G-type AF order, where Fe^{3+} ions are surrounded by six neighboring Fe^{3+} ions with spin antiparallel to the central ion [8]. However, BFO exhibits weak FM properties due to the spiral magnetic spin cycloid. Therefore, the cross-coupling between these ferroic parameters in BFO is weak and is poorly understood due to weak FM.

Although FE properties are prominent, the major drawbacks for the BFO-related devices are high leakage, a tendency towards fatigue [9], and weak FM. Ion modification with rare-earth cations has been demonstrated to be a successful method for the reduction in leakage and the improvement of magnetic properties at once [10]. FE and FM enhancements were also found in A-doped BFO ceramics and films (A = Sm, La, Pr, Gd, Dy, and Ca), due to the suppressed oxygen vacancies in BFO caused by the proper A substitution [11–14]. Among rare earth elements, Gd has higher exchange interaction and large magnetic moment [15]. Gd addition into BFO ceramics could improve FE and FM properties [16], but the MF properties are too low for practical applications up to now. Although the investigations related to A-site-doped BFO were reported, Gd-doped BFO thin films are still lacking. Besides, it is desirable and important to understand mechanical properties of the films, since most practical applications of functional devices are fabricated with thin films.

Therefore, $Bi_{1-x}Gd_xFeO_3$ (x = 0.00, 0.05, 0.10, and 0.15) (BGFO) polycrystalline films are prepared on the glass substrates by pulsed laser deposition in this work. A systematic investigation of the structural evolution, surface morphology, microstructure, leakage mechanisms, ferroelectric, and magnetic properties of BGFO films is reported.

2. Experiment

The solid-state reaction method was used to prepare $Bi_{1-x}Gd_xFeO_3$ targets (x = 0.00–0.15), as described elsewhere [17]. A PLD method was employed to deposit BGFO thin films on Pt buffered Corning's code 1737 glass substrates. A 20-nm-thick Pt layer was grown on a glass substrate as a bottom electrode by sputtering with a power of 60 W within Ar atmosphere of 10 mTorr at *RT*. A pulsed Nd:YAG laser with a laser wavelength of 355 nm was used to deposit 300-nm-thick BGFO films on the Pt layer at a substrate temperature of 300–500 °C and an O_2 pressure of 30 mTorr. Pulsed laser energy of 2.5 mJ with a repetition rate of 5 Hz was used for BGFO deposition.

The chemical composition of the BGFO thin films was determined using X-ray fluorescence analysis, and double-checked by an energy-dispersive X-ray spectroscopy analysis. X-ray diffractometry (XRD; PANalytical X'Pert PRO, Almelo, The Netherlands) with cupper K_α radiation was used to determine the phase and crystallographic orientations. The XRD pattern was refined and structural parameters were determined using the HighScore Plus version 3.0 software. The inorganic crystal structure database was used to infer pseudo-cubic (space group: *Pm3m*), rhombohedral (space group: *R3c*), orthorhombic (space group: *Pnma*), and orthorhombic $Bi_2Fe_4O_9$ (space group: *Pbam*) structures [18–21]. The surface and grain morphologies were observed by atomic force microscopy (AFM; Force Genie AFM, Taiwan, China). The microstructure was directly observed by transmission electron microscopy (TEM; JEOL JEM-2100, Tokyo, Japan). Sputtering was used to deposit a circular-shaped Pt layer with a diameter of 200 µm as a top electrode for following ferroelectric measurements. FE behavior and leakage test were analyzed using an TF-2000 Analyzer FE-Module ferroelectric test device (axiACCT Co., Aachen, Germany). Magnetization–magnetic field (*M-H*) curves of BGFO films were measured using a vibrating sample magnetometer.

The nanomechanical properties of BGFO films were measured using a nanoindenter with a Berkovich tip, and the hardness was defined as the applied indentation load divided by the projected contact area as follows:

$$H = \frac{P_{max}}{A_p} \quad (1)$$

where A_p is the projected contact area between the indenter and the sample surface at the maximum indentation load, P_{max}. For a perfectly sharp Berkovich indenter, the projected area A_p was given by $A_p = 24.56 h_c^2$ with h_c being the true contact depth. In general, the indentation depth should never exceed 30% of the film thickness to avoid the substrate effect on hardness measurements [22].

3. Results and Discussion

At first, the phase composition of BGFO films (x = 0.00–0.15) at 400 °C is analyzed and the XRD patterns are shown in Figure 1. A strong Pt(111) diffraction peak found in all samples reveals a flat and well-crystallized Pt(111) underlayer, which could help to grow following BGFO films. The BFO film consists of a rhombohedral structure, $R3c$ space group, and no other phase is detected. For Gd substitution for Bi in BGFO films, the phase transformation is observed. When x is increased from 0 to 0.05, BFO(104) and BFO(110) peaks merge into a single peak of BGFO(110). It indicates that the rhombohedral structure is transformed into a pseudo-cubic structure with Gd substitution, in good accordance with previous literature [23]. With increasing x from 0.05 to 0.10, the texture of pseudo-cubic structure changes from isotropic orientation to (110). For the sample with a higher x = 0.15, the orthorhombic structure with a space group of $Pnma$ is found to coexist with the pseudo-cubic structure. The presence of the orthorhombic phase for higher Gd content is consistent with the result of BGFO bulks [24]. Additionally, the shift of diffraction peaks for the perovskite phase toward a higher angle with the increase in Gd content x in Figure 1 reveals the entry of substituted Gd^{3+} ion into the perovskite BGFO phase to occupy A site due to smaller ionic radius of Gd^{3+} than Bi^{3+}.

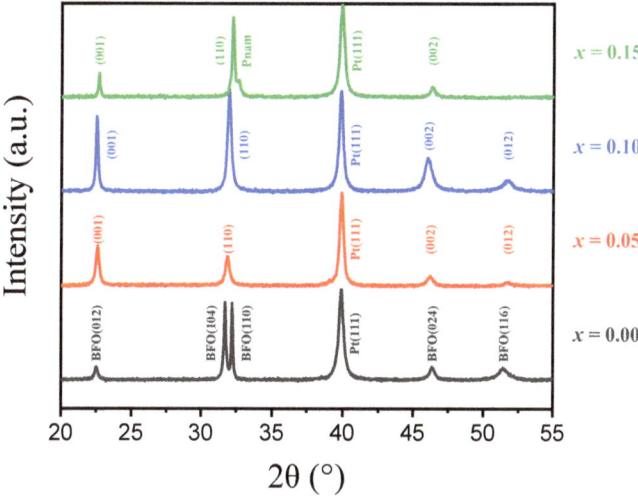

Figure 1. XRD patterns of $Bi_{1-x}Gd_xFeO_3$ (x = 0.00, 0.05, 0.10, and 0.15) films.

TEM images of $Bi_{1-x}Gd_xFeO_3$ with x = 0.00, 0.05, 0.10, and 0.15 are shown in Figure 2a–d. The average grain size, estimated through Image J software, is refined from 63 to 26 nm when Gd content x is increased from 0.00 to 0.15. The size distribution of the grains is also analyzed, and the distribution histograms by fitting with Gaussian distribution are shown in Figure 2e–h. The singlet distribution for x = 0.00–0.10 and dual size distribution for x = 0.15 are found. When x is increased from 0.00 to 0.10, the size and distribution of the grain becomes gradually small and narrow, shown in Figure 2e–g. As x is further increased to 0.15, there are two distributions for grain size in Figure 2h. According to XRD analysis, smaller grains belong to the orthorhombic phase for the sample with x = 0.15. Change in grain size with Gd substitution and the content x might be related to the evolution and texture of the structure.

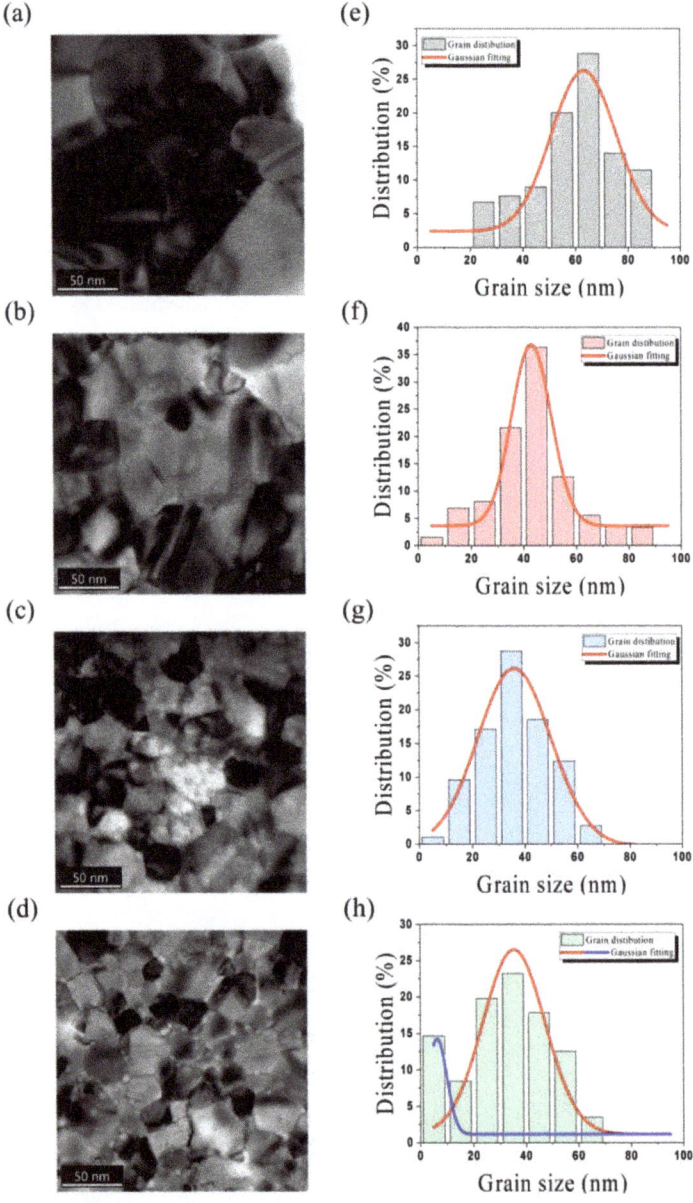

Figure 2. TEM images of $Bi_{1-x}Gd_xFeO_3$ films: (**a**) x = 0.00, (**b**) 0.05, (**c**) 0.10, and (**d**) 0.15. Grain size distribution of the films: (**e**) x = 0, (**f**) 0.05, (**g**) 0.10, and (**h**) 0.15.

AFM is used to observe the surface morphology and their AFM images are shown in Figure 3. Flat surface with the root-mean-square-roughness (R_{rms}) in the range of 6.7–13.4 nm is found. R_{rms} is decreased from 13.4 to 6.7 nm as x is increased from 0.00 to 0.15. The flattened surface with Gd substitution is possibly related to grain refinement, shown in Figure 2.

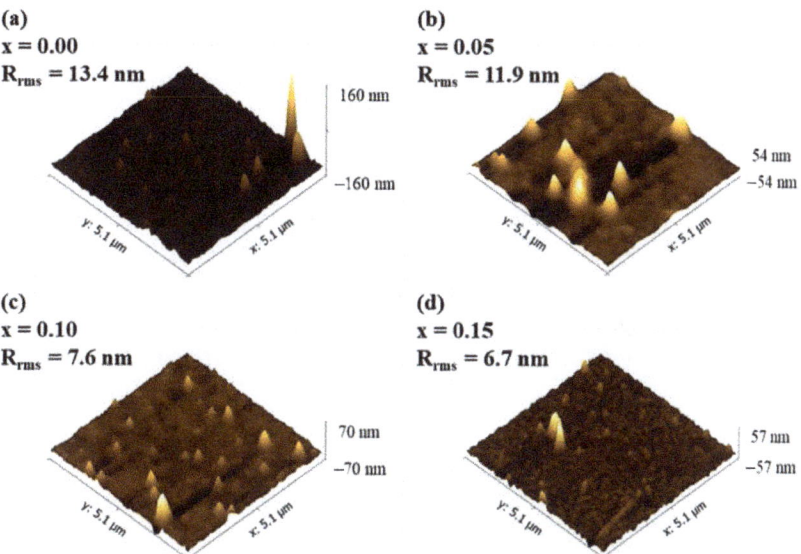

Figure 3. AFM images of $Bi_{1-x}Gd_xFeO_3$ films: (**a**) x = 0.00, (**b**) 0.05, (**c**) 0.10, and (**d**) 0.15.

For $Bi_{1-x}Gd_xFeO_3$ films with x = 0.00, 0.05, 0.10, and 0.15 films, polarization electric (P-E) field curves are shown in Figure 4a–d. All films exhibit a typical FE behavior with good FE properties. With increasing x, the remanent polarization ($2P_r$) increases from 41 µC/cm^2 for x = 0.00 to 72 µC/cm^2 for x = 0.05, and $2P_r$ enhancement may be due to the suppressed vacancies [25]. As x is increased to 0.10, $2P_r$ further increases up to a maximum value of 120 µC/cm^2, related to the texture transformation, form isotropy to (110). Finally, $2P_r$ decreases to 38 µC/cm^2 for higher Gd content of x = 0.15, related to the appearance of the paraelectric orthorhombic phase. On the other hand, with increasing Gd content, the coercive field (E_c) monotonically decreases from 320 kV/cm for x = 0.00 to 248 kV/cm for x = 0.15. The reduction in E_c with x may result from two facts: phase transformation and grain refinement, which decrease the energy barrier for the switching of FE polarization [26].

Figure 5a plots the leakage current density (*J*) versus the external electric field (*E*) for $Bi_{1-x}Gd_xFeO_3$ films. Large *J* is found for BFO film, which might be related to the existence of oxygen vacancies and rougher interfaces. For Gd-substituted films, a dramatic decrease in more than one order of magnitude in leakage is found. The suppressed leakage, possibly related to phase transformation and flattened surface, improves FE properties. However, due to the appearance of the paraelectric orthorhombic phase, the FE properties is decreased even though the leakage is further reduced at x = 0.15 in this study.

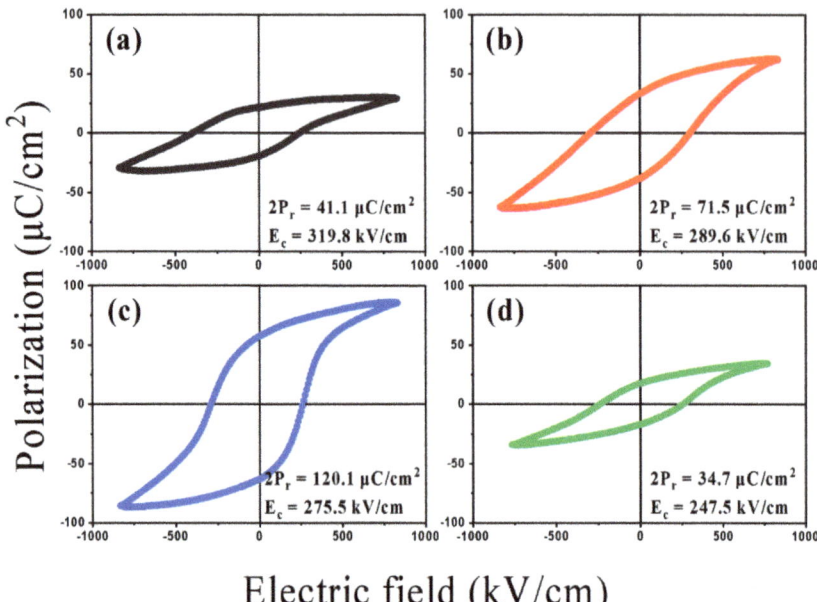

Figure 4. P-E curves of $Bi_{1-x}Gd_xFeO_3$ films: (**a**) x = 0.00, (**b**) 0.05, (**c**) 0.10, and (**d**) 0.15.

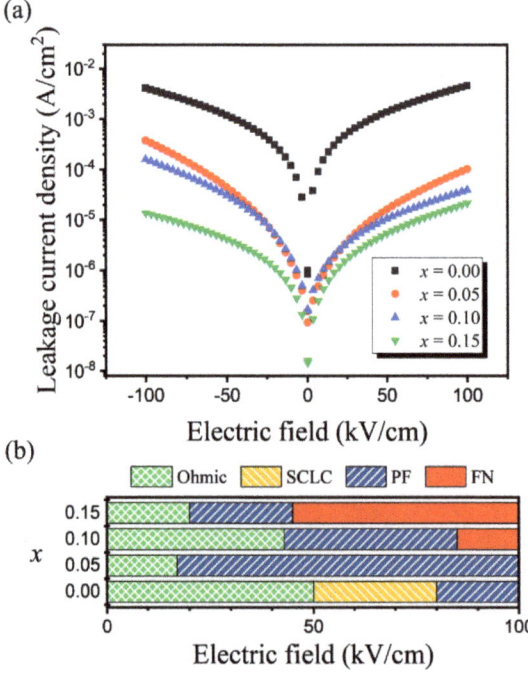

Figure 5. (**a**) Leakage current density and (**b**) leakage mechanism of $Bi_{1-x}Gd_xFeO_3$ (x = 0.00, 0.05, 0.10, and 0.15) films.

In order to further understand the leakage mechanisms of this studied films, five conduction mechanisms adopted to fit are as follows [27,28]:

For Ohmic Conduction

$$J_{Ohmic} = nq\mu E \Rightarrow log(J) \propto log(E), \qquad (2)$$

where n is the number of charge carriers, q is the electronic charge, and μ is the carrier mobility.

For Space Charge Limited Conduction (SCLC)

$$J_{SCLC} = \frac{9\mu\varepsilon E^2}{8d} \Rightarrow log(J) \propto 2log(E) \qquad (3)$$

where ε is the dielectric constant, and d is the film thickness.

For Poole–Frenkel Emission (PF)

$$J_{PF} = AE exp\left[-q\left(\frac{\varphi_T - \sqrt{\frac{qE}{\pi\varepsilon_r\varepsilon_o}}}{kT}\right)\right] \Rightarrow log\left(\frac{J}{E}\right) \propto \frac{\sqrt{\frac{q^3}{4\pi\varepsilon_r\varepsilon_o}}}{kT} E^{\frac{1}{2}} \qquad (4)$$

where A is a constant, φ_T is the trap ionization energy, ε_r is the dynamic (or optical) dielectric permittivity, ε_o is the permittivity of free space, k is the Boltzmann constant and T is the thermodynamic temperature.

For Schottky Emission

$$J_{Schottky} = BT^2 exp\left[-q\left(\frac{\varphi_B - \sqrt{\frac{qE}{4\pi\varepsilon_r\varepsilon_o}}}{kT}\right)\right] \Rightarrow log(J) \propto \frac{\sqrt{\frac{q^3}{4\pi\varepsilon_r\varepsilon_o}}}{kT} E^{\frac{1}{2}} \qquad (5)$$

where B is a constant, and φ_B is the interface potential barrier.

For Fowler–Nordheim (FN) Tunneling

$$J_{FN} = C\frac{E^2}{\varphi_B}\left[\frac{D}{E}\varphi_B^{\frac{3}{2}}\right] \Rightarrow log\left(\frac{J}{E^2}\right) \propto \frac{1}{E} \qquad (6)$$

where C is a constant.

The analyses will make assess if the leakage current follows Ohmic, SCLC, Schottky, PF emission, or FN tunneling, as described in Equations (2)–(6). Whether ε_r, obtained from the slopes, is within the range of 6.25–9.61 [29] for the developed BFO films is adopted to determine the mechanism dominated by Schottky or PF emission.

The fitting results and the evolution of leakage mechanism with E are summarized in Figure 5b. For all samples, Ohmic mechanism dominates at low E, and PF mechanism, which electrons can thermally emit into the conduction band from trapping centers, governs at larger E. In addition to Ohmic and PF mechanisms, SCLC mechanism is also found for BFO film, but disappears for BGFO films. The disappearance of the SCLC mechanism with Gd substitution, which may be a result of the reduced oxygen vacancy, is helpful in improving FE properties. Besides, for higher Gd content, FN tunneling occurs at larger E. FN tunneling currents which generally occur in the relatively insulating situation, for instance, from a solid surface into a vacuum or any weakly conducting dielectric. In this case, the exhibited FN tunneling mechanism proves the improved insulating ability with Gd substitution.

Especially near the morphotropic phase boundaries (MPBs), the structure in BFO could be easily affected by external electric field, stress, and growth temperature [30]. To optimize the FE characteristics and to explore the relationship between structures and growth temperature, structure and FE properties of the films with $x = 0.10$ at various growth temperatures (T_g) in range of 300–500 °C are also studied. XRD patterns are shown in Figure 6. At lower T_g, no diffraction peak belonging to the perovskite structure is found because heat energy is too low to form the perovskite phase. With increasing T_g from 350 to 450 °C, the peaks belonging to pseudo-cubic structure appear to become stronger, and besides, the orientation of crystal is changed from almost isotropy to BGFO(110) texture. At higher T_g = 500 °C, the texture of BGFO films is changed to (001), and an extra weak peak, belonging to the $Bi_2Fe_4O_9$ phase, appears possibly related to the volatilization of Bi at higher T_g. For the film grown at T_g = 500 °C, (110) and (002) diffraction peaks are split into (110), (121)$_o$, (002)$_o$ and (002), (202)$_o$, (040)$_o$, respectively, where (121)$_o$, (002)$_o$, (202)$_o$, and (040)$_o$ belong to the diffraction peaks of the orthorhombic phase. The appearance and coexistence of the orthorhombic phase for BGFO with x = 0.10 at higher T_g is consistent with phase diagram reported in the previous research of Kan et al. [30].

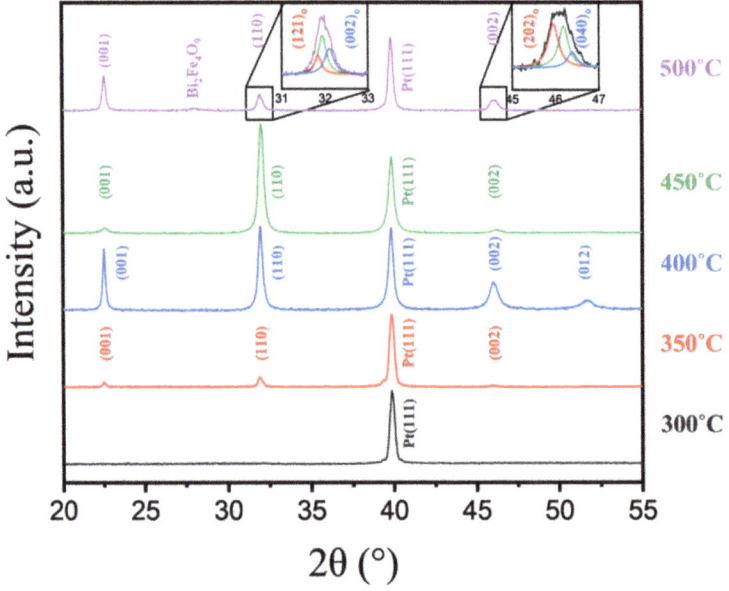

Figure 6. XRD patterns of $Bi_{0.9}Gd_{0.1}FeO_3$ (T_g = 300, 350, 400, 450, and 500 °C) films.

For $Bi_{0.9}Gd_{0.1}FeO_3$ films with T_g = 350–500 °C, electrical polarization electric (P-E) field curves are shown in Figure 7a–d. At lower $T_{g'}$ = 300 °C, no FE behavior is found due to the absence of the perovskite phase. Good P-E properties are attained at $T_{g'}$ = 350–500 °C. As T_g is increased from 350 to 500 °C, $2P_r$ is increased from 69.6 µC/cm^2 at T_g = 350 °C to a maximum $2P_r$ = 133.5 µC/cm^2 at T_g = 450 °C, and finally decreased to 58.8 µC/cm^2 at T_g = 500 °C. The improved FE properties with increasing T_g is related to the promoted crystallinity and texture constitution, and importantly, the highest $2P_r$ could be mainly attributed to BGFO (110) texture at 450 °C. At higher T_g = 500 °C, due to low breakdown voltage caused by high leakage, the saturation hysteresis curve cannot be obtained.

The curves of the leakage current density (J) versus the external electric field (E) of $Bi_{0.9}Gd_{0.1}FeO_3$ films at T_g = 350–500 °C are shown in Figure 8. Low leakage, observed for BGFO at T_g = 350–500 °C, contributes to good FE properties. The slight increase in leakage with increasing T_g is presumably related to texture evolution, and the appearance of 2:4:9 phase at higher T_g = 500 °C.

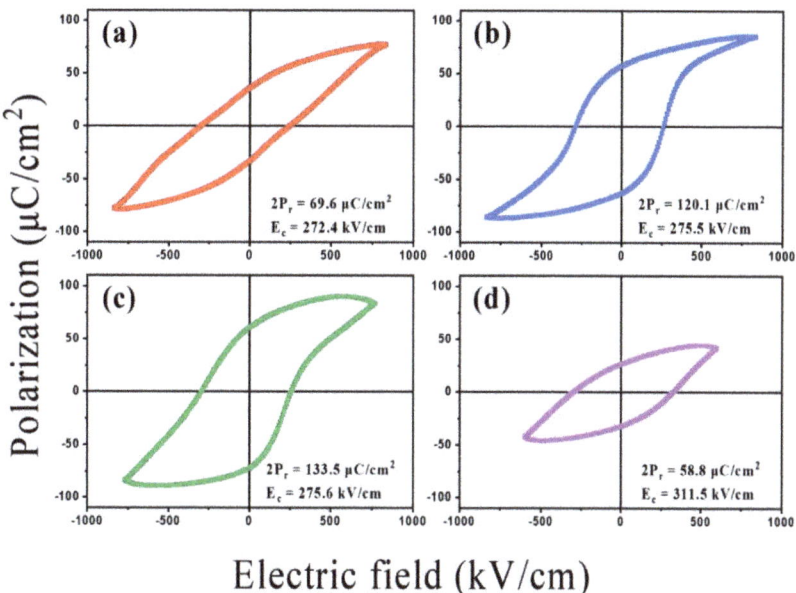

Figure 7. P-E curves of $Bi_{0.9}Gd_{0.1}FeO_3$ films: (**a**) T_g = 350 °C, (**b**) 400 °C, (**c**) 450 °C, and (**d**) 500 °C.

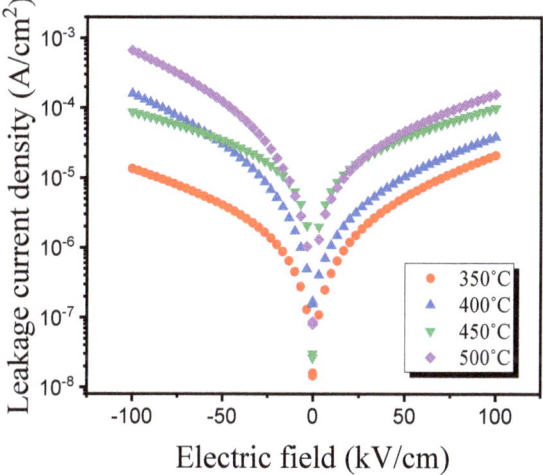

Figure 8. Leakage current density of $Bi_{0.9}Gd_{0.1}FeO_3$ (T_g = 350, 400, 450, and 500 °C) films.

Additionally, the magnetic properties of $Bi_{1-x}Gd_xFeO_3$ films are investigated, too. Their *M-H* curves are depicted in Figure 9. All BGFO films exhibit ferromagnetic behavior and exhibit enhanced magnetic properties. The saturation magnetization (M_s) increases dramatically from 4.9 to 23.9 emu/cm^3, but the magnetic coercive field (H_c) drops from 502 to 350 Oe as the Gd content increases from 0.00 to 0.15. The magnetization enhancement due to Gd substitution is presumably related to the distorted Fe-O-Fe bond, influencing the Fe-O-Fe exchange way, and the magnetic moment of Gd^{3+} ion [31].

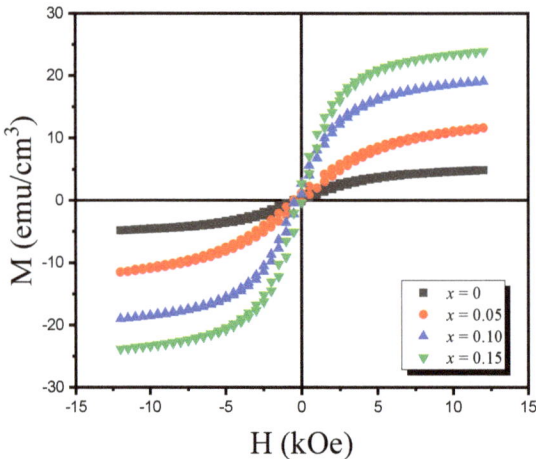

Figure 9. M-H curves of $Bi_{1-x}Gd_xFeO_3$ (x = 0.00, 0.05, 0.10, and 0.15) films.

The curves of hardness versus load penetration depth are illustrated in Figure 10a. Above a depth of 120 nm, the hardness of all samples reduces to a constant value. Generally, the relatively high hardness at a small depth results from the transition from fully elastic to elastic/plastic contact. The hardness subsequently decreases to a stable value. At this stage, the hardness value obtained is considered to be the inherent hardness for BGFO films. The hardness is therefore determined based on average values of hardness at depths of 60 to 90 nm. This range of penetration depth was chosen intentionally to be deep enough for observing plastic deformation during indentation, yet shallow enough to avoid the complications arising from the effects of surface roughness [22] and substrate [32]. With increasing Gd content, the hardness increases from 5.9 GPa for x = 0.00 to 8.3 GPa for x = 0.10, and then decreases to 8.0 GPa for x = 0.15. The change in hardness with x might be related to the phase and texture constitutions and grain refinement. With increasing x from 0.00 to 0.05, the denser pseudo-cubic structure forms and replaces rhombohedral, where the densities of rhombohedral and pseudo-cubic structure, estimated from XRD, are 8.33 and 8.47 g/cm^3, respectively. With further increasing x to 0.10, the change in texture for pseudo-cubic structure from isotropy to (110) might also help the increase in hardness. In addition to evolutions of the phase and texture, grain size refinement with x, demonstrating the grain boundary's effectiveness in impeding dislocation movement, also contributes to hardness enhancement. When x is further increased to 0.15, the slight decrease in hardness might be related to the presence of the orthorhombic phase.

MF and mechanical characteristics of the Gd-doped BFO bulks and films, which have been developed up to now [33–37], are listed in Table 1. First of all, BGFO bulks exhibit poor ferroelectric and weak ferromagnetic properties. Besides, the significant improvement in ferroelectric properties is found for BGFO polycrystalline films, especially for $Bi_{0.95}Gd_{0.05}FeO_3$ with the increased $2P_r$ of 90 µC/cm^2 [36]. In general, simultaneously improving ferroelectric and ferromagnetic properties is extremely difficult. In this study, BGFO films deposited on the refined Pt(111) underlayer at the reduced temperature of 400–450 °C show superior FE and FM properties. Furthermore, $2P_r$ of 133.5 µC/cm^2 and M_s of 19.1 emu/cm^3 attained for the $Bi_{0.9}Gd_{0.1}FeO_3$ thin film suggests that Gd-substituted BFO thin films on a Pt electrode buffered glass substrate at a low deposition temperature become a useful multiferroic material.

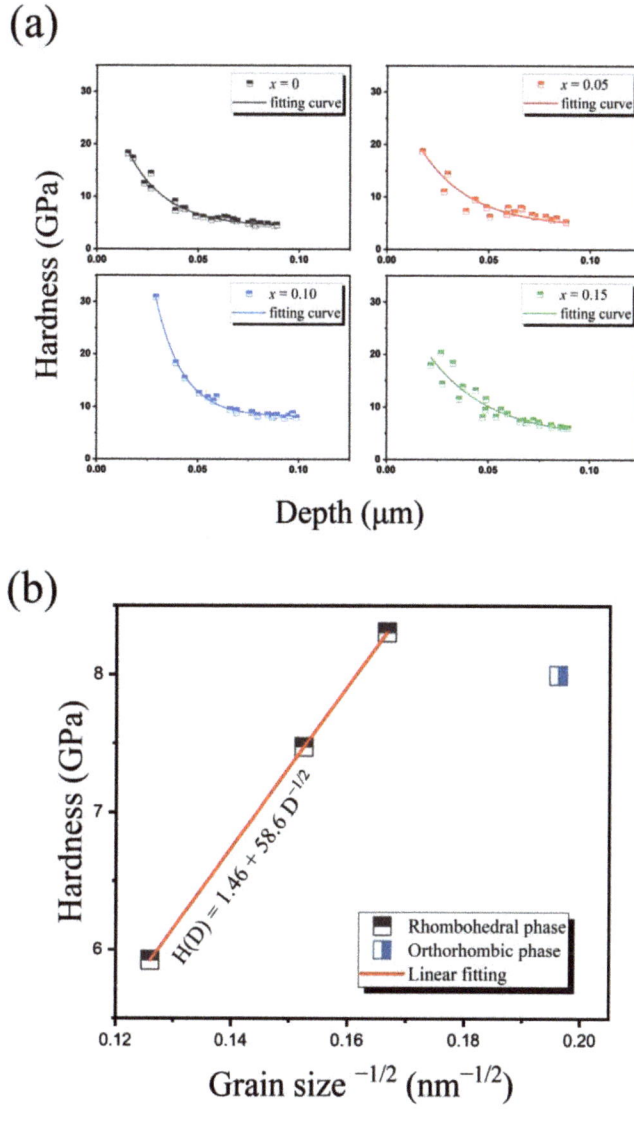

Figure 10. (a) Hardness versus depth curves and (b) Hardness versus grain size $^{-1/2}$ of $Bi_{1-x}Gd_xFeO_3$ films.

Table 1. Comparison of the developed Gd-doped BFO bulks and films: the multiferroic and nanomechanical characteristics.

Composition	Sample Form	$2P_r$ ($\mu C/cm^2$)	E_c (kV/cm)	M_s (emu/cm^3)	H_c (kOe)	Hardness (GPa)	Ref.
0.05	Bulk	25.6	107.0	~0.13	1.5	-	[33]
0.15	Bulk	-	-	~1.4	~5.0	-	[23]
0.10	Bulk	15	50	0.05	~2.0	-	[34]
0.10	Polycrystalline Films	72.6	196.5	-	-	-	[35]
0.05	Polycrystalline Films	~90	~300	-	-	-	[36]
0.10	Polycrystalline Films	80	5.0	8.33	<10.0	-	[37]
0.05	Polycrystalline Films	71.5	289.6	11.6	502	7.5	This work
0.10	Polycrystalline Films	133.5	275.6	19.1	395	8.3	This work
0.15	Polycrystalline Films	34.7	247.5	23.9	350	8.0	This work

4. Conclusions

The effects of Gd content on the structure, microstructure, MF, and nanomechanical properties of BGFO thin films deposited by PLD on 20-nm-thick Pt underlayer buffered glass substrates at reduced substrate temperatures of 400–450 °C were reported in this work. The perovskite phase in BGFO films with x = 0.00–0.15 was verified. The phase transition from a rhombohedral for x = 0.00 to a pseudo-cubic for x = 0.05–0.10, and an additional orthorhombic phase for higher x = 0.15 was found. The surface roughness and grain size of BGFO decreased as the Gd content increases. The hardness of BGFO thin films ranged from 5.9 to 8.3 GPa, and nanomechanical properties were strongly dependent on the phase and texture constitutions and grain size. Furthermore, it was found that microstructure, surface morphology, and Gd content all have a strong influence on ferroelectric properties. For BGFO films with x = 0.05–0.15, ferroelectric properties improved to $2P_r$ = 72–133.5 C/cm^2 and E_c of 248–312 kV/m, possibly because of the suppressed leakage current which resulted from the suppressed oxygen vacancy, the flattened interface, and microstructure refinement. The enhanced M_s of 11.6–23.9 emu/cm^3 seen in BGFO films due to Gd substitution was caused by the magnetic moment of the Gd^{3+} ion and distortion of the Fe-O-Fe bond angle. This study indicates that Gd-substituted BFO thin films on a Pt electrode buffered glass substrate at the reduced deposition temperature may be a useful multiferroic material.

Author Contributions: Conceptualization, H.-W.C.; methodology, H.-W.C.; software, T.-K.L.; validation, H.-W.C.; formal analysis, T.-K.L.; investigation, T.-K.L. and W.-C.C.; resources, H.-W.C., C.-R.W., D.-H.W., C.-S.T. and P.-Y.C.; data curation, T.-K.L.; writing—original draft preparation, T.-K.L.; writing—review and editing, H.-W.C.; visualization, T.-K.L. and W.-C.C.; supervision, H.-W.C.; project administration, H.-W.C.; funding acquisition, H.-W.C. and C.-R.W. All authors have read and agreed to the published version of the manuscript.

Funding: This research was financially supported by Ministry of Science and Technology of Taiwan, under Grant No. MOST-109-2112-M-194-006 and MOST-109-2112-M-029-006.

Institutional Review Board Statement: Not applicable.

Informed Consent Statement: Not applicable.

Data Availability Statement: Not applicable.

Conflicts of Interest: The authors declare no conflict of interest.

References

1. Tabares-Munoz, C.; Rivera, J.-P.; Schmid, H. Ferroelectric domains, birefringence and absorption of single crystals of BiFeO$_3$. *Ferroelectrics* **1984**, *55*, 235–238. [CrossRef]
2. Wang, J.; Neaton, J.B.; Zheng, H.; Nagarajan, V.; Ogale, S.B.; Liu, B.; Viehland, D.; Vaithyanathan, V.; Schlom, D.G.; Waghmare, U.V.; et al. Epitaxial BiFeO$_3$ multiferroic thin film heterostructures. *Science* **2003**, *299*, 1719–1722. [CrossRef]

3. Smolenskiĭ, G.A.; Chupis, I.E. Ferroelectromagnets. *Sov. Phys. Usp.* **1982**, *25*, 475. [CrossRef]
4. Li, J.; Wang, J.; Wuttig, M.; Ramesh, R.; Wang, N.; Ruette, B.; Pyatakov, A.P.; Zvezdin, A.K.; Li, D.V.J.; Wang, J.; et al. Dramatically enhanced polarization in (001), (101), and (111) BiFeO$_3$ thin films due to epitaxial-induced transitions. *Appl. Phys. Lett.* **2004**, *84*, 5261. [CrossRef]
5. Teague, J.R.; Gerson, R.; James, W.J. Dielectric hysteresis in single crystal BiFeO$_3$. *Solid State Commun.* **1970**, *8*, 1073–1074. [CrossRef]
6. Shvartsman, V.V.; Kleemann, W.; Haumont, R.; Kreisel, J. Large bulk polarization and regular domain structure in ceramic BiFeO$_3$. *Appl. Phys. Lett.* **2007**, *90*, 475. [CrossRef]
7. Neaton, J.B.; Ederer, C.; Waghmare, U.V.; Spaldin, N.A.; Rabe, K.M. First-principles study of spontaneous polarization in multiferroic BiFeO$_3$. *Phys. Rev. B* **2005**, *71*, 14113. [CrossRef]
8. Catalan, G.; Scott, J.F. Physics and applications of bismuth ferrite. *Adv. Mater.* **2009**, *21*, 2463–2485. [CrossRef]
9. Jang, H.W.; Baek, S.H.; Ortiz, D.; Folkman, C.M.; Eom, C.B.; Chu, Y.H.; Shafer, P.; Ramesh, R.; Vaithyanathan, V.; Schlom, D.G. Epitaxial (001) BiFeO$_3$ membranes with substantially reduced fatigue and leakage. *Appl. Phys. Lett.* **2008**, *92*, 1719. [CrossRef]
10. Lahmar, A.; Habouti, S.; Dietze, M.; Solterbeck, C.-H.; Es-Souni, M. Effects of rare earth manganites on structural, ferroelectric, and magnetic properties of BiFeO$_3$ thin films. *Appl. Phys. Lett.* **2009**, *94*, 141106. [CrossRef]
11. Lin, T.K.; Chang, H.W.; Wang, C.R.; Wei, D.H.; Tu, C.S. Multiferroic and nanomechanical properties of Bi$_{1-x}$R$_x$FeO$_3$ polycrystalline films (R = La, Pr, Sm, and Ho; x = 0–0.15). *J. Alloys Compd.* **2020**, *846*, 156080. [CrossRef]
12. Murashov, V.A.; Rakov, D.N.; Ionov, V.M.; Dubenko, I.S.; Titov, Y.V.; Goreli, V.S. Magnetoelectric (Bi, Ln)FeO3 compounds: Crystal growth, structure and properties. *Ferroelectrics* **1994**, *162*, 11. [CrossRef]
13. Gabbasova, Z.V.; Kuz'min, M.D.; Zvezdin, A.K.; Dubenko, I.S.; Murashov, V.A.; Rakov, D.N.; Krynetsky, I.B. Bi$_{1-x}$R$_x$FeO$_3$ (R = rare earth): A family of novel magnetoelectrics. *Phys. Lett. A* **1991**, *158*, 491–498. [CrossRef]
14. Chang, H.W.; Yuan, F.T.; Tu, K.T.; Lin, S.Y.; Wang, C.R.; Tu, C.S. Multiferroic properties of Bi$_{1-x}$A$_x$FeO$_3$ polycrystalline films on glass substrates (A = Ca, Sr, Ba and x = 0.05–0.15). *J. Alloys Compd.* **2016**, *683*, 427–432. [CrossRef]
15. Khmelevskyi, S.; Khmelevska, T.; Ruban, A.V.; Mohn, P. Magnetic exchange interactions in the paramagnetic state of hcp Gd. *J. Phys. Condens. Matter* **2007**, *19*, 326218. [CrossRef]
16. Hu, W.W.; Chen, Y.; Yuan, H.M.; Li, G.H.; Qiao, Y.; Qin, Y.Y.; Feng, S.H. Structure, magnetic, and ferroelectric properties of Bi$_{1-x}$Gd$_x$FeO$_3$ nanoparticles. *J. Phys. Chem. C* **2011**, *115*, 8869. [CrossRef]
17. Tu, C.S.; Chen, C.S.; Chen, P.Y.; Wei, H.H.; Schmidt, V.H.; Lin, C.Y.; Anthoniappen, J.; Lee, J.M. Enhanced photovoltaic effects in A-site samarium doped BiFeO$_3$ ceramics: The roles of domain structure and electronic state. *J. Eur. Ceram. Soc.* **2016**, *36*, 1149. [CrossRef]
18. Song, D.; He, H.; Duan, Y.; Li, J. Crystal structure, electronic structure, and magnetic properties of bismuth-strontium ferrites. *J. Alloys Compd.* **2001**, *315*, 259–264.
19. Murashov, V.A.; Fischer, P.; Przenioslo, R.; Sosnowska, I. Neutron diffraction studies of the crystal and magnetic structures of BiFeO$_3$ and Bi$_{0.93}$La$_{0.07}$FeO$_3$. *J. Magn. Magn. Mater.* **1996**, *160*, 384–385.
20. Eibschuetz, M.; Coppens, P. Determination of the crystal structure of yttrium orthoferrite and refinement of gadolinium orthoferrite. *Acta Crystallogr.* **1965**, *19*, 524.
21. Markin, V.N.; Tutov, A.G. The X-ray structural analysis of the antiferromagnetic Bi$_2$Fe$_4$O$_9$ and the isotypical combinations Bi$_2$Ga$_4$O$_9$ and Bi$_2$Al$_4$O$_9$. *Neorg. Mater.* **1970**, *6*, 2014.
22. Li, X.D.; Gao, H.; Murphy, C.J.; Caswell, K.K. Nanoindentation of silver nanowires. *Nano Lett.* **2003**, *11*, 1495–1498. [CrossRef]
23. Ablat, A.; Wu, R.; Mamat, M.; Li, J.; Muhemmed, E.; Si, C.; Wu, R.; Wang, J.O.; Qian, H.J.; Ibrahim, K. Structural analysis and magnetic properties of Gd doped BiFeO$_3$ ceramics. *Ceram. Int.* **2014**, *40*, 14083–14089. [CrossRef]
24. Khomchenko, V.A.; Kiselev, D.A.; Bdikin, I.K.; Shvartsman, V.V.; Borisov, P.; Kleemann, W.; Vieira, J.M.; Kholkin, A.L. Crystal structure and multiferroic properties of Gd-substituted BiFeO$_3$. *Appl. Phys. Lett.* **2008**, *93*, 759. [CrossRef]
25. Jena, A.K.; Chelvane, J.A.; Mohanty, J. Evidence for dielectric suppression in non-magnetic modified multiferroic bismuth ferrite. *J. Appl. Phys.* **2019**, *126*, 184101. [CrossRef]
26. Shelke, V.; Mazumdar, D.; Srinivasan, G.; Kumar, A.; Jesse, S.; Kalinin, S.; Baddorf, A.; Gupta, A. Reduced coercive field in BiFeO$_3$ thin films through domain engineering. *Adv. Mater.* **2011**, *23*, 669–672. [CrossRef] [PubMed]
27. Liu, Y.; Tan, G.Q.; Ren, X.X.; Li, J.C.; Xue, M.T.; Ren, H.J.; Xia, A.; Liu, W.L. Electric field dependence of ferroelectric stability in BiFeO$_3$ thin films co-doped with Er and Mn. *Ceram. Int.* **2020**, *46*, 18690. [CrossRef]
28. Ranjith, R.; Prellier, W.; Cheah, J.W.; Wang, J.; Wu, T. Dc leakage behavior and conduction mechanism in (BiFeO$_3$)m(SrTiO$_3$)m superlattices. *Appl. Phys. Lett.* **2008**, *92*, 232905. [CrossRef]
29. Pabst, G.W.; Martin, L.W.; Chu, Y.-H.; Ramesh, R. Leakage mechanisms in BiFeO$_3$ thin films. *Appl. Phys. Lett.* **2007**, *90*, 1719. [CrossRef]
30. Kan, D.; Palova, L.; Anbusathaiah, V.; Cheng, C.-J.; Fujino, S.; Nagarajan, V.; Rabe, K.M.; Takeuchi, I. Universal behavior and electric-field-induced structural transition in rare-earth-substituted BiFeO$_3$. *Adv. Funct. Mater.* **2010**, *20*, 1108–1115. [CrossRef]
31. Nayek, C.; Tamilselvan, A.; Thirmal, C.; Murugavel, P.; Balakumar, S. Origin of enhanced magnetization in rare earth doped multiferroic bismuth ferrite. *J. Appl. Phys.* **2014**, *115*, 73902. [CrossRef]
32. Biswas, S.K.; Pethica, J.B. Effect of roughness on the measurement of nanohardness—A computer simulation study. *Appl. Phys. Lett.* **1997**, *71*, 1059–1061.

33. Yao, Y.; Liu, W.; Chan, Y.; Leung, C.; Mak, C. Studies of rare-earth-doped BiFeO$_3$ ceramics. *Int. J. Appl. Ceram. Technol.* **2011**, *8*, 1246. [CrossRef]
34. Pradhan, S.K.; Das, J.; Rout, P.P.; Das, S.K.; Mishra, D.K.; Sahu, D.R.; Pradhan, A.K.; Srinivasu, V.V.; Nayak, B.B.; Verma, S.; et al. Defect driven multiferroicity in Gd doped BiFeO$_3$ at room temperature. *J. Magn. Magn. Mater.* **2010**, *322*, 3614–3622. [CrossRef]
35. Wang, X.Z.; Yan, B.W.; Dai, Z.G.; Liu, M.F.; Liu, H.R. Enhanced ferroelectric properties of Gd- substitution BiFeO$_3$ thin films prepared by sol-gel process. *Ferroelectrics* **2010**, *410*, 96–101. [CrossRef]
36. Wu, J.; Wang, J.; Xiao, D.; Zhu, J.J. Effect of (Bi,Gd)FeO$_3$ layer thickness on the microstructure and electrical properties of BiFeO$_3$ thin films. *Am. Ceram. Soc.* **2011**, *94*, 4291. [CrossRef]
37. Cheng, Z.X.; Wang, X.L.; Dou, S.X.; Kimura, H.; Ozawa, K. Enhancement of ferroelectricity and ferromagnetism in rare earth element doped BiFeO$_3$. *J. Appl. Phys.* **2008**, *104*, 116109. [CrossRef]

Article

Effects of Substrate Temperature on Nanomechanical Properties of Pulsed Laser Deposited Bi_2Te_3 Films

Hui-Ping Cheng [1], Phuoc Huu Le [2,*], Le Thi Cam Tuyen [3,*], Sheng-Rui Jian [1,4,5,*], Yu-Chen Chung [1], I-Ju Teng [6], Chih-Ming Lin [7] and Jenh-Yih Juang [8]

1. Department of Materials Science and Engineering, I-Shou University, Kaohsiung 84001, Taiwan; jasonping87@gmail.com (H.-P.C.); ricky12579@gmail.com (Y.-C.C.)
2. Department of Physics and Biophysics, Faculty of Basic Sciences, Can Tho University of Medicine and Pharmacy, Can Tho City 94000, Vietnam
3. Department of Chemical Engineering, College of Engineering Technology, Can Tho University, 3/2 Street, Can Tho City 900000, Vietnam
4. Department of Fragrance and Cosmetic Science, College of Pharmacy, Kaohsiung Medical University, 100 Shi-Chuan 1st Road, Kaohsiung 80708, Taiwan
5. Department of Mechanical and Automation Engineering, I-Shou University, Kaohsiung 84001, Taiwan
6. College of Engineering, National Taiwan University of Science and Technology, Taipei 106335, Taiwan; eru7023@gmail.com
7. Department of Physics, National Tsing Hua University, Hsinchu 30013, Taiwan; cm_lin@phys.nthu.edu.tw
8. Department of Electrophysics, National Yang Ming Chiao Tung University, Hsinchu 30010, Taiwan; jyjuang@nycu.edu.tw
* Correspondence: lhuuphuoc@ctump.edu.vn (P.H.L.); ltctuyen@ctu.edu.vn (L.T.C.T.); srjian@gmail.com (S.-R.J.); Tel.: +886-7-6577711-3130 (S.-R.J.)

Abstract: The correlations among microstructure, surface morphology, hardness, and elastic modulus of Bi_2Te_3 thin films deposited on c-plane sapphire substrates by pulsed laser deposition are investigated. X-ray diffraction (XRD) and transmission electron microscopy are used to characterize the microstructures of the Bi_2Te_3 thin films. The XRD analyses revealed that the Bi_2Te_3 thin films were highly (00l)-oriented and exhibited progressively improved crystallinity when the substrate temperature (T_S) increased. The hardness and elastic modulus of the Bi_2Te_3 thin films determined by nanoindentation operated with the continuous contact stiffness measurement (CSM) mode are both substantially larger than those reported for bulk samples, albeit both decrease monotonically with increasing crystallite size and follow the Hall—Petch relation closely. Moreover, the Berkovich nanoindentation-induced crack exhibited trans-granular cracking behaviors for all films investigated. The fracture toughness was significantly higher for films deposited at the lower T_S; meanwhile, the fracture energy was almost the same when the crystallite size was suppressed, which indicated a prominent role of grain boundary in governing the deformation characteristics of the present Bi_2Te_3 films.

Keywords: Bi_2Te_3 thin films; XRD; SEM; nanoindentation; pop-in; hardness

1. Introduction

Bismuth telluride, Bi_2Te_3, is a 3D topological insulator and an excellent thermoelectric (TE) material that works well at room temperature [1,2]. TE materials are of interest for heat pump and power generator applications. The performance of TE materials is quantified by a dimensionless figure of merit (ZT), expressed as $ZT = S^2\sigma T/\kappa$, where S, σ, κ, and T are the Seebeck coefficient, electrical conductivity, thermal conductivity, and absolute temperature, respectively. Generally, thin films can be fabricated by various methods, such as aerosol-assisted chemical vapor deposition [3,4], dip-coating [5], pulsed laser deposition (PLD) [6], etc. PLD offers advantages such as a higher instantaneous deposition rate, relatively high reproducibility, and low costs.

In addition to improving the properties of TE materials, mechanical characterizations are of critical importance when the reliability of TE devices is concerned [2,7,8]. For example, the performances of TE devices can be significantly degraded due to contact loading during operation. In addition, inhomogeneous thermal expansion may occur in TE generators because they are regularly subjected to the cyclic temperature gradient during processing. Consequently, the inhomogeneous thermal expansion/contraction induces repetitive expansion/shrinkage and the corresponding stress/strain in TE materials to possibly cause fatigue cracking, performance degradation, and even failure of the TE generators. Therefore, it is important to have a comprehensive understanding of the mechanical properties of TE materials (i.e., Bi_2Te_3) to provide vital information for fabricating efficient and endurable Bi_2Te_3-based devices.

Nanoindentation has become a widely used technique for extracting prominent mechanical properties, namely the hardness and elastic modulus, as well as to unveil the dislocation-mediated plastic deformation and the fracture behaviors of a wide variety of nanostructures [9–11] and oxide thin films [12–14]. The relationship between the microstructure and nanomechanical characterizations of the Bi_2Te_3/Al_2O_3 (001) thin films deposited at the various T_S by means of PLD is systematically investigated in this study. Al_2O_3 (001) is used because it is an insulating–popular substrate and has moderate lattice mismatch between Bi_2Te_3 and Al_2O_3 (8.7%). It is found that T_S evidently introduces drastic modification in the film's microstructure, crystallinity, and crystallite size, which in turn manipulates the mechanical characterizations, such as hardness, elastic modulus, fracture toughness, and fracture energy of the Bi_2Te_3 films.

2. Materials and Methods

In this study, a 99.99%-pure tellurium-excessive target (Bi_2Te_8) was used to deposit the Bi_2Te_3 films on Al_2O_3 (001) substrates by PLD. The reason for choosing the tellurium-excessive target was to overcome the issues of high-doping carriers in the Te-deficient non-stoichiometric Bi_2Te_3 phase as well as to avoid the formation of unwanted phases. A KrF excimer pulsed laser was used to ablate the target. The energy density and repetition rate of the laser pulses were 5.7 J/cm^2 and 10 Hz, respectively. The vacuum chamber was evacuated to a base pressure of 2×10^{-6} Torr. Prior to loading into the chamber, the Al_2O_3 (001) substrates with a size of 4 mm $\times$ 4 mm were sequentially ultrasonically cleaned in acetone, methanol, and deionized water baths for 30 min. Then, helium gas was introduced into the chamber, with the pressure being maintained at 200 mTorr throughout the entire deposition process. The number of laser pulses was 15,000, and the film thickness was 1154–1428 nm. The substrate temperature (T_S) was controlled at 225 °C, 250 °C, and 300 °C. We selected these T_S because they were known as the suitable temperatures for growing high-quality Bi_2Te_3 films with excellent thermoelectric properties [6].

The crystal structure of the Bi_2Te_3 thin films was determined by X-ray diffraction (XRD; Bruker D2, Billerica, MA, USA) using Cu Kα radiation (wavelength of 1.5406 Å) in the θ–2θ configuration. The surface morphology and thickness of the films were examined using a field-emission scanning electron microscope (SEM, JEOL JSM-6500, Pleasanton, CA, USA) at an applied voltage of 15 kV, working distance of 10.5 mm, and magnification of 30,000. The composition of the films was analyzed using an Oxford energy-dispersive X-ray spectroscope (EDS) attached to the SEM. For EDS analyses, the accelerating voltage of the electron beam was set at 15 kV, and the dead time and collection time were 22–30% and 60 s, respectively. Digital images from a high-resolution transmission electron microscope (HRTEM; Tecnai F20, ThermoFisher, Waltham, MA, USA) operated at 200 kV were recorded using a Gatan 2 k $\times$ 2 k CCD camera system to obtain detailed film-structure information. The HRTEM specimens were prepared using a standard mechanical thinning and Ar ion milling procedure.

An MTS NanoXP® system (MTS Corporation, Nano Instruments Innovation Center, Oak Ridge, TN, USA) with a load force resolution of 50 nN and a displacement resolution of 0.1 nm was used for conducting the nanoindentation tests at room temperature. The indentation depth of the Berkovich diamond indenter was 55 nm, and the strain rate varied from 0.01 to 1 s^{-1}. Additional harmonic modulation was superimposed simultaneously onto the indenter when continuous stiffness measurements (CSM) were performed [15]. The modulation amplitude and frequency were set at 2 nm and 45 Hz, respectively. Special care was taken to ensure that the thermal drift was less than 0.01 nm/s before each test was conducted. We performed at least 20 indents for each sample to ensure the statistical significance of the results.

The projected contact area between the indenter tip and films surface, A_p, and the maximum indentation loading, P_m, are used to define the hardness as $H = P_m/A_p$. The Berkovich indenter tip, A_p, and the contact depth, h_c, are correlated as $A_p = 24.56 h_c^2$. Following Sneddon's analysis [16], the elastic modulus of the film (E_f) is given by $S_c = 2\beta E_r(\sqrt{A_p}/\sqrt{\pi})$, where S_c is the contact stiffness of the thin film and β is a geometric constant, with $\beta \approx 1$ for the Berkovich indenter tip. The reduced elastic modulus (E_r) is further utilized to determine E_f using $1/E_r = [(1-\nu_f^2)/E_f + (1-\nu_i^2)/E_i]$, with ν being the Poisson's ratio and the subscripts, f and i, denoting the parameters for the film and the indenter tip, respectively. For the diamond indenter tip used here, $E_i = 1141$ GPa and $\nu_i = 0.07$ [17], and $\nu_f = 0.25$ is assumed.

3. Results and Discussion

Figure 1a shows the XRD patterns of films, which can be unambiguously indexed as (003), (006), and (0015) of the rhombohedral Bi$_2$Te$_3$ phase (JCPDS card No. 82-0358). It is worth noting that the intensity ratios between the (0 0 6) and (1 0 16) diffraction peaks were 6.7, 7.8, and 20.2 for films deposited at 225 °C, 250 °C, and 300 °C, respectively, which were all substantially higher than the value of 1 obtained from the standard powder diffraction data file (JCPDS card No. 82-0358), confirming the highly (00l)-oriented characteristic of the present Bi$_2$Te$_3$ films. In addition, the preferred in-plane orientation was not found by the XRD phi-scan. Therefore, the Bi$_2$Te$_3$ films are polycrystalline and highly c-axis-oriented (or textured films). Noticeably, the crystal structure of Bi$_2$Te$_3$ is rhombohedral with a space group D_{3d}^5 ($R\bar{3}m$), which can be represented by a hexagonal primitive cell consisting of three quintuple-layers (QL). Each QL is about 1 nm-thick with 5-atomic-layer stacking in sequence, namely –(Te$^{(1)}$–Bi–Te$^{(2)}$–Bi–Te$^{(1)}$)–, along the c-axis, as depicted schematically in Figure 1b. The bonding between the QLs is the Van der Waals (VdW) Te$^{(1)}$–Te$^{(1)}$ bond, which is significantly weaker than the ionic–covalent Bi-Te bonds within the QLs [6,18].

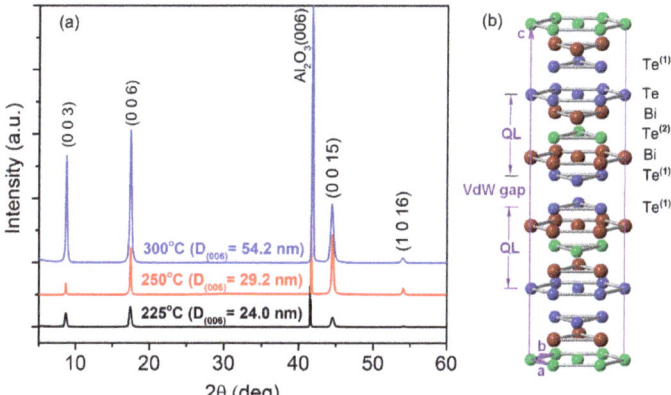

Figure 1. (a) XRD patterns of the Bi$_2$Te$_3$ thin films deposited on the c-plane sapphire substrates at the various T_S of 225 °C, 250 °C, and 300 °C, respectively. $D_{(006)}$ is the calculated grain size of the films using the Scherrer equation and Bi$_2$Te$_3$ (006) peaks. (b) The schematics depicts the crystal structure of Bi$_2$Te$_3$.

It is also evident from Figure 1a that the diffraction peaks exhibit higher intensity and narrower width with increasing T_S, indicating that the crystallinity of the films is progressively improved when T_S increases. We used Scherrer's formula [19], $D_s = 0.9\lambda/(\beta\cos\theta)$, to estimate the crystallite size (D_s) of the Bi_2Te_3 thin films, wherein λ, β, and θ are the X-ray wavelength, the full-width-at-half-maximum (FWHM) of the selected diffraction peak, and the corresponding Bragg diffraction angle, respectively. Here, we chose the (006) peak to calculate crystallite sizes, and the sizes are 24.0 nm, 29.2 nm, and 54.2 nm for the films deposited at T_S of 225 °C, 250 °C, and 300 °C, respectively. These results clearly indicate that T_S imposes a marked effect on the crystallite size. This observation can be understood as the following. At a lower T_S, due to the higher extent of supersaturation and the lower surface diffusion rate, the reduced critical size and the nucleation energy barrier for the nuclei are reduced, which leads to an increased number of nucleation sites and an eventually smaller grain size [6].

Furthermore, by using the Williamson—Hall (WH) equation [20,21], the effects of the crystalline size-induced broadening and strain-induced broadening of XRD results for the Bi_2Te_3 thin films at various T_S can be determined. The WH equation is as follows:

$$\beta\cos\theta = (0.9\lambda/D_{WH}) + 4\varepsilon \cdot \sin\theta \tag{1}$$

where ε is the microstrain. The WH equation represents a straight line between $4\sin\theta$ (X-axis) and $\beta\cos\theta$ (Y-axis). The slope of the line gives the value of the microstrain. Both the crystalline size (D_{WH}) and microstrain (ε) contribute to the broadening of the XRD spectra of the Bi_2Te_3 thin films and are listed in Table 1. It is noted that the incorporation of the microstrain effect is attributed to the different values of the crystalline size significantly. The values of the crystalline size of the T_S-treated Bi_2Te_3 thin films calculated using Scherrer's equation are noticeably smaller than those calculated using the WH equation that gave rise to the influence of the microstrain on the XRD results. Furthermore, both the values of D_s and D_{WH} are enlarged, indicating that the overall crystallinity of the Bi_2Te_3 thin films was remarkably improved by the increased T_S. The ε increases monotonically from 0.21% to 0.29% with increasing T_S from 225 °C to 300 °C, which can be attributed to the newly created interfaces associated with the T_S-dependent crystallite size and evolving grain shapes.

Table 1. The structural and mechanical properties of the Bi_2Te_3 thin films.

Bi_2Te_3	D_s (nm)	D_{WH} (nm)	ε (%)	H (GPa)	E_f (GPa)	τ_c (GPa)	K_c (MPa·m$^{1/2}$)	G_c (Jm^{-2})
Bulk [22]	—	—	—	1.6 ± 0.2	32.4 ± 2.9	—	—	—
Thin films [23] (helium gas pressure) 2×10^{-5}–2×10^{-3} Torr	11–20	—	—	2.9–4.0	106.3–127.5	0.9–1.3	—	—
Thin films [#] T_S = 225 °C	24.0	32.6	0.21	5.2 ± 0.3	125.2 ± 6.9	2.2	1.42	0.15
Thin films [#] T_S = 250 °C	29.2	40.2	0.24	4.0 ± 0.1	98.3 ± 2.1	1.4	1.21	0.14
Thin films [#] T_S = 300 °C	54.2	62.7	0.29	3.4 ± 0.2	62.5 ± 1.4	1.0	0.88	0.12

[#]: this work.

Figure 2a shows a low-magnification TEM image Bi_2Te_3 film deposited at the T_S of 300 °C, in which some crystallites are presented by the dotted areas. It can be seen that the crystallite sizes are approximately 50–60 nm, agreeing well with the estimated crystallite size obtained from the XRD results. Moreover, an HRTEM image of a crystallite clearly presents the projected periods of 0.51 nm along the c-axis (Figure 2b), which corresponds to the lattice spacing of the Bi_2Te_3 (006) planes. These TEM results are further confirmed by the crystallite sizes and gain insight into the crystal structure of the textured Bi_2Te_3 films in this study.

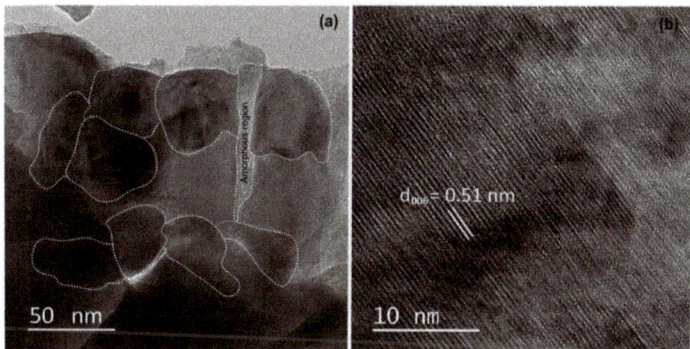

Figure 2. TEM images of the Bi_2Te_3 film deposited at the T_S of 300 °C. The dotted areas in (**a**) are used to guide the eyes of some crystallites; (**b**) the magnified HRTEM image obtained from (**a**).

Figure 3 depicts the SEM photographs, revealing the surface morphology of the Bi_2Te_3 films grown at different T_S from 225 to 300 °C. The SEM images show that all films exhibit the typical polygonal granular morphology of a polycrystalline microstructure, and the grain size increases with increasing T_S. This result is in line with the T_S-dependent crystallite size tendency obtained from the XRD results (Figure 1). It is noted that the grain size observed from the SEM images shown in Figure 3 appears to be larger than the crystallite size estimated using the XRD data or observed directly from the HRTEM images because a "grain" in the SEM image may be composed of agglomerated grains and/or even include amorphous regions [24]. In addition, the EDS results confirm that all films have stoichiometry very close to that of the Bi_2Te_3 phase, which is necessary for obtaining the pure Bi_2Te_3 phase for all the investigated films. Noticeably, the Te composition reduces slightly from 60.12 at.% for the 225 °C film to 59.48 at.% for the 300 °C film. This could be due to the re-evaporation of Te from the heated substrates being much faster than that of Bi at elevated temperatures. The cross-sectional SEM images in Figure 3 show that the films had a thickness in the 1154–1428 nm range and layered structures.

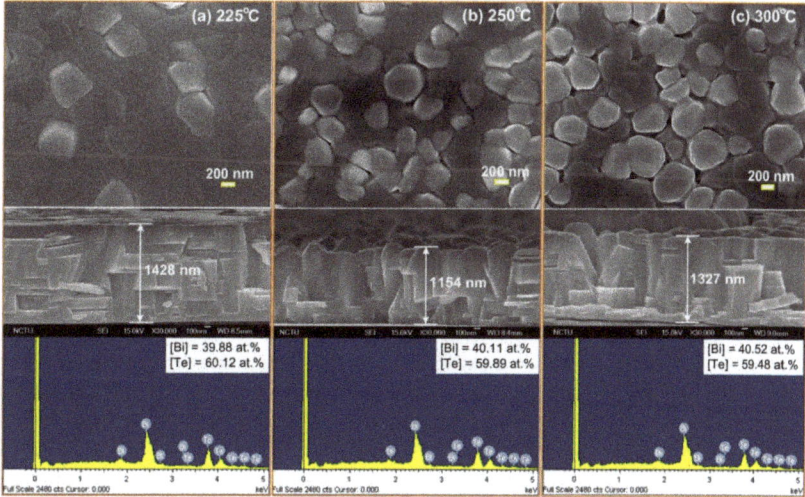

Figure 3. SEM images (upper: top view and middle: cross section) and EDS spectra of the Bi_2Te_3 thin films deposited at various substrate temperatures of 225, 250, and 300 °C.

As discussed above, the T_S evidently showed significant effects on the film microstructures. The next prominent question is how it affects the nanomechanical properties of the films. Figure 4 shows the typical load—depth curves (Ph-curves) obtained from the nanoindentation CSM measurements on the Bi_2Te_3 films at various T_S of 225 °C, 250 °C, and 300 °C. We kept the total penetration depth to within approximately 55 nm, which is well below the 30% criterion (film thickness of 1154–1428 nm) suggested by Li et al. [25], in which the indentation depth should never exceed 30% of the films' thickness or the dimension of the nanostructures to avoid any complications from the substrate.

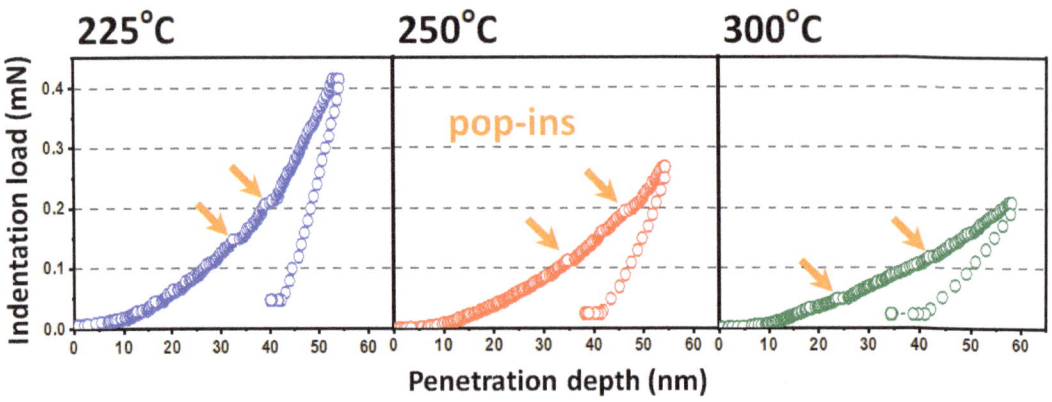

Figure 4. CSM nanoindentation Ph-curves of the Bi_2Te_3 thin films deposited on the c-plane sapphire substrates at the various T_S from 225 to 300 °C.

The hardness and Young's modulus of the Bi_2Te_3 thin films are directly determined by the Ph-curves obtained from CSM measurements following the Oliver and Pharr method [17]. The results are shown in Figure 5. Briefly, the values of hardness (H) of the Bi_2Te_3 films are 5.2 ± 0.3, 4.0 ± 0.1, and 3.4 ± 0.2 GPa for the films deposited at 225 °C, 250 °C, and 300 °C, respectively. Similarly, the values of Young's modulus (E_f) are 125.2 ± 6.9, 98.3 ± 2.1, and 62.5 ± 1.4 GPa for the Bi_2Te_3 films grown at 225 °C, 250 °C, and 300 °C, respectively. This means that both the values of H and E_f of the Bi_2Te_3 thin films monotonically decrease with increasing T_S. Noticeably, the H and E_f of the present Bi_2Te_3 thin films are significantly larger than that reported for the bulk Bi_2Te_3 ($H = 1.6 \pm 0.2$ GPa and $E = 32.4 \pm 2.9$ GPa) [22]. The reason for the apparent discrepancy is not clear at present. Nevertheless, by comparing the results of the films in this study with that reported in Ref. [23], the stoichiometric levels and crystallite orientation of the films may have intimate correlations with the H and E_f results. We found that the H and E_f values of the close stoichiometric Bi_2Te_3 films grown using the Bi_2Te_8 target are significantly larger than those of the Te-deficient Bi_2Te_3 films grown using the Bi_2Te_3 target [23]. Similar behaviors have also been reported in the Bi_2Se_3 thin films [26]. Notably, for the present Bi_2Te_3 films, the intensity of the (006)-diffraction peak is dominantly higher than that of the (0015)-diffraction peak, whereas the Bi_2Te_3 films in Ref. [23] showed reversed behavior for the intensity of the two peaks, presumably due to the different experimental conditions. We found that the mechanical properties of the PLD-grown Bi_2Te_3 films on the Al_2O_3 (001) substrates substantially were enhanced when they were grown using a Te-rich target (e.g., Bi_2Te_8) under a relatively high helium gas (e.g., 200 mTorr) pressure and at a moderately low T_S (e.g., 225 °C). Additionally, compared with the previous studies [22,23], the structural and mechanical properties of the Bi_2Te_3 thin films are listed in Table 1.

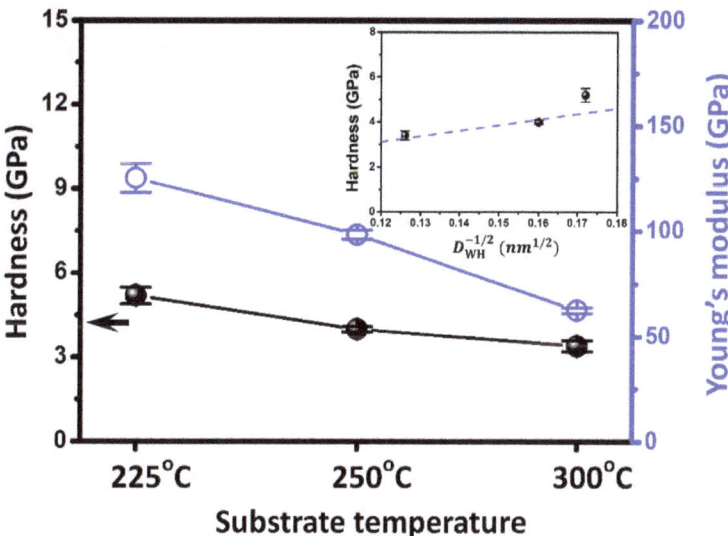

Figure 5. The hardness as a function of grain size for the Bi_2Te_3 thin films. The dash lines are the fitting using the Hall—Petch equation.

On the other hand, as displayed in the inset of Figure 5, the hardness of the films as a function of the crystallite size can be described satisfactorily by the empirical Hall—Petch relation [27], where $H(D_{WH}) = H_0 + kD_{WH}^{-1/2}$ (H_0 is the lattice friction stress and k is the Hall-Petch constant). The fitting yields $H(D_{WH}) = 24.7 D_{WH}^{-1/2} + 0.2$. Since the dislocation motion is recognized to play the primary role in giving rise to the phenomena describable by the Hall—Petch relation, it can also explain the pop-ins in the Ph-curves for the Bi_2Te_3 thin films by linking the observed pop-ins event to the abrupt plastic flow associated with the massive dislocation activities during nanoindentation (as shown in Figure 4). As is evident from the above results, higher T_S apparently has led to a microstructure with larger crystallite size and better crystallinity for the Bi_2Te_3 thin films associated with the increased T_S that could reduce the capability of hindering the dislocation movement, hence leading to decreases in the H and E_f values.

As indicated by the arrows shown in Figures 4 and 6, along the loading segment of the Ph-curves, clear discontinuities reflecting the pop-ins phenomena are observed. Such behavior, in fact, has been ubiquitously observed in single crystal [22] and thin films [23] of Bi_2Te_3, when similar nanoindentation tests were undertaken. The fact that it occurs in a vast variety of loading segments associated with a wide range of corresponding strain rates during the test indicates that the phenomena, especially the first pop-in event, are not activated thermally. Instead, the phenomena are often explained in terms of dislocation nucleation and/or propagation [28,29], or development of induced micro-cracks [30,31] during nanoindentation. The possibility of a phase pressure-induced transition, however, can be ruled out. Due to the in situ high pressure XRD experiments carried out on Bi_2Te_3 [32–34], the magnitude of the applied pressure required to induce the phase transitions is orders of magnitude higher than the apparent room-temperature hardness obtained for the present Bi_2Te_3 films. Moreover, the absence of "pop-out" discontinuities along the unloading segment of the Ph-curves (Figures 4 and 6) also support that, unlike that observed in nanoindented Si [35,36], the phase transition is not involved here. Consequently, we believe that the predominant deformation mechanism prevailing in the present case must mainly associate with the nucleation and subsequent propagation of dislocations.

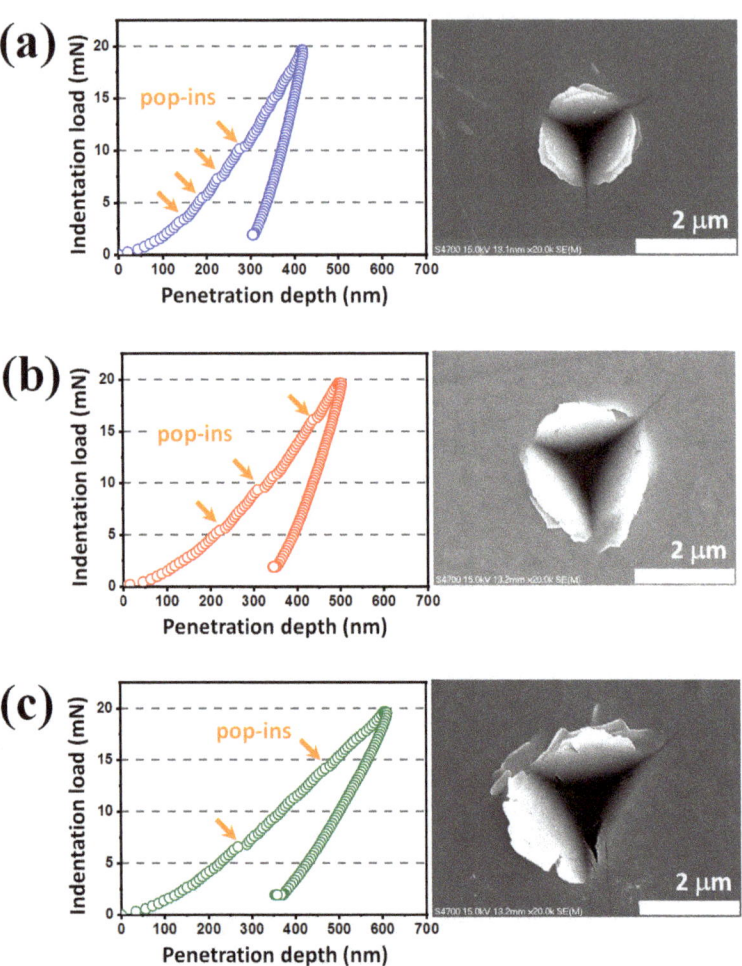

Figure 6. Berkovich nanoindentation on the Bi_2Te_3 thin films at various T_S: (**a**) 225 °C, (**b**) 250 °C, and (**c**) 300 °C. The corresponding SEM indentations are displayed to the right of the load-depth curves.

For the context of the dislocation activity-mediated scenario, the first "pop-in" event can be attributed to the transition of deformation behaviors. Namely, the pop-in is the onset of plasticity, reflecting the indentation load at which the system switches from elastic to plastic deformation due to the movement of dislocations. Based on the above discussion, one can further calculate the corresponding critical shear stress (τ_c), at which the dislocation movement is initiated using the following equation: $\tau_c = (0.31/\pi)[6P_c(E_r/R)^2]^{1/3}$ [37], where P_c is the load at which the load-depth discontinuity occurs, R is the radius of the tip of nanoindenter, and E_r is the reduced elastic modulus defined in the Materials and Method section. The obtained values for τ_c are approximately 2.2, 1.4, and 1.0 GPa for the Bi_2Te_3 films deposited at the T_S of 225, 250, and 300 °C, respectively. Alternatively, τ_c may also be regarded as the stress responsible for massive homogeneous nucleation of the dislocations within the region deformed underneath the tip.

Figure 6 shows the phenomena of Berkovich nanoindentation-induced cracking and the pile-up along the corners and edges of the residual indent clearly. Fracture toughness (K_c) is another prominent mechanical property of materials in nanoindentation, which can be determined by [38] $K_c = \alpha \cdot (P_m/c^{3/2}) \cdot (E_f/H)^{1/2}$, where α is an empirical constant depending solely on the geometry of the indenter, which is taken to be 0.016 for the Berkovich indenter, and c is the trace length of the radial crack appearing on the material surface at a maximum indentation loading (P_m) of 20 mN. The K_c values of the Bi$_2$Te$_3$ thin films thus obtained are 1.42, 1.21, and 0.88 MPa·m$^{1/2}$ for films deposited at the T_S of 225 °C, 250 °C, and 300 °C, respectively. Moreover, the fracture energy (G_c) of the Bi$_2$Te$_3$ thin films is estimated based on the elastic modulus and fracture toughness using the equation [39] $G_c = K_c^2 \cdot (1 - \nu^2 / E_f)$, where the G_c values of the Bi$_2$Te$_3$ thin films are 0.15, 0.14, and 0.12 Jm^{-2} for the films deposited at the T_S of 225 °C, 250 °C, and 300 °C, respectively. The values of Kc and Gc of the Bi$_2$Te$_3$ thin films are also listed in Table 1. Accordingly, as is evident from the SEM photographs shown in Figure 3, the cracks propagate in a straight line and exhibit a trans-granular cracking behavior, which confirms that the grain boundaries have effectively obstructed the inter-granular crack propagation.

It is worth noting that the Bi$_2$Te$_3$ film grown at the T_S of 225 °C can be considered the optimal film because of its excellent mechanical properties of $H = 5.2 \pm 0.3$ GPa and $E_f = 125.2 \pm 6.9$ GPa for thermoelectric applications [40].

4. Conclusions

In summary, the Bi$_2$Te$_3$ thin films were grown on c-plane sapphire substrates at various T_S from 225 to 300 °C under a helium ambient pressure of 200 mTorr using a Bi$_2$Te$_8$ target. The T_S dependence of the structural, morphological, compositional, and nanomechanical properties of the Bi$_2$Te$_3$ films was systematically studied using XRD, TEM, SEM, EDS, and nanoindentation methods. As a result, all the films exhibited the Bi$_2$Te$_3$ phase, highly c-axis preferred orientation, granular morphology, and good stoichiometry. Moreover, the crystallite size of the films monotonically increased with increasing T_S from 225 to 300 °C. The hardness (Young's modulus) of the Bi$_2$Te$_3$ thin films decreased from 5.2 GPa (125.2 GPa) to 3.4 GPa (62.5 GPa) when T_S increased from 225 to 300 °C. The T_S-dependent hardness and Young's modulus is associated with the variation in crystallite size, which can be explained by the dislocation-mediated mechanism underlying the Hall—Petch relation. The calculated values of K_c and G_c of the Bi$_2$Te$_3$ thin films were in the ranges of 0.88–1.42 MPa·m$^{1/2}$ and 0.12–0.15 Jm^{-2}, and their values were systematically decreased with increasing T_S.

Author Contributions: H.-P.C., P.H.L., L.T.C.T. and Y.-C.C. contributed to the experiments and analyses. P.H.L., L.T.C.T., S.-R.J., I.-J.T., C.-M.L. and J.-Y.J. contributed to the discussion on materials characterizations. P.H.L., L.T.C.T., S.-R.J. and J.-Y.J. designed the project of experiments and drafted the manuscript. All authors have read and agreed to the published version of the manuscript.

Funding: The research was funded by the Ministry of Science and Technology, Taiwan, under Contract Nos. MOST 110-2221-E-214-013, MOST 109-2112-M-009-014-MY2, and Vietnam National Foundation for Science and Technology Development (NAFOSTED) under Grant No. 103.02-2019.374.

Institutional Review Board Statement: Not applicable.

Informed Consent Statement: Not applicable.

Data Availability Statement: Not applicable.

Acknowledgments: We acknowledge the support of time and facilities from Can Tho University of Medicine and Pharmacy for this study, and thank the MANALAB at ISU.

Conflicts of Interest: The authors declare no conflict of interest.

References

1. Luo, X.; Sullivan, M.B.; Quek, S.Y. First-principles investigations of the atomic, electronic, and thermoelectric properties of equilibrium and strained Bi$_2$Se$_3$ and Bi$_2$Te$_3$ including van der Waals interactions. *Phys. Rev. B* **2012**, *86*, 184111. [CrossRef]
2. Rowe, D.M. *CRC Handbook of Thermoelectrics*; CRC Press Inc.: New York, NY, USA, 1995.
3. Mansoor, M.A.; Ismail, A.; Yahya, R.; Arifin, Z.; Tiekink, E.R.T.; Weng, N.S.; Mazhar, M.; Esmaeili, A.R. Perovskite-structured PbTiO$_3$ thin films grown from a single-source precursor. *Inorg. Chem.* **2013**, *52*, 5624–5626.

4. Munawar, K.; Mansoor, M.A.; Olmstead, M.M.; Zaharinie, T.; Zubir, M.N.M.; Haniffa, M.; Wan Jefrey Basirun, W.J.; Mazhar, M. Fabrication of Ag–ZnO composite thin films for plasmonic enhanced water splitting. *Mater. Chem. Phys.* **2020**, *255*, 123220. [CrossRef]
5. Liaqat, R.; Mansoor, M.A.; Iqbal, J.; Jilani, A.; Shakir, S.; Kalam, A.; Wageh, S. Fabrication of metal (Cu and Cr) incorporated nickel oxide films for electrochemical oxidation of methanol. *Crystals* **2021**, *11*, 1398. [CrossRef]
6. Le, P.H.; Liao, C.-N.; Luo, C.W.; Leu, J. Thermoelectric properties of nanostructured bismuth–telluride thin films grown using pulsed laser deposition. *J. Alloys Compd.* **2014**, *615*, 546–552. [CrossRef]
7. Nandihalli, N. Thermoelectric films and periodic structures and spin Seebeck effect systems: Facets of performance optimization. *Mater. Today Energy* **2022**, *25*, 100965. [CrossRef]
8. Nandihalli, N.; Liu, C.-J.; Mori, T. Polymer based thermoelectric nanocomposite materials and devices: Fabrication and characteristics. *Nano Energy* **2020**, *78*, 105186. [CrossRef]
9. Bao, L.H.; Xu, Z.H.; Li, R.; Li, X.D. Catalyst-free synthesis and structural and mechanical characterization of single crystalline $Ca_2B_2O_5 \cdot H_2O$ nanobelts and stacking faulted $Ca_2B_2O_5$ nanogrooves. *Nano Lett.* **2010**, *10*, 255–262. [CrossRef]
10. Sun, Y.; Liu, J.; Blom, D.; Koley, G.; Duan, Z.; Wang, G.; Li, X.D. Atomic-scale imaging correlation on the deformation and sensing mechanisms of SnO_2 nanowire. *Appl. Phys. Lett.* **2014**, *105*, 243105. [CrossRef]
11. Jiang, T.; Khabaz, F.; Marne, A.; Wu, C.; Gearba, R.; Bodepudi, R.; Bonnecaze, R.T.; Liechti, K.M.; Korgel, B.A. Mechanical properties of hydrogenated amorphous silicon (a-Si:H) particles. *J. Appl. Phys.* **2019**, *126*, 204303. [CrossRef]
12. Wiatrowski, A.; Obstarczyk, A.; Mazur, M.; Kaczmarek, D.; Wojcieszak, D. Characterization of HfO_2 optical coatings deposited by mf magnetron sputtering. *Coatings* **2019**, *9*, 106. [CrossRef]
13. Wang, S.H.; Jian, S.R.; Chen, G.J.; Cheng, H.Z.; Juang, J.Y. Annealing-driven microstructural evolution and its effects on the surface and nanomechanical properties of Cu-doped NiO thin films. *Coatings* **2019**, *9*, 107. [CrossRef]
14. Bayansal, F.; Şahin, O.; Çetinkara, H.A. Mechanical and structural properties of Li-doped CuO thin films deposited by the successive ionic layer adsorption and reaction method. *Thin Solid Films* **2020**, *697*, 137839. [CrossRef]
15. Li, X.D.; Bhushan, B. A review of nanoindentation continuous stiffness measurement technique and its applications. *Mater. Charact.* **2002**, *48*, 11–36. [CrossRef]
16. Sneddon, I.N. The relation between load and penetration in the axisymmetric boussinesq problem for a punch of arbitrary profile. *Int. J. Eng. Sci.* **1965**, *3*, 47–57. [CrossRef]
17. Oliver, W.C.; Pharr, G.M. An improved technique for determining hardness and elastic modulus using load and displacement sensing indentation experiments. *J. Mater. Res.* **1992**, *7*, 1564–1583. [CrossRef]
18. Chen, Y.L.; Analytis, J.G.; Chu, J.H.; Liu, Z.K.; Mo, S.K.; Qi, X.L.; Zhang, H.J.; Lu, D.H.; Dai, X.; Fang, Z.; et al. Experimental realization of a three-dimensional topological insulator, Bi_2Te_3. *Science* **2009**, *325*, 178–181. [CrossRef]
19. Cullity, B.D.; Stock, S.R. *Element of X-ray Diffraction*; Prentice Hall: Upper Saddle River, NJ, USA, 2001; p. 170.
20. Goh, K.H.; Haseeb, A.S.M.A.; Wong, Y.H. Effect of oxidation temperature on physical and electrical properties of Sm_2O_3 thin-film gate oxide on Si substrate. *J. Electron. Mater.* **2016**, *45*, 5302–5312. [CrossRef]
21. Venkatewarlu, K.; Bose, A.C.; Rameshbabu, N. X-ray peak broadening studies of nanocrystalline hydroxyapatite by Williamson–Hall analysis. *Physica B* **2010**, *405*, 4256–4261. [CrossRef]
22. Lamuta, C.; Cupolillo, A.; Politano, A.; Aliev, Z.S.; Babanly, M.B.; Chulkov, E.V.; Alfano, M.; Pagnotta, L. Nanoindentation of single-crystal Bi_2Te_3 topological insulators grown with the Bridgman–Stockbarger method. *Phys. Status Solidi B* **2016**, *253*, 1082–1086. [CrossRef]
23. Tsai, C.H.; Tseng, Y.C.; Jian, S.R.; Liao, Y.Y.; Lin, C.M.; Yang, P.F.; Chen, D.L.; Chen, H.J.; Luo, C.W.; Juang, J.Y. Nanomechanical properties of Bi_2Te_3 thin films by nanoindentation. *J. Alloys Compd.* **2015**, *619*, 834–838.
24. Tuyen, L.T.C.; Jian, S.R.; Tien, N.T.; Le, P.H. Nanomechanical and material properties of fluorine-doped tin oxide thin films prepared by ultrasonic spray pyrolysis: Effects of F-doping. *Materials* **2019**, *12*, 1665. [CrossRef] [PubMed]
25. Li, X.D.; Gao, H.S.; Murphy, C.J.; Gou, L.F. Nanoindentation of Cu_2O nanocubes. *Nano Lett.* **2004**, *4*, 1903. [CrossRef]
26. Hwang, Y.M.; Pan, C.T.; Le, P.H.; Uyen, N.N.; Tuyen, L.T.C.; Nguyen, V.; Luo, C.W.; Juang, J.Y.; Leu, J.; Jian, S.R. Effects of stoichiometry on structural, morphological and nanomechanical properties of Bi_2Se_3 thin films deposited on InP(111) substrates by pulsed laser deposition. *Coatings* **2020**, *10*, 958. [CrossRef]
27. Greer, J.R.; De Hosson, J.T.M. Plasticity in small-sized metallic systems: Intrinsic versus extrinsic size effect. *Prog. Mater. Sci.* **2011**, *56*, 654–724. [CrossRef]
28. Jian, S.R.; Juang, J.Y.; Lai, Y.S. Cross-sectional transmission electron microscopy observations of structural damage in $Al_{0.16}Ga_{0.84}N$ thin film under contact loading. *J. Appl. Phys.* **2008**, *103*, 033503. [CrossRef]
29. Jian, S.R. Cathodoluminescence rosettes in c-plane GaN films under Berkovich nanoindentation. *Opt. Mater.* **2013**, *35*, 2707–2709. [CrossRef]
30. Yen, C.Y.; Jian, S.R.; Lai, Y.S.; Yang, P.F.; Liao, Y.Y.; Jang, J.S.C.; Lin, T.H.; Juang, J.Y. Mechanical properties of the hexagonal $HoMnO_3$ thin films by nanoindentation. *J. Alloys Compd.* **2010**, *508*, 523–527. [CrossRef]
31. Smolik, J.; Kacprzyńska-Gołacka, J.; Sowa, S.; Piasek, A. The analysis of resistance to brittle cracking of tungsten doped TiB_2 coatings obtained by magnetron sputtering. *Coatings* **2020**, *10*, 807. [CrossRef]
32. Zhu, L.; Wang, H.; Wang, Y.; Lv, J.; Ma, Y.; Cui, Q.; Ma, Y.; Zou, G. Substitutional alloy of Bi and Te at high pressure. *Phys. Rev. Lett.* **2011**, *106*, 145501. [CrossRef]

33. Einaga, M.; Ohmura, A.; Nakayama, A.; Ishikawa, F.; Yamada, Y.; Nakano, S. Pressure-induced phase transition of Bi_2Te_3 to a bcc structure. *Phys. Rev. B* **2011**, *83*, 092102. [CrossRef]
34. Manjón, F.J.; Vilaplana, R.; Gomis, O.; Pérez-González, E.; Santamaría-Pérez, D.; Marín-Borrás, V.; Segura, A.; González, J.; Rodríguez-Hernández, P.; Muñoz, A.; et al. High-pressure studies of topological insulators Bi_2Se_3, Bi_2Te_3, and Sb_2Te_3. *Phys. Status Solidi B* **2013**, *250*, 669–676. [CrossRef]
35. Bradby, J.E.; Williams, J.S.; Wong-Leung, J.; Swain, M.V.; Munroe, P. Transmission electron microscopy observation of deformation microstructure under spherical indentation in silicon. *Appl. Phys. Lett.* **2000**, *77*, 3749. [CrossRef]
36. Jian, S.R.; Chen, G.J.; Juang, J.Y. Nanoindentation-induced phase transformation in (1 1 0)-oriented Si single-crystals. *Curr. Opin. Solid State Mater. Sci.* **2010**, *14*, 69–74. [CrossRef]
37. Johnson, K.L. *Contact Mechanics*; Cambridge University Press: Cambridge, UK, 1985.
38. Casellas, D.; Caro, J.; Molas, S.; Prado, J.M.; Valls, I. Fracture toughness of carbides in tool steels evaluated by nanoindentation. *Acta Mater.* **2007**, *55*, 4277–4286. [CrossRef]
39. Rafiee, M.A.; Rafiee, J.; Srivastava, I.; Wang, Z.; Song, H.; Yu, Z.Z.; Koratkar, N. Fracture and Fatigue in Graphene Nanocomposites. *Small* **2010**, *6*, 179–183. [CrossRef]
40. Le, P.H.; Liao, C.N.; Luo, C.W.; Lin, J.Y.; Leu, J. Thermoelectric properties of bismuth-selenide films with controlled morphology and texture grown using pulsed laser deposition. *Appl. Surf. Sci.* **2013**, *285*, 657–663. [CrossRef]

Article

Effects of Stoichiometry on Structural, Morphological and Nanomechanical Properties of Bi₂Se₃ Thin Films Deposited on InP(111) Substrates by Pulsed Laser Deposition

Yeong-Maw Hwang [1], Cheng-Tang Pan [1], Bo-Syun Chen [1], Phuoc Huu Le [2,*], Ngo Ngoc Uyen [2], Le Thi Cam Tuyen [3], Vanthan Nguyen [4], Chih-Wei Luo [5], Jenh-Yih Juang [5], Jihperng Leu [6] and Sheng-Rui Jian [7,8,*]

[1] Department of Mechanical and Electro-Mechanical Engineering, National Sun Yat-Sen University, Kaohsiung 804, Taiwan; ymhwang@mail.nsysu.edu.tw (Y.-M.H.); pan@mem.nsysu.edu.tw (C.-T.P.); bosyun815@gmail.com (B.-S.C.)
[2] Department of Physics and Biophysics, Faculty of Basic Sciences, Can Tho University of Medicine and Pharmacy, 179 Nguyen Van Cu Street, Can Tho City 94000, Vietnam; nnuyen@ctump.edu.vn
[3] Faculty of Basic Sciences, Nam Can Tho University, 168 Nguyen Van Cu (Ext) Street, Can Tho City 94000, Vietnam; ltctuyen89@gmail.com
[4] Department of Electrical and Electronic Engineering, Faculty of Automotive Engineering, Ngo Quyen University, Thu Dau Mot City 820000, Vietnam; nguyenvanthan1010@gmail.com
[5] Department of Electrophysics, National Chiao Tung University, Hsinchu 300, Taiwan; cwluo@mail.nctu.edu.tw (C.-W.L.); jyjuang@g2.nctu.edu.tw (J.-Y.J.)
[6] Department of Materials Science and Engineering, National Chiao Tung University, Hsinchu 300, Taiwan; jimleu@mail.nctu.edu.tw
[7] Department of Materials Science and Engineering, I-Shou University, Kaohsiung 840, Taiwan
[8] Department of Fragrance and Cosmetic Science, Kaohsiung Medical University, 100 Shin-Chuan 1st Road, Kaohsiung 80782, Taiwan
* Correspondence: lhuuphuoc@ctump.edu.vn (P.H.L.); srjian@gmail.com (S.-R.J.);
Tel.: +886-7-657-7711 (ext. 3130) (S.-R.J.)

Received: 7 August 2020; Accepted: 30 September 2020; Published: 5 October 2020

Abstract: In the present study, the structural, morphological, compositional, nanomechanical, and surface wetting properties of Bi_2Se_3 thin films prepared using a stoichiometric Bi_2Se_3 target and a Se-rich Bi_2Se_5 target are investigated. The Bi_2Se_3 films were grown on InP(111) substrates by using pulsed laser deposition. X-ray diffraction results revealed that all the as-grown thin films exhibited were highly c-axis-oriented Bi_2Se_3 phase with slight shift in diffraction angles, presumably due to slight stoichiometry changes. The energy dispersive X-ray spectroscopy analyses indicated that the Se-rich target gives rise to a nearly stoichiometric Bi_2Se_3 films, while the stoichiometric target only resulted in Se-deficient and Bi-rich films. Atomic force microscopy images showed that the films' surfaces mainly consist of triangular pyramids with step-and-terrace structures with average roughness, R_a, being ~2.41 nm and ~1.65 nm for films grown with Bi_2Se_3 and Bi_2Se_5 targets, respectively. The hardness (Young's modulus) of the Bi_2Se_3 thin films grown from the Bi_2Se_3 and Bi_2Se_5 targets were 5.4 GPa (110.2 GPa) and 10.3 GPa (186.5 GPa), respectively. The contact angle measurements of water droplets gave the results that the contact angle (surface energy) of the Bi_2Se_3 films obtained from the Bi_2Se_3 and Bi_2Se_5 targets were 80° (21.4 mJ/m²) and 110° (11.9 mJ/m²), respectively.

Keywords: Bi_2Se_3 thin films; nanoindentation; hardness; pop-in; surface energy

1. Introduction

Bismuth selenide (Bi_2Se_3) is of great interest owing to its intriguing physical properties as a three-dimensional topological insulator [1–5], and potential applications in spintronics [6], optoelectronics [7] and quantum computation [8]. In addition, Bi_2Se_3 possesses excellent thermoelectric properties at room-temperature [9,10] and low temperature regimes [11]. For fundamental studies and application purposes, it is essential to grow Bi_2Se_3 thin films with high-quality and to have comprehensive characterizations of their physical properties, including the mechanical properties [12–14].

Nanoindentation is a versatile technique ubiquitously used to obtain the basic mechanical parameters, such as the hardness and elastic modulus, as well as to delineate the deformation mechanisms, creep and fracture behaviors of various nanostructured materials [15–18] and thin films [19–23] with very high sensitivity and excellent resolution. On the other hand, wettability is an important property of a solid surface, which is intimately related to the chemical compositions and morphology of the surface [24]. The peculiar wetting behaviors exhibited on the surface of two-dimensional and van der Waals layered materials have been receiving dramatically increased interest in recent years [25–27]. It implies that specific water–substrate interaction features are relevant to the atomic and electronic structures of the layered materials. In particular, the hydrophobic surface (water contact angle, $\theta_{CA} > 90°$) can be used in many applications of self-cleaning surfaces and antifogging [28,29]. Consequently, how to control the behavior of hydrophobicity or hydrophilicity of films' surfaces is also of great importance in realizing the designed functionality for device applications.

Because of the high volatility of selenium (Se), Bi_2Se_3 tends to form Se vacancies or antisites that serve as donors to result in a sufficiently high carrier concentration and low carrier mobility [30,31]. When severe loss of Se-atoms occurs during the thin-film growth at elevated substrate temperatures, pure phase Bi_2Se_3 film is usually not achieved, and the obtained films may present impurity phases or even turn into another phase [32]. Thus, to overcome this problem and obtain high-quality stoichiometric Bi_2Se_3 thin-films, a Se-rich environment is necessary during films' growth. Indeed, this strategy has been employed to grow high-quality Bi_2Se_3 thin films by creating a Se-rich environment with a Se:Bi flux ratio ranging from 10:1 to 20:1 using molecular beam epitaxy (MBE) [33,34]. Pulsed laser deposition (PLD) offers a high instantaneous deposition rate, relatively high reproducibility, and low costs. The PLD has been used for growing epitaxial and polycrystalline Bi_2Se_3 thin films [9,30,35–37]. In 2011, Onose et al. [35] successfully grew epitaxial Bi_2Se_3 thin-films on InP(111) substrates using a designed target with an atomic ratio of Bi:Se of 2:8. Yet, systematic investigations on the effects of target composition, and hence the resultant films' stoichiometry, on the properties of Bi_2Se_3 thin films have been relatively scarce.

Herein, we conducted comprehensive characterizations of the structural, compositional, morphological, nanomechanical, and wetting properties of Bi_2Se_3 thin films grown on InP(111) substrates by PLD. In particular, two different targets (i.e., a stoichiometric target of Bi_2Se_3 and a Se-rich target of Bi_2Se_5) were deliberately used to tune the stoichiometry of the resultant Bi_2Se_3 films and to unveil its effects on the surface wettability and nanomechanical properties, since both characteristics are of pivotal importance for their practical applications in Bi_2Se_3 thin film-based microelectronic and spintronic devices.

2. Materials and Methods

In order to study the effects of film stoichiometry, two targets with different composition effects were used. One is stoichiometric Bi_2Se_3 and another is a Se-rich target with a nominal composition of Bi_2Se_5. The targets were purchased from Ultimate Materials Technology Co., Ltd. (Ping-Tung City, Taiwan). Noticeably, though having differences in Se/Bi atomic ratios of 3/2 and 5/2, both Bi_2Se_3 and Bi_2Se_5 targets were polycrystalline with the right Bi_2Se_3 phase. Bi_2Se_3 thin films were deposited on InP(111) substrates using PLD at a substrate temperature of 350 °C in vacuum at a base pressure of 4×10^{-6} Torr (~0.53 mPa). For the PLD process, ultraviolet (UV) pulses (20-ns duration) from a KrF excimer laser (λ = 248 nm, repetition: 1 Hz) were focused on the polycrystalline Bi_2Se_3 or Bi_2Se_5 target at a fluence of 5.5 J/cm². The target-to-substrate distance was 40 mm. The target was ablated for approximately 5 min in order to clean its surface before every deposition. The deposition time was 25 min, which resulted in an average Bi_2Se_3 film thickness of approximately 191 nm (the growth rate of approximately 1.27 Å/pulse).

The crystal structure and surface morphology of the Bi_2Se_3 thin films were characterized by X-ray diffraction (XRD; Bruker D8, CuKα radiation, λ = 1.5406 Å, Bruker, Billerica, MA, USA) and field emission scanning electron microscopy (SEM, JEOL JSM-6500, JEOL, Pleasanton, CA, USA) operated at an accelerating voltage of 15 kV, respectively. Film compositions were analyzed through Oxford energy-dispersive X-ray spectroscopy (EDS, Inca X-sight 7558, Oxford Instruments plc., Oxfordshire, UK) equipped with the SEM instrument at an accelerating voltage of 15 kV, dead time of 22–30%, and collection time of 60 s. The atomic percentage of each film was determined by averaging the values measured in 5 or more distinct 14 × 20 μm² areas on the surface of films. Moreover, the surface morphology and roughness of the thin films were examined using atomic force microscopy (AFM; Veeco Escope, Veeco, New York, USA).

The nanoindentation was performed on a Nanoindenter MTS NanoXP® system (MTS Cooperation, Nano Instruments Innovation Center, Oak Ridge, TN, USA) with a pyramid-shaped Berkovich diamond tip. The nanomechanical properties of the Bi_2Se_3 thin films were measured by nanoindentation with a continuous contact stiffness mode (CSM) [38]. At least 20 indentations were performed on each sample and the distance between the adjacent indents was kept at least 10 μm apart to avoid mutual interferences. We also followed the analytic method proposed by Oliver and Pharr [39] to determine the hardness and Young's modulus of measured materials from the load–displacement results. Thus, the hardness (H) and Young's modulus (E) of the Bi_2Se_3 thin films are obtained and the results are listed in Table 1. Moreover, the surface wettability of the Bi_2Se_3 thin films under ambient conditions was monitored using a Ramehart Model 200 contact angle goniometer (Ramé-hart, Succasunna, NJ, USA) with deionized water as the liquid.

Table 1. The microstructural parameters, nanomechanical properties, contact angle and surface energy of Bi_2Se_3 thin films. The mechanical properties of InP(111) are also listed.

Sample	D (nm)	R_a (nm)	H (GPa)	E (GPa)	τ_{max} (GPa)	θ_{CA}	$(\gamma^d)_s$ (mJ/m²)
Bi_2Se_3 thin film on InP(111) substrate (Bi_2Se_3 target)	29.7	2.41	5.4	110.2	1.8	80°	21.4
Bi_2Se_3 thin film on InP(111) substrate (Bi_2Se_5 target)	26.0	1.65	10.3	186.5	3.4	110°	11.9
Bi_2Se_3 thin film on sapphire substrate [14]	34.2	8.5	~2.1	~58.6	~0.7	—	—
Single-crystal Bi_2Se_3 [13]	—	—	~0.4–0.9	~2–9	—	—	—
Single-crystal InP(111) [40]	—	—	~5	72.4–76.2	1.96	—	—

3. Results and Discussion

3.1. Structural and Morphological Properties

Bi_2Se_3 has a rhombohedral structure with a space group $D_{3d}^5(R\overline{3}m)$ that can be described by a hexagonal primitive cell with three five-atomic-layer thick lamellae of $-(Se^{(1)}-Bi-Se^{(2)}-Bi-Se^{(1)})-$, in which the atomic layers are stacked in sequence along the c-axis [9]. The XRD patterns of the Bi_2Se_3 thin films obtained from the Bi_2Se_3 and Bi_2Se_5 targets are shown in Figure 1. As is evident from Figure 1, besides the diffraction peaks of InP substrates at 26.3° and 54.1° (JCPDS PDF#00-032-0452), the films exhibited highly c-axis-preferred orientation with (006), (0015), and (0021) diffraction peaks of the Bi_2Se_3 phase (JCPDS PDF#33-0214). However, minor diffraction peaks belonging to the BiSe phase (PDF#29-0246) can be identified. It is noticed that, although both of the as-grown films exhibit highly c-axis preferred orientation of the Bi_2Se_3 phase, a slight relative shift in diffraction angles indicative of modification of the c-axis parameter is observed. Indeed, by using the dominant Bi_2Se_3 (006) and Bi_2Se_3 (0015) peaks and the hexagonal unit cell relationship [32], the average c-axis lattice constant of the Bi$_2$Se$_3$ thin films prepared using Bi$_2$Se$_3$ and Bi$_2$Se$_5$ targets were 28.39 Å and 28.25 Å, respectively, whose values were slightly smaller the c-axis lattice constant of 28.63 Å from the database of Bi$_2$Se$_3$ powder (JCPDS PDF#33-0214). This could be due to the difference in the internal stress built up during the deposition.

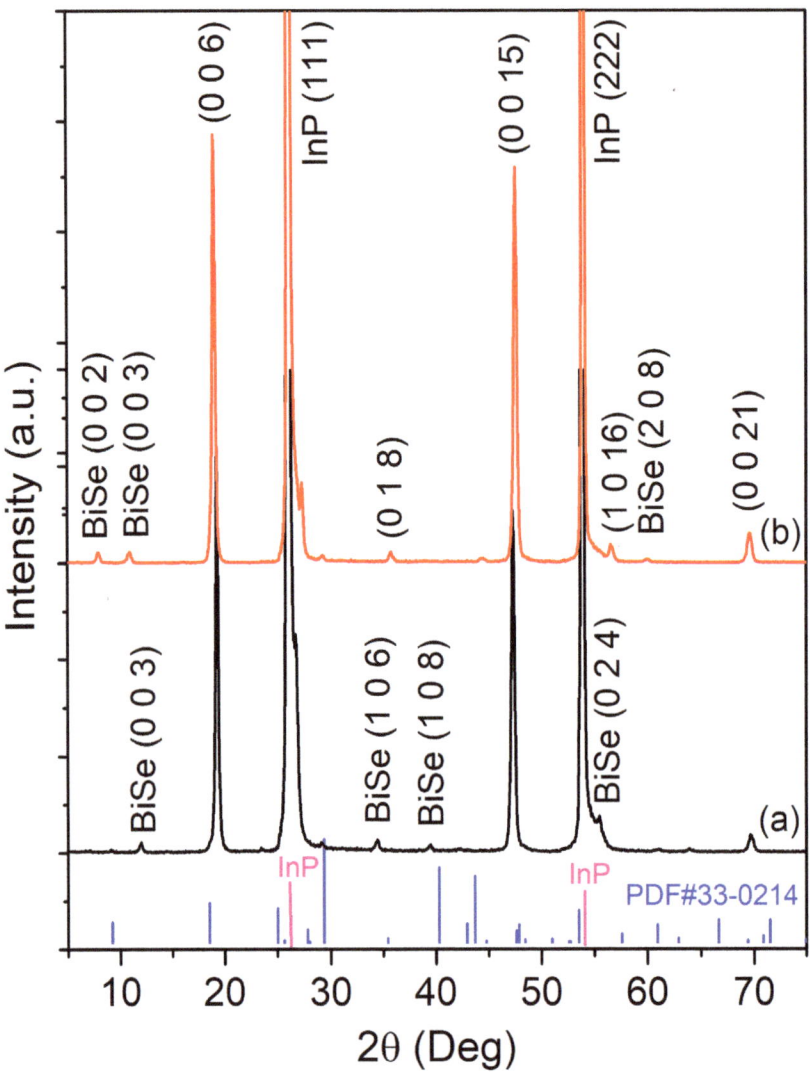

Figure 1. XRD patterns of Bi$_2$Se$_3$ thin films grown on InP (111) substrates from two different targets of Bi$_2$Se$_3$ (**a**) and Bi$_2$Se$_5$ (**b**) using pulsed laser deposition.

The grain sizes (D) of the Bi$_2$Se$_3$ films were estimated using the Scherrer equation $D = 0.9\lambda/\beta\cos\theta$, where λ, β, and θ are the X-ray wavelength, full width at half maximum of the Bi$_2$Se$_3$ (006)-oriented peak, and Bragg diffraction angle, respectively. The estimated D values of the Bi$_2$Se$_3$ thin films prepared using Bi$_2$Se$_3$ target and Bi$_2$Se$_5$ target were 29.7 nm and 26.0 nm, respectively.

Figure 2 shows the AFM and SEM-EDS results of Bi$_2$Se$_3$ thin films prepared using the Bi$_2$Se$_3$ and Bi$_2$Se$_5$ target, respectively. As shown in Figure 2a,b, the films mainly consist of triangular pyramids with features of step-and-terrace structures. This is a clear indication that the films are growing along the [0001] direction, which is consistent with XRD results displayed in Figure 1. The films also exhibit highly smooth surfaces with the centerline average roughness R_a being ~2.41 nm and ~1.65 nm for films grown from the Bi_2Se_3 target and from the Bi_2Se_5 target, respectively. In addition, the films grown from the Bi_2Se_5 target also show clearer step-and-terrace structures with fewer large particle-like outgrowth defects on the surface as compared to the film grown from the Bi_2Se_3 target (see 3D images), indicating that these films are closer to the stoichiometric composition and, thus, are less defective.

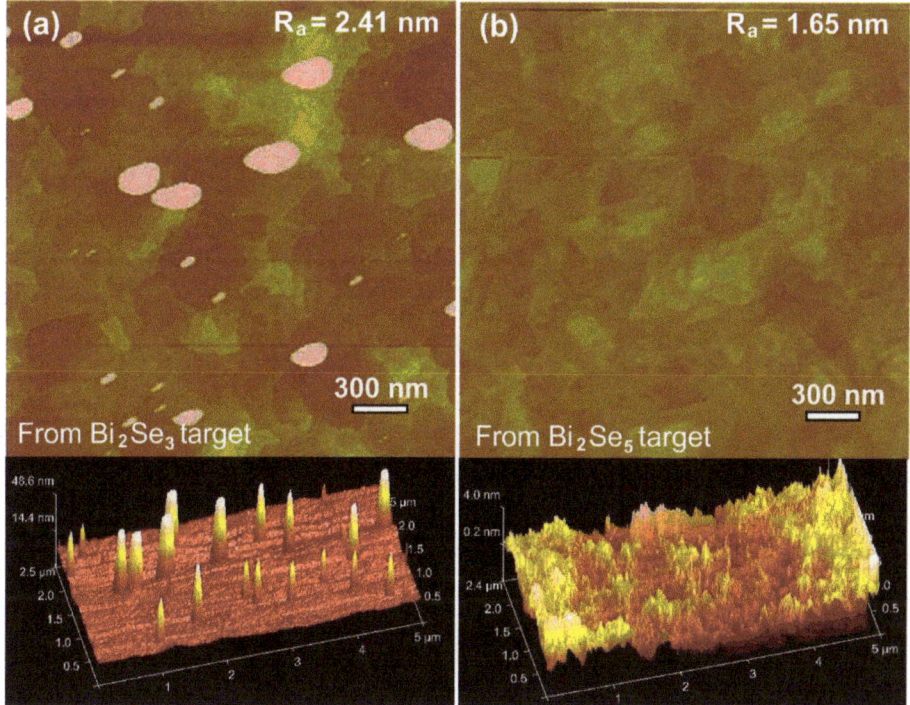

Figure 2. 2D and 3D AFM images of the Bi_2Se_3 thin films deposited from (**a**) Bi_2Se_3 target and (**b**) Bi_2Se_5 target.

The top-view SEM images displayed in Figure 3a,b further confirmed the aforementioned surface morphology. The cross-sectional view images shown at the bottom of Figure 3a,b indicate that the films are rather uniform with their thickness being in the range of 185~197 nm. Furthermore, as is evident from the EDS results displayed in the insets of Figure 3a,b and the typical EDS spectra of the corresponding thin films shown in Figure 3c, the composition of the film prepared from the Bi_2Se_3 target clearly showed a substantial Se-deficiency of about 4.4 at.%, while the film prepared from the Bi_2Se_5 target is nearly stoichiometric, which is consistent with the conjectures discussed above. Intuitively, it is rather straightforward to explain why the Bi_2Se_3 target would lead to Bi-rich (or Se-deficient) film by recognizing that the re-evaporation of Se from the heated substrate (~350 °C) is much faster than Bi owing the much higher vapor pressure of Se [9,41]. The present results also suggest that to obtain stoichiometric Bi_2Se_3 films, a Se-excessive target is essential. We note that stoichiometric Bi_2Se_3 and Bi_2Te_3 films have been shown to exhibit reduced carrier concentration and increased carrier mobility, which led to the enhanced thermoelectric properties and provided suitable conditions for investigating the topological surface states [9,30,42].

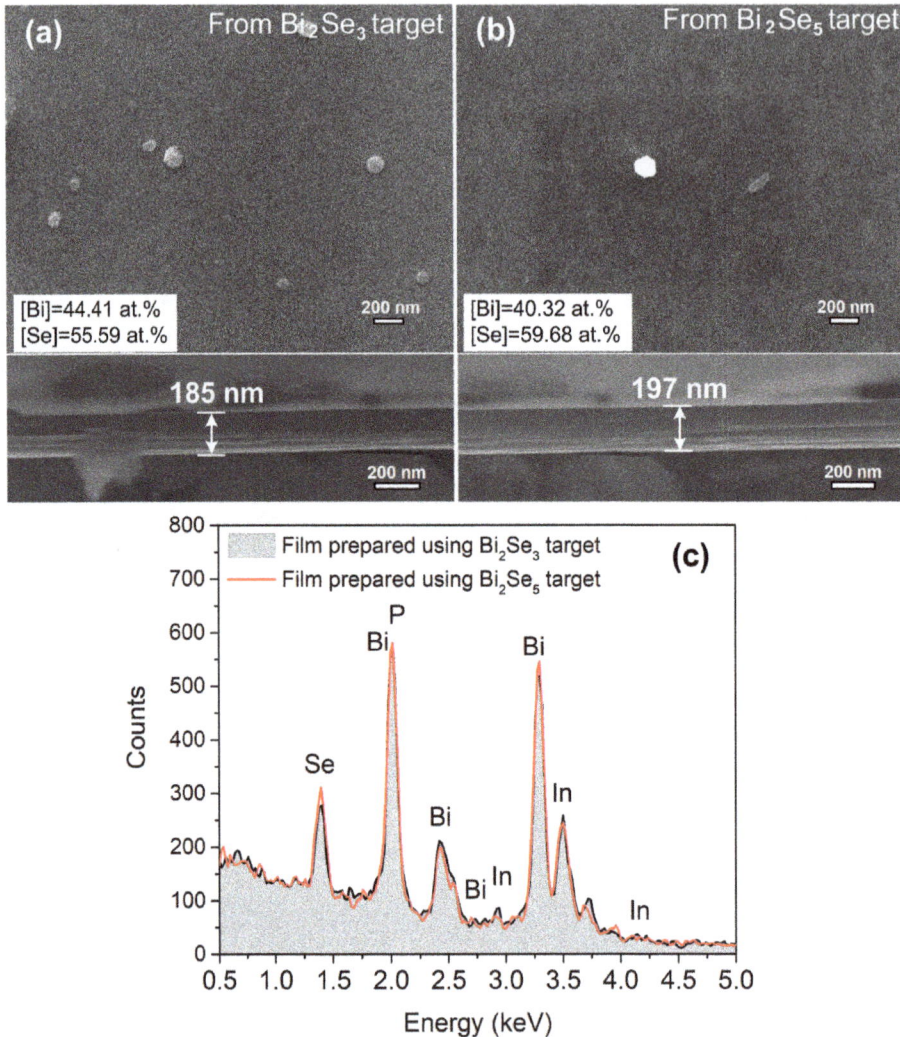

Figure 3. Top-view and cross-sectional SEM images of the Bi$_2$Se$_3$ thin films deposited from (**a**) Bi$_2$Se$_3$ target and (**b**) Bi$_2$Se$_5$ target. (**c**) EDS spectra of the corresponding Bi$_2$Se$_3$ thin films.

3.2. Nanomechanical Properties

The typical nanoindentation load–displacement curves of Bi$_2$Se$_3$ thin film deposited on InP(111) substrates are shown in Figure 4a. The hardness and Young's modulus of Bi$_2$Se$_3$ thin films were calculated from the load–displacement curves [39]; the Poisson's ratio of Bi$_2$Se$_3$ films is set to 0.25 in this study. Figure 4b,c present the penetration depth dependence of hardness and Young's modulus are obtained using the CSM method. In 2004, Li et al. [15] indicated that nanoindentation depth should never exceed 30% of the film's thickness. In this work, the CSM technique system is applied to record stiffness data along with load and displacement data dynamically, making it possible to calculate the hardness and Young's modulus at every data point and get their average values during the indentation experiment [15,39]. The mechanical properties obtained under nanoindentation exhibit a convergent manner and are steady with a rational tolerance around penetrating depths of 40~60nm, reflecting that the material properties obtained are intrinsic and the substrate effect on the present thin films for hardness and modulus tests is negligible. The obtained values of hardness (*H*) and Young's modulus (*E*) are listed in Table 1 together with those reported in the literature for Bi$_2$Se$_3$ single crystals and thin films deposited on sapphire substrates.

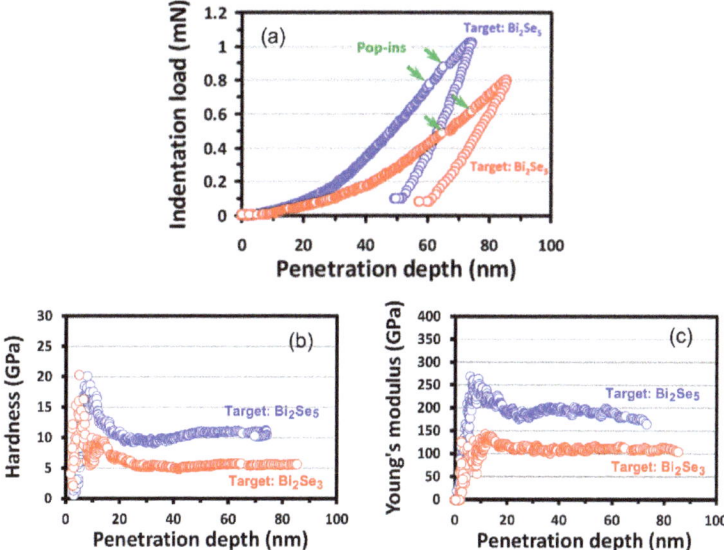

Figure 4. (**a**) The load–displacement curves of Bi$_2$Se$_3$ thin films deposited on InP(111) substrates using two different target compositions. A clear single "pop-in" behavior is displayed in both curves during loading. (**b**) A hardness—displacement curve and (**c**) a Young's modulus–displacement curve for a Bi$_2$Se$_3$ thin films deposited using Bi$_2$Se$_3$ and Bi$_2$Se$_5$ targets.

From Table 1, it is somewhat surprising to observe that the values of hardness and Young's modulus of the Bi$_2$Se$_3$ thin films are much larger than those of single crystals. The reason for this peculiar observation, especially the very low values for single crystals, is not clear at present. However, by comparing the results for films, the two prominent mechanical property parameters appear to have intimate correlations with the grain size (*D*) and surface roughness (*R$_a$*). For films grown on InP(111) substrate, as in the present case, the lattice mismatch between the Bi$_2$Se$_3$ thin films and substrate is about 0.2% [35], which, in turn, consistently resulted in films with better crystallinity, as indicated by the narrower full width at half maximum of the diffraction peaks, namely ~0.3° for films grown on InP(111) as compared to that of ~0.5° for the films grown on sapphire substrate [14]. Moreover, when comparing the results for the films grown with different targets, it further indicates that stoichiometry of the film can play an even more prominent role in determining the mechanical properties. Namely, the hardness and Young's modulus of the stoichiometric Bi$_2$Se$_3$ thin films are both about two times larger than that of Se-deficient films, which are again about two times larger than that grown on sapphire substrate. The enhancement of *H* and *E* values can be explained by considering the film crystallinity and surface roughness. It has been reported that the crystallinity of Bi$_2$Se$_3$ thin films deposited on InP(111) substrate was better than those deposited on Al$_2$O$_3$ and Si substrates [35]. In general, better film crystallinity often results in superior nanomechanical properties [43,44]. Therefore, compared with those reported in [14], the larger values of hardness and Young's modulus of the present Bi$_2$Se$_3$ thin films could be attributed to their better crystallinity. Furthermore, the film surface roughness can also be an important factor. Jian et al. [45] reported that the nanomechanical properties of ZnO thin films were significantly enhanced as the film surfaces became smoother. Even for AISI 316L stainless steel, the mechanical properties were found to decrease with increasing surface roughness [46]. Since the surface roughness of the present films are all below 2.41 nm, it is reasonable to account, at least partially, for the enhanced H and E values.

Turning to the deformation behaviors during nanoindentation, it is evident that there are several pop-ins occurring along the loading segment for both load–displacement curves shown in Figure 4a. It is noted that similar phenomena were found in the previous studies [13,14], where the pop-ins were also observed in nanoindented Bi_2Se_3 single-crystal and thin films, despite the fact that the loads at which the pop-ins took place varied in each individual measurement. Moreover, it is noted that there is no sign of reverse discontinuity in the unloading portion of the load–displacement curves (the so-called "pop-out" event) being observed. The reverse discontinuity is commonly ascribed to the pressure-induced phase transformation that has been observed in Si or Ge single crystals [47,48]. The absence of these incidences indicates that pressure-induced phase transition did not occur for the Bi_2Se_3 films in the pressure range applied in this study. In fact, Yu et al. [49] have reported that the pressure-induced phase transition in Bi_2Se_3 occurred at pressures of 35.6 and 81.2 GPa as revealed, respectively, by Raman spectroscopy and synchrotron XRD experiments conducted in a diamond anvil cell. These values are much higher than the room-temperature hardness of the present hexagonal Bi_2Se_3 thin films. On the other hand, the pop-in behaviors during nanoindentation have been reported previously in other hexagonal structured materials, such as sapphire [50] and ZnO single crystals [51], as well as GaN thin films [52–54] by using the Berkovich indenter tip. It is generally conceived that the nanoindentation-induced deformation mechanism in these hexagonal-structured materials were primarily dominated by the nucleation and/or propagation of dislocations. Thus, it is plausible to believe that similar mechanisms must have been prevailing in the present Bi_2Se_3 thin films. Reasonably, it can be seen from Table 1 that the hardness of Bi_2Se_3 thin films increases when D value decreases, partially due to grain boundary hardening.

Within the context of the dislocation-mediated deformation scenarios, the first pop-in event may reflect the transition from perfectly elastic to plastic deformation. Namely, it is the onset of plasticity in Bi_2Se_3 thin films. Under this circumstance, the corresponding critical shear stress (τ_{max}) under the Berkovich indenter at an indentation load, P_c, where the load–displacement discontinuity occurs, can be determined by using the following relation [55]:

$$\tau_{max} = 0.31 \left(\frac{6 P_c E^2}{\pi^3 R^2} \right)^{1/3} \tag{1}$$

where R is the radius of the tip of nanoindenter. The obtained τ_{max} values are 1.8 and 3.4 GPa for Bi_2Se_3 thin films grown using Bi_2Se_3 and Bi_2Se_5 targets, respectively. The τ_{max} is responsible for the homogeneous dislocation nucleation within the deformation region underneath the indenter tip.

3.3. Wettability Behavior

The surface wettability of the Bi_2Se_3 thin films was examined by water contact angle measurements. If the contact angle (θ_{CA}) is greater than 90°, it is said to be hydrophobic, otherwise it is hydrophilic. In Figure 5, the values of θ_{CA} for films are 80° and 110° for films grown using the Bi_2Se_3 target and the Bi_2Se_5 target, respectively.

As described above, the surface roughness measured by the AFM indicated that the Bi_2Se_3 thin film grown using the Bi_2Se_5 target have smaller surface roughness, suggesting that the wettability behavior of the surface was significantly affected by the surface morphology of the films [56]. Alternatively, the atomic arrangements and existence of surface defects might also play a role in the eventual surface energy. In general, the surface wettability is a measurement of surface energy and is most commonly quantified by θ_{CA} [57]. The surface energy for Bi_2Se_3 thin films was calculated by means of the Fowkes–Girifalco–Good (FGG) theory [58]. According to the analysis of the FGG method, the considered critical interaction is the dispersive force or the van der Waals force across the interface existing between the water droplet and the solid surface. The FGG equation is given as:

$$\gamma_{ls} = \gamma_s + \gamma_l - 2\sqrt{(\gamma^d)_s + (\gamma^d)_l} \qquad (2)$$

where $(\gamma^d)_s$ and $(\gamma^d)_l$ are the dispersive portions of surface tension for the solid and liquid surfaces, respectively. By combining Young's equation [56] with Equation (2) and taking the nonpolar liquid deionized water as the testing liquid and set $(\gamma^d)_l = \gamma_l$, the Girifalco–Good–Fowkes–Young equation becomes as: $(\gamma^d)_s = \gamma_l(cos\theta_{CA} + 1)/4$, where $(\gamma^d)_s$ is the surface energy of measured materials. Using γ_l = 72.8 mJ/m^2, the values of surface energy obtained were 21.4 mJ/m^2 and 11.9 mJ/m^2 for films grown with the Bi$_2$Se$_3$ target and Bi$_2$Se$_5$ target, respectively. The lower surface energy gives rise to higher hydrophobicity. It is noted that the θ_{CA} of 110° for the present stoichiometric Bi$_2$Se$_3$ thin films deposited on InP(111) substrates using PLD is even larger than that (θ_{CA}~98.4°) of Bi$_2$Se$_3$ thin films deposited on SrTiO$_3$(111) substrate by MBE [59]. In any case, the present study suggests that both the hydrophobic/hydrophilic transition behavior and nanomechanical properties of the Bi$_2$Se$_3$ thin films can be manipulated by controlling the target compositions.

Figure 5. Contact angle test: the images of water droplets on the Bi$_2$Se$_3$ thin film surfaces.

4. Conclusions

The present study evidently illustrated that stoichiometry, which can be manipulated by tuning the target composition, can give rise to significant effects on the microstructural, morphological, compositional, nanomechanical and surface wetting properties of the Bi$_2$Se$_3$/InP (111) thin films. The Bi$_2$Se$_3$ thin films were grown using PLD from a stoichiometric Bi$_2$Se$_3$ target and a Se-rich Bi$_2$Se$_5$ target at a substrate temperature of 350 °C in a vacuum with a base pressure of ~4 × 10^{-6} Torr. The films were highly (00l)-oriented with smooth surfaces consisting mainly of triangular step-and-terrace structures, which is the common feature of epitaxial Bi$_2$Se$_3$ thin films. Compared to the films grown from the Bi$_2$Se$_3$ target, using the Bi$_2$Se$_5$ target is more favorable for obtaining stoichiometric films with larger hardness and Young's modulus. In addition, the contact angle (surface energy) of the Bi$_2$Se$_3$ films deposited from the Bi$_2$Se$_3$ and Bi$_2$Se$_5$ targets were 80° (21.4 mJ/m^2) and 110° (11.9 mJ/m^2), respectively. These results suggest that, in addition to the usual factors such as surface roughness and grain morphology, stoichiometry as well as defect chemistry originated from Se-deficiency may also play important roles in determining the eventual nanomechanical and wettability properties of Bi$_2$Se$_3$ thin films.

Author Contributions: Data curation, Y.-M.H., C.-T.P., B.-S.C., P.H.L., L.T.C.T., N.N.U. and V.N.; Formal analysis, Y.-M.H., C.-T.P., B.-S.C., L.T.C.T., N.N.U. and V.N.; Funding acquisition, J.-Y.J.; Resources, P.H.L., C.-W.L., J.-Y.J.,

J.L. and S.-R.J.; Writing—original draft, P.H.L. and S.-R.J.; Writing—review & editing, J.-Y.J. All authors have read and agreed to the published version of the manuscript.

Funding: This research was funded by the Ministry of Science and Technology, Taiwan under Contract Nos. MOST 109-2221-E-214-016.

Acknowledgments: The authors would like to thank T.-C. Lin for her technical support in the nanoindentation experiments.

Conflicts of Interest: The authors declare no conflict of interest.

References

1. Zhang, H.; Liu, C.X.; Qi, X.L.; Dai, X.; Fang, Z.; Zhang, S.C. Topological insulators in Bi_2Se_3, Bi_2Te_3 and Sb_2Te_3 with a single Dirac cone on the surface. *Nat. Phys.* **2009**, *5*, 438–442. [CrossRef]
2. Moore, J.E. The birth of topological insulators. *Nature* **2010**, *464*, 194–198. [CrossRef]
3. Xia, Y.; Qian, D.; Hsieh, D.; Wray, L.; Pal, A.; Lin, H.; Bansil, A.; Grauer, D.; Hor, Y.S.; Cava, R.J.; et al. Observation of a large-gap topological-insulator class with a single Dirac cone on the surface. *Nat. Phys.* **2009**, *5*, 398–402. [CrossRef]
4. Wiedmann, S.; Jost, A.; Fauqué, B.; van Dijk, J.; Meijer, M.J.; Khouri, T.; Pezzini, S.; Grauer, S.; Schreyeck, S.; Brüne, C.; et al. Anisotropic and strong negative magnetoresistance in the three-dimensional topological insulator Bi_2Se_3. *Phys. Rev. B* **2016**, *94*, 081302. [CrossRef]
5. Le, P.H.; Luo, C.W. *Ultrafast Dynamics in Topological Insulators*; InTech: London, UK, 2018.
6. Yazyev, O.V.; Moore, J.E.; Louie, S.G. Spin polarization and transport of surface states in the topological insulators Bi_2Se_3 and Bi_2Te_3 from first principles. *Phys. Rev. Lett.* **2010**, *105*, 266806. [CrossRef] [PubMed]
7. Min, W.L.; Betancourt, A.P.; Jiang, P.; Jiang, B. Bioinspired broadband antireflection coatings on GaSb. *Appl. Phys. Lett.* **2008**, *92*, 141109. [CrossRef]
8. Qi, X.L.; Zhang, S.C. Topological insulators and superconductors. *Rev. Mod. Phys.* **2011**, *83*, 1057–1110. [CrossRef]
9. Le, P.H.; Liao, C.N.; Luo, C.W.; Lin, J.Y.; Leu, J. Thermoelectric properties of bismuth-selenide films with controlled morphology and texture grown using pulsed laser deposition. *Appl. Surf. Sci.* **2013**, *285*, 657. [CrossRef]
10. Le, P.H.; Luo, C.W. *Thermoelectric and Topological Insulator Bismuth Chalcogenide Thin Films Grown Using Pulsed Laser Deposition*; InTech: London, UK, 2016.
11. Hor, Y.; Richardella, A.; Roushan, P.; Xia, Y.; Checkelsky, J.; Yazdani, A.; Hasan, M.; Ong, N.; Cava, R. p-type Bi_2Se_3 for topological insulator and low-temperature thermoelectric applications. *Phys. Rev. B* **2009**, *79*, 195208. [CrossRef]
12. Wang, E.; Ding, H.; Fedorov, A.V.; Yao, W.; Li, Z.; Lv, Y.F.; Zhao, K.; Zhang, L.G.; Xu, Z.; Schneeloch, J.; et al. Fully gapped topological surface states in Bi_2Se_3 films induced by a d-wave high-temperature superconductor. *Nat. Phys.* **2013**, *9*, 621–625. [CrossRef]
13. Gupta, S.; Vijayan, N.; Krishna, A.; Thukral, K.; Maurya, K.K.; Muthiah, S.; Dhar, A.; Singh, B.; Bhagavannarayana, G. Enhancement of thermoelectric figure of merit in Bi_2Se_3 crystals through a necking process. *J. Appl. Crystallogr.* **2015**, *48*, 533–541. [CrossRef]
14. Lai, H.D.; Jian, S.R.; Tuyen, L.T.C.; Le, P.H.; Luo, C.W.; Juang, J.Y. Nanoindentation of Bi_2Se_3 Thin Films. *Micromachines* **2018**, *9*, 518. [CrossRef] [PubMed]
15. Li, X.D.; Gao, H.; Murphy, C.J.; Gou, L. Nanoindentation of Cu_2O nanocubes. *Nano Lett.* **2004**, *4*, 1903–1907. [CrossRef]
16. Bao, L.; Xu, Z.H.; Li, R.; Li, X. Catalyst-free synthesis and structural and mechanical characterization of single crystalline $Ca_2B_2O_5 \cdot H_2O$ nanobelts and stacking faulted $Ca_2B_2O_5$ nanogrooves. *Nano Lett.* **2010**, *10*, 255–262. [CrossRef]
17. Jiang, Y.; Hor, J.L.; Lee, D.; Turner, K.T. Toughening nanoparticle films via polymer infiltration and confinement. *ACS Appl. Mater. Interfaces* **2018**, *10*, 44011–44017. [CrossRef]
18. Ma, Y.; Huang, X.; Song, Y.; Hang, W.; Zhang, T. Room-temperature creep behavior and activation volume of dislocation nucleation in a $LiTaO_3$ single crystal by nanoindentation. *Materials* **2019**, *12*, 1683. [CrossRef]
19. Zaman, A.; Meletis, E.I. Microstructure and mechanical properties of TaN thin films prepared by reactive magnetron sputtering. *Coatings* **2017**, *7*, 209. [CrossRef]

20. Azizpour, A.; Hahn, R.; Klimashin, F.F.; Wojcik, T.; Poursaeidi, E.; Mayrhofer, P.H. Deformation and cracking mechanisms in CrN/TiN multilayer coatings. *Coatings* **2019**, *9*, 363. [CrossRef]
21. Wiatrowski, A.; Obstarczyk, A.; Mazur, M.; Kaczmarek, D.; Wojcieszak, D. Characterization of HfO_2 optical coatings deposited by MF magnetron sputtering. *Coatings* **2019**, *9*, 106. [CrossRef]
22. Tuyen, L.T.C.; Jian, S.R.; Tien, N.T.; Le, P.H. Nanomechanical and material properties of Fluorine-doped tin oxide thin films prepared by ultrasonic spray pyrolysis: Effects of F-doping. *Materials* **2019**, *12*, 1665. [CrossRef]
23. Hwang, Y.M.; Pan, C.T.; Lu, Y.X.; Jian, S.R.; Chang, H.W.; Juang, J.Y. Influence of post-annealing on the structural and nanomechanical properties of Co thin films. *Micromachines* **2020**, *11*, 180. [CrossRef] [PubMed]
24. Quere, D. Wetting and Roughness. *Ann. Rev. Mater. Res.* **2008**, *38*, 71–99. [CrossRef]
25. Rafiee, J.; Mi, X.; Gullapalli, H.; Thomas, A.V.; Yavari, F.; Shi, Y.; Ajayan, P.M.; Koratkar, N.A. Wetting transparency of graphene. *Nat. Mater.* **2012**, *11*, 217–222. [CrossRef] [PubMed]
26. Gaur, A.P.S.; Sahoo, S.; Ahmadi, M.; Dash, S.P.; Guinel, M.J.-F.; Katiyar, R.S. Surface energy engineering for tunable wettability through controlled synthesis of MoS_2. *Nano Lett.* **2014**, *14*, 4314–4321. [CrossRef] [PubMed]
27. Chow, P.K.; Singh, E.; Viana, B.C.; Gao, J.; Luo, J.; Li, J.; Lin, Z.; Arriaga, L.E.; Shi, Y.; Wang, Z.; et al. Wetting of mono and few-layered WS_2 and MoS_2 films supported on Si/SiO_2 substrates. *ACS Nano* **2015**, *9*, 30023–30031. [CrossRef] [PubMed]
28. Fujishima, A.; Rao, T.N.; Tryk, D.A. Titanium dioxide photocatalysis. *J. Photochem. Photobiol. C Photochem. Rev.* **2000**, *1*, 1–21. [CrossRef]
29. Guo, Z.; Liu, W.; Su, B.L. Superhydrophobic surfaces: From natural to biomimetic to functional. *J. Colloid Interface Sci.* **2011**, *353*, 335–355. [CrossRef]
30. Le, P.H.; Wu, K.H.; Luo, C.W.; Leu, J. Growth and characterization of topological insulator Bi_2Se_3 thin films on $SrTiO_3$ using pulsed laser deposition. *Thin Solid Film.* **2013**, *534*, 659–665. [CrossRef]
31. Richardella, A.; Zhang, D.M.; Lee, J.S.; Koser, A.; Rench, D.W.; Yeats, A.L.; Buckley, B.B.; Awschalom, D.D.; Samarth, N. Coherent heteroepitaxy of Bi_2Se_3 on GaAs (111)B. *Appl. Phys. Lett.* **2010**, *97*, 262104. [CrossRef]
32. Tuyen, L.T.C.; Le, P.H.; Luo, C.W.; Leu, J. Thermoelectric properties of nanocrystalline Bi_3Se_2Te thin films grown using pulsed laser deposition. *J. Alloy. Compd.* **2016**, *673*, 107–114. [CrossRef]
33. Wang, Z.Y.; Guo, X.; Li, H.D.; Wong, T.L.; Wang, N.; Xie, M.H. Superlattices of Bi_2Se_3/In_2Se_3: Growth characteristics and structural properties. *Appl. Phys. Lett.* **2011**, *99*, 023112. [CrossRef]
34. Taskin, A.A.; Sasaki, S.; Segawa, K.; Ando, Y. Manifestation of topological protection in transport properties of epitaxial Bi_2Se_3 thin films. *Phys. Rev. Lett.* **2012**, *109*, 066803. [CrossRef] [PubMed]
35. Onose, Y.; Yoshimi, R.; Tsukazaki, A.; Yuan, H.; Hidaka, T.; Iwasa, Y.; Kawasaki, M.; Tokura, Y. Pulsed laser deposition and ionic liquid gate control of epitaxial Bi_2Se_3 thin films. *Appl. Phys. Express.* **2011**, *4*, 083001. [CrossRef]
36. Lee, Y.F.; Punugupati, S.; Wu, F.; Jin, Z.; Narayan, J.; Schwartz, J. Evidence for topological surface states in epitaxial Bi_2Se_3 thin film grown by pulsed laser deposition through magneto-transport measurements. *Curr. Opin. Solid State Mater. Sci.* **2014**, *18*, 279–285. [CrossRef]
37. Orgiani, P.; Bigi, C.; Das, P.K.; Fujii, J.; Ciancio, R.; Gobaut, B.; Galdi, A.; Sacco, C.; Maritato, L.; Torelli, P.; et al. Structural and electronic properties of Bi_2Se_3 topological insulator thin films grown by pulsed laser deposition. *Appl. Phys. Lett.* **2017**, *110*, 171601. [CrossRef]
38. Li, X.; Bhushan, B. A review of nanoindentation continuous stiffness measurement technique and its applications. *Mater. Charact.* **2002**, *48*, 11–36. [CrossRef]
39. Oliver, W.C.; Pharr, G.M. An improved technique for determining hardness and elastic modulus using load and displacement sensing indentation experiments. *J. Mater. Res.* **1992**, *7*, 1564–1583. [CrossRef]
40. Chrobak, D.; Chrobak, A.; Nowak, R. Effect of doping on nanoindentation induced incipient plasticity in InP crystal. *AIP Adv.* **2019**, *9*, 125323. [CrossRef]
41. Noro, H.; Sato, K.; Kagechika, H. The thermoelectric properties and crystallography of Bi-Sb-Te-Se thin films grown by ion beam sputtering. *J. Appl. Phys.* **1993**, *73*, 1252–1260. [CrossRef]
42. Le, P.H.; Liao, C.N.; Luo, C.W.; Lin, J.Y.; Leu, J. Thermoelectric properties of nanostructured bismuth–telluride thin films grown using pulsed laser deposition. *J. Alloy. Compd.* **2014**, *615*, 546–552. [CrossRef]

43. Wang, S.H.; Jian, S.R.; Chen, G.J.; Cheng, H.Z.; Juang, J.Y. Annealing-driven microstructural evoluation and its effects on the surface and nanomechanical properties of Cu-doped NiO thin films. *Coatings* **2019**, *9*, 107. [CrossRef]
44. Lou, B.S.; Moirangthem, I.; Lee, J.W. Fabrication of tungsten nitride thin films by superimposed HiPIMS and MF system: Effects of nitrogen flow rate. *Surf. Coat. Technol.* **2020**, *393*, 125743. [CrossRef]
45. Jian, S.R.; Teng, I.J.; Yang, P.F.; Lai, Y.S.; Lu, J.M.; Chang, J.G.; Ju, S.P. Surface morphological and nanomechanical properties of PLD-derived ZnO thin films. *Nanoscale Res. Lett.* **2008**, *3*, 186–193. [CrossRef]
46. Walter, C.; Antretter, T.; Daniel, R.; Mitterer, C. Finite element simulation of the effect of surface roughness on nanoindentation of thin films with spherical indenters. *Surf. Coat. Technol.* **2007**, *202*, 1103–1107. [CrossRef]
47. Jian, S.R.; Chen, G.J.; Juang, J.Y. Nanoindentation-induced phase transformation in (11 0)-oriented Si single-crystals. *Curr. Opin. Solid State Mater. Sci.* **2010**, *14*, 69–74. [CrossRef]
48. Jang, J.; Lance, M.J.; Wen, S.; Pharr, G.M. Evidence for nanoindentation-induced phase transformations in germanium. *Appl. Phys. Lett.* **2009**, *86*, 131907. [CrossRef]
49. Yu, Z.; Wang, L.; Hu, Q.; Zhao, J.; Yan, S.; Yang, K.; Sinogeikin, S.; Gu, G.; Mao, H.-K. Structural phase transitions in Bi_2Se_3 under high pressure. *Sci. Rep.* **2015**, *5*, 1–9. [CrossRef]
50. Mao, W.G.; Shen, Y.G.; Lu, C. Nanoscale elastic-plastic deformation and stress distributions of the C plane of sapphire single crystal during nanoindentation. *J. Eur. Ceram. Soc.* **2011**, *31*, 1865–1871. [CrossRef]
51. Jian, S.R. Pop-in effects and dislocation nucleation of c-plane single-crystal ZnO by Berkovich nanoindentation. *J. Alloy. Compd.* **2015**, *644*, 54–58. [CrossRef]
52. Chien, C.H.; Jian, S.R.; Wang, C.T.; Juang, J.Y.; Huang, J.C.; Lai, Y.S. Cross-sectional transmission electron microscopy observations on the Berkovich indentation-induced deformation microstructures in GaN thin films. *J. Phys. D Appl. Phys.* **2007**, *40*, 3985. [CrossRef]
53. Jian, S.R. Mechanical deformation induced in Si and GaN under Berkovich nanoindentation. *Nanoscale Res. Lett.* **2008**, *3*, 6–13. [CrossRef]
54. Jian, S.R. Cathodoluminescence rosettes in c-plane GaN films under Berkovich nanoindentation. *Opt. Mater.* **2013**, *35*, 2707–2709. [CrossRef]
55. Johnson, K.L. *Contact Mechanics*; Cambridge University Press: Cambridge, UK, 1985.
56. Ottone, C.; Lamberti, A.; Fontana, M.; Cauda, V. Wetting Behavior of Hierarchical Oxide Nanostructures: TiO_2 Nanotubes from Anodic Oxidation Decorated with ZnO Nanostructures. *J. Electrochem. Soc.* **2014**, *164*, D484. [CrossRef]
57. Angelo, M.S.; McCandless, B.E.; Birkmire, R.W.; Rykov, S.A.; Chen, J.G. Contact wetting angle as a characterization for processing CdTe/CdS solar cells. *Pro. Photovolt.* **2007**, *15*, 93–111. [CrossRef]
58. Mahadik, D.B.; Rao, A.V.; Parale, V.G.; Kavale, M.S.; Wagh, P.B.; Ingale, S.V.; Gupta, S.C. Effect of surface composition and roughness on the apparent surface free energy of silica aerogel materials. *Appl. Phys. Lett.* **2011**, *99*, 104104. [CrossRef]
59. Zhao, P.; Huang, Y.; Shen, Y.; Yang, S.; Chen, L.; Wu, K.; Li, H.; Meng, S. A modified Wenzel model for water wetting on van der Waals layered materials with topographic surfaces. *Nanoscale* **2017**, *9*, 3843–3849. [CrossRef] [PubMed]

© 2020 by the authors. Licensee MDPI, Basel, Switzerland. This article is an open access article distributed under the terms and conditions of the Creative Commons Attribution (CC BY) license (http://creativecommons.org/licenses/by/4.0/).

Article

Finite Element Analysis of Nanoindentation Responses in Bi$_2$Se$_3$ Thin Films

Shu-Wei Cheng [1,†], Bo-Syun Chen [2,†], Sheng-Rui Jian [1,3,4,*], Yu-Min Hu [3,*], Phuoc Huu Le [5,*], Le Thi Cam Tuyen [6], Jyh-Wei Lee [7] and Jenh-Yih Juang [8]

[1] Department of Materials Science and Engineering, I-Shou University, Kaohsiung 84001, Taiwan
[2] Department of Mechanical and Electro-Mechanical Engineering, National Sun Yat-sen University, Kaohsiung 80424, Taiwan
[3] Department of Applied Physics, National University of Kaohsiung, Kaohsiung 81148, Taiwan
[4] Department of Fragrance and Cosmetic Science, College of Pharmacy, Kaohsiung Medical University, Kaohsiung 80708, Taiwan
[5] Department of Physics and Biophysics, Faculty of Basic Sciences, Can Tho University of Medicine and Pharmacy, Can Tho City 94000, Vietnam
[6] Department of Chemical Engineering, College of Engineering Technology, Can Tho University, Can Tho City 900000, Vietnam
[7] Department of Materials Engineering, Ming Chi University of Technology, New Taipei City 24301, Taiwan
[8] Department of Electrophysics, National Yang Ming Chiao Tung University, Hsinchu 30010, Taiwan
* Correspondence: srjian@gmail.com (S.-R.J.); ymhu@nuk.edu.tw (Y.-M.H.); lhuuphuoc@ctump.edu.vn (P.H.L.)
† These authors contributed equally to this work.

Abstract: In this study, the nanoindentation responses of Bi$_2$Se$_3$ thin film were quantitatively analyzed and simulated by using the finite element method (FEM). The hardness and Young's modulus of Bi$_2$Se$_3$ thin films were experimentally determined using the continuous contact stiffness measurements option built into a Berkovich nanoindenter. Concurrently, FEM was conducted to establish a model describing the contact mechanics at the film/substrate interface, which was then used to reproduce the nanoindentation load-depth and hardness-depth curves. As such, the appropriate material parameters were obtained by correlating the FEM results with the corresponding experimental load-displacement curves. Moreover, the detailed nanoindentation-induced stress distribution in the vicinity around the interface of Bi$_2$Se$_3$ thin film and c-plane sapphires was mapped by FEM simulation for three different indenters, namely, the Berkovich, spherical and flat punch indenters. The results indicated that the nanoindentation-induced stress distribution at the film/substrate interface is indeed strongly dependent on the indenter's geometric shape.

Keywords: Bi$_2$Se$_3$ thin film; nanoindentation; finite element method

1. Introduction

With its unique quintuple layer structure, Bi$_2$Se$_3$ behaves like a narrow band gap semiconductor with excellent thermoelectric properties near room temperature, as well as a 3D topological insulator with a large bulk band gap (0.3 eV) and topologically protected surface state [1,2]. Such rich emergent physical properties automatically invite tremendous research interest due to its potential applications in a wide range of next generation devices [3,4]. However, while most of the research has been focusing on the thermoelectric [5] and transport properties [6], research on the mechanical characterizations has not drawn equal attention. Since for most device fabrication processes, contact-induced damage may significantly affect the properties of the films upon which devices are made, which in turn would substantially influence the performance of the devices, a comprehensive understanding of the mechanical properties of Bi$_2$Se$_3$, especially how the film reacts when under localized compressive stress, is indispensable for fabricating efficient and endurable devices.

Nanoindentation is a popular technique being widely adopted to obtain prominent mechanical property parameters, such as the Young's modulus and hardness, with various materials, especially for film/substrate systems [7–10]. However, nanoindentation itself does not provide information about the mechanism of nanoindentation-induced deformation mechanisms and stress distribution within the film/substrate system in a direct manner, which, from a practical point of view, is more relevant to stress-induced deteriorations during device processing. In this respect, finite element modeling (FEM) might serve in a complementary role not only in revealing the nanomechanical properties of thin films [11,12], but also in unveiling the stress distribution within the film and at the interface during nanoindentation [13,14], or even in explaining the crack formation and delamination phenomenon [15,16]. Nevertheless, simulating the nanoindentation process is often a highly complicated task due to the nonlinear behavior of the process. For instance, Lichinchi et al. [17] reported the results of combining FEM simulations with nanoindentation using a Berkovich indenter tip on TiN thin film deposited on high-speed steel and concluded that no apparent differences are observed between the experimental load-displacement (P-h) curves and those obtained from FEM using the 2D axisymmetric model with a conical indenter and/or the 3D pyramidal model. The results suggest the feasibility of combining FEM simulation and actual indentation measurements to extract the prominent parameters for a more detailed understanding of the contact-induced mechanistic behaviors.

In this study, a combination of experiments and 2D axisymmetric FEM analysis on the nanoindentation responses of Bi_2Se_3 thin films deposited on c-plane sapphires is investigated. The nanomechanical properties, e.g., hardness, Young's modulus, as well as the P-h curves, are experimentally measured. The FEM analysis is carried out to simulate the experimentally measured P-h and hardness-depth curves. Moreover, the effects of the indenters' geometries, for Berkovich, flat punch and spherical indenters, on the interfacial stress distribution of the Bi_2Se_3 thin film/c-plane sapphire system during the nanoindentation processes are also discussed. The fact that the present FEM simulations are able to replicate the main features of actual nanoindentation experiments evidently validates the feasibility of reliably developing the mechanical deformation of Bi_2Se_3 thin films deposited on c-plane sapphire substrate using FEM simulation, which has been largely missing in previous investigations. Moreover, the comparisons performed on indenters with various geometries may also provide an efficient means to evaluate the contact-induced deformation encountered in practical device fabrication processes, wherein the shape of the contact tip is often case dependent.

2. Materials and Methods

The Bi_2Se_3 thin films investigated in this work were grown on c-plane sapphire substrates by using pulsed laser deposition method with an average thickness of about 360 nm. The details of growth procedures in preparing these Bi_2Se_3 thin films can be found elsewhere [7].

The nanoindentation measurements were carried out on a Nanoindenter MTS NanoXP® system (MTS Cooperation, Nano Instruments Innovation Center, Oak Ridge, TN, USA) with a diamond pyramid-shaped Berkovich-type indenter tip, whose radius of curvature is 50 nm. The measurements were performed with a continuous stiffness mode [18] and a constant nominal strain rate of 0.05 s^{-1}. The hardness (H) and Young's modulus (E) of Bi_2Se_3 thin films, calculated from the P-h curves based on the analytic method developed by Oliver and Pharr [19], are about 1.8 GPa and 70 GPa, respectively.

To investigate the nanoindentation-induced deformation behaviors of Bi_2Se_3 thin films on the c-plane sapphire substrate, the film–substrate structure was modelled by FEM. In particular, not only the P-h curves, but also the distribution of stress fields, strain fields, and the profile of the indentation are also analyzed using FEM under nanoindentation. In this model, the diameter and thickness of the sapphire substrate are both assumed to be 10,000 nm and the thickness of a Bi_2Se_3 thin film is taken as 360 nm. FEM simulations are made with the 2D axisymmetric conical indenter, which is equivalent to the Berkovich indenter, and the indentation curves are evaluated. The actual indenter was constructed from a diamond with a height of 2200 nm. To define an axis symmetrical model, an equivalent conical indenter with a semi apical angle of 70.3° having the same contact area as the Berkovich indenter was used [20]. In addition, simulations with flat punch and spherical indenters are also included for comparison, as depicted schematically in Figure 1. The geometrical dimensions of the flat punch indenter tip are: 1200 nm in height and 1600 nm in diameter [21]. For the spherical indenter, the radius of the tip is 1500 nm [22].

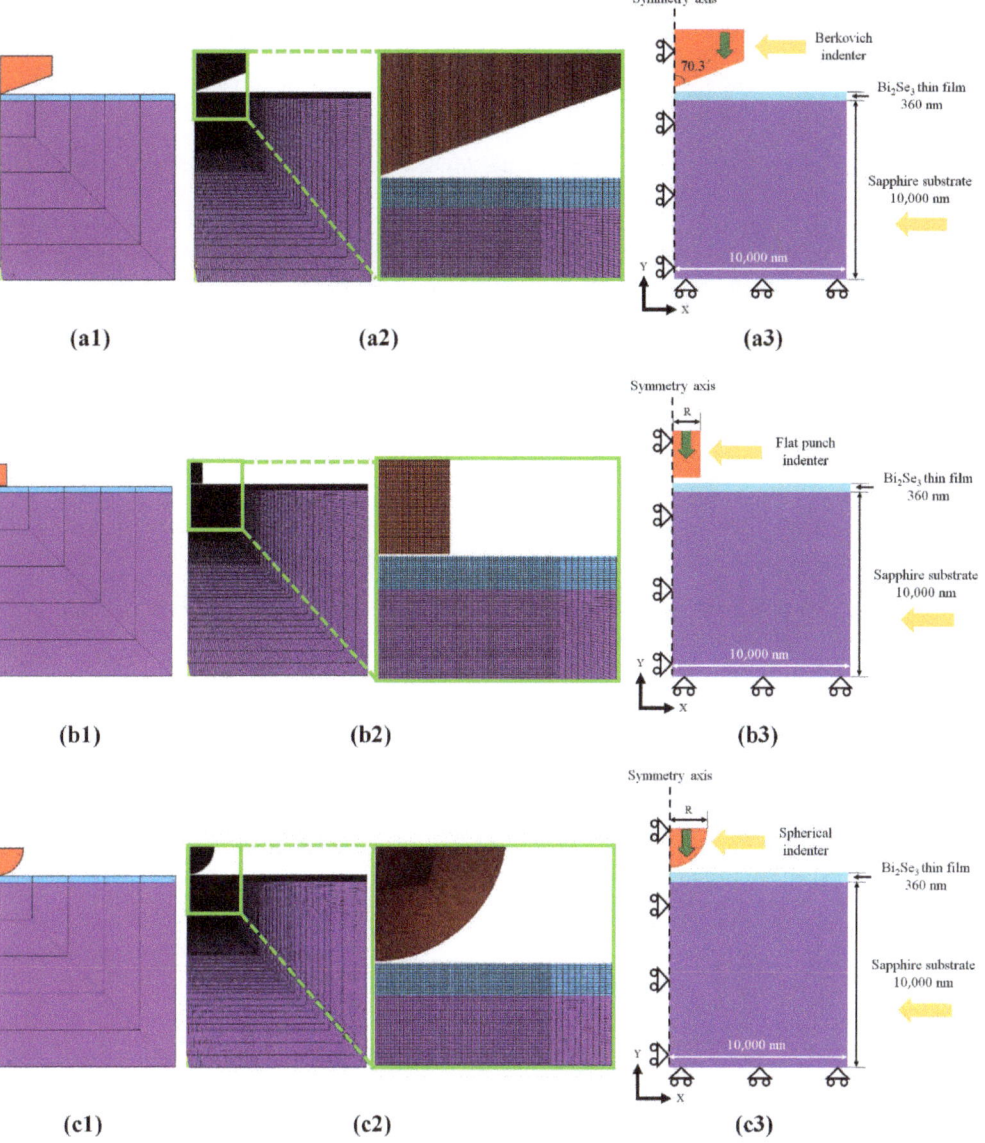

Figure 1. The entire specimen with (**a1**) Berkovich, (**b1**) flat punch and (**c1**) spherical indenters for the 2D model before meshing. Mesh of the entire specimen with (**a2**) Berkovich, (**b2**) flat punch and (**c2**) spherical indenters for the 2D model. The geometry, boundary, and load conditions of (**a3**) Berkovich, (**b3**) flat punch and (**c3**) spherical indenters.

By using an applied force of 0.12 mN, the maximum depth can be obtained for the analysis of mesh convergence. The displacement is measured along the Y-axis in Figure 2. The result is the displacement of the center of the sample, which is used to verify the maximum depth of the experiment. When the boundary conditions apply the same load, the increase of the mesh does not affect the maximum displacement, which means that the number of meshes in the model reaches a convergent state. A high-density mesh is performed in this critical area to confirm the accuracy of the simulation analysis. From the maximum curve, it is known that the number of elements is more than 20,000 after reaching the convergence state, as shown in Figure 2. We increase the mesh density near the contact point of the indenter tip and the film to observe the changes in the vicinity of this point, as shown in Figure 1(a2,b2,c2). The model uses a mapped mesh, not a free mesh, in order to be able to control the number and quality of elements. Material properties of Bi_2Se_3 used in the model were obtained from the experimental data. The simplest relationship between hardness (H) and yield stress (σ_y) is approximated as $H \approx 3\sigma_y$ according to Gupta et al. [14] and is used as the presumed value. Material properties of the indenter and substrate required for modeling are listed in Table 1. It is noted that the bi-linear model requires the Young's modulus to represent the elastic phase and the tangent modulus to describe the plastic phase. Nevertheless, from the nanoindentation experiments, we can only obtain the Young's modulus and hardness, not the tangent modulus, for the simulation input. Because of the lack of the tangent modulus, our simulations became sort of elastic perfect-plastic, although our material model was not only considering the linear flexible part. This is believed to be the primary reason why the obtained stress contours were not homogeneous (see below).

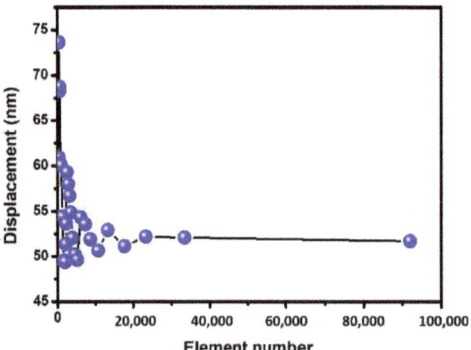

Figure 2. The element number convergence analysis, indicating the mesh independence above element number of 20,000.

Table 1. Mechanical properties of diamond indenter, Bi_2Se_3 film and sapphire substrate.

Component	Materials	Properties
Indenter	Diamond [19]	Elastic modulus: 1140 GPa Poisson's ratio: 0.07
Thin film	Bi_2Se_3 [this work]	Elastic modulus: 70 GPa Poisson's ratio: 0.25 Initial yield stress: 0.6 GPa
Substrate	Sapphire [23]	Elastic modulus: 450 GPa Poisson's ratio: 0.23 Initial yield stress: 8.7 GPa

Additional assumptions considered for the simulation are outlined as follows:
(a) The material is considered completely homogeneous, isotropic, and defect-free.

(b) The residual stress in the deposited film is assumed to be negligible.
(c) During the simulation the material undergoes elastic deformation perfectly prior to the set in of plastic deformation.
(d) The contact between film and substrate is considered as perfectly bonded so that there is no delamination during the indentation process and the contact between film and the indenter is considered to be frictionless.

By using the 2D models, the plastic deformation behaviors of Bi_2Se_3 thin film on the sapphire substrate during nanoindentation can be observed. Since the system is considered to be symmetrical, only half of the test vehicle is displayed. The 2D axis symmetry finite element model with Berkovich (33,568 elements and 100,824 nodes), flat punch (18,628 elements and 55,848 nodes) and spherical (26,689 elements and 80,131 nodes) indenters are shown in Figure 1(a2, b2,c2), respectively. The PLANE 183 element type was used. The Bi_2Se_3 thin films, sapphire substrate and pressure indenter employ a quadrilateral element, each defined by eight nodes. The four-node quadrilateral element is used, and high mesh refinement is adopted in the vicinity of the indent because of relatively larger deformation. The contact elements CONTA 172/TARGE 169 were used in the contact area. The interactions between the indenter and specimen are modeled as a contact pair with no friction. The behavior of the indenter, which is usually made of high-hardness and high-yield stress materials such as diamond, is assumed to be rigid. A finer mesh is adopted to increase the numerical accuracy. However, this also implies that much larger computational time is needed. Thus, one needs to perform a convergence study to get a mesh that balances the accuracy and available computing resources. We did an element convergence analysis for the element model plotted in Figure 2. To improve the accuracy of the modeling and save computational time, it is important to check the mesh density and refine them accordingly. The deformation is primarily concentrated underneath the indenter and around the indented region. Thus, denser meshes around the region are used, as shown in Figure 1(a2,b2,c2). A static analysis including large deflection was performed using the commercial finite element software package ANSYS. In the constraint condition, the Y axis of the model is the axis of symmetry, and the nodes are set to be fixed in the X direction. In particular, the nodes on the bottom of the sample are set to be fixed in the Y direction. The load was simulated by applying a displacement to the indenter during the nanoindentation simulation, as indicated by the green arrows shown in Figure 1(a3,b3,c3).

The elastic/plastic properties of thin films have been successfully studied by FEM simulation and applied to various material systems [24,25]. Based on these methods, we propose the detailed procedures described below for simulating the material system. During loading, the indenter is controlled by displacement or force and the indenter is pressed to a certain maximum displacement or loading force. During unloading, the indenter tip returns to its original position at the same rate. Here, we used displacement control patterns to characterize indentation behavior. Both loading and unloading separate 5 nm at a step depth to ensure stable convergence. The P-h curve is obtained by the displacement of the vertical reaction force and the rigid indenter. Both the film and substrate are assumed to be homogeneous, isotropic and elastic/plastic. However, the plastic deformation does not allow full recovery, and actual recovery is related to the relaxation of elastic strain. The contact between the film and substrate is assumed to be perfect during indentation. A perfect bonding condition between the film and substrate is also assumed, and the stress is used to determine the extent of the plastic state in the materials. The equivalent yield stress is calculated by the stress tensor and the material's equivalent stress begins to yield when it reaches the yield stress. The von Mises or equivalent strain ε_e is computed as:

$$\varepsilon_e = \frac{1}{1+\nu}\left(\frac{1}{2}\left[(\varepsilon_1-\varepsilon_2)^2+(\varepsilon_2-\varepsilon_3)^2+(\varepsilon_3-\varepsilon_1)^2\right]\right)^{\frac{1}{2}} \quad (1)$$

where ν = effective Poisson's ratio.

3. Results and Discussion

As displayed in Figure 3, the presented FEM simulation evidently reproduces the main features of the P-h curve obtained from the Berkovich tip indented Bi_2Se_3 thin films grown on c-plane sapphire substrate, despite a slight offset of the indentation load value for the first pop-in event. In particular, FEM simulation clearly mimics the multiple-step feature on the loading segment of the P-h curve seen in real indentation measurements. Such a feature is known to closely relate to the elastic–plastic deformation and associated dislocation activities. In any case, it is apparent that good agreement between the nanoindentation measurement and FEM results can be obtained, which, in turn, allows one to extract the corresponding equivalent stresses from the nodes of the FEM P-h curve.

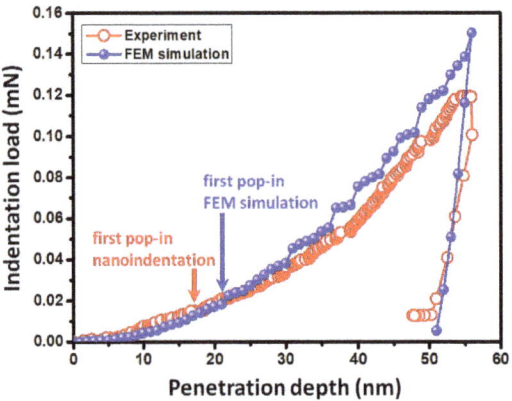

Figure 3. The comparsion of the experimental load-displacement curve with that obtained from FEM simulation during nanoindnetation.

Moreover, as shown in Figure 4, the behavior of penetration-depth-dependent hardness also exhibits very good agreement between the FEM simulation and the nanoindentation experimental data. It is noted that in this case the maximum indentation depth of the FEM result is 56 nm, which is only 15% of the film thickness (~360 nm) and, hence, well within the substrate-effect-free criterion suggested by Li et al. [26]. Following the Oliver-Pharr model [19], the values of H and E for the present Bi_2Se_3 thin film calculated from the FEM simulated P-h curve are about 2.1 GPa and 86.4 GPa, respectively, which are close to the experimentally recorded values of H~1.8 GPa and E~ 70 GPa.

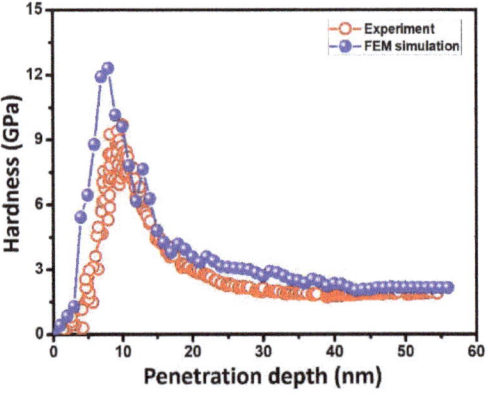

Figure 4. The comparsion of the experimental penetration-depth-dependent hardness and that obtained from FEM simulation during nanoindnetation.

Despite the consistency mentioned above, it is noted that there exist some subtle discrepancies between the experimental P-h curve and that obtained from FEM simulation. Firstly, as seen in Figure 3, the indentation load of FEM simulation starts to deviate and becomes slightly higher than the experimental value as the indentation depth is beyond 20 nm. This could be due to the fact that the regions underneath the indenter may have transformed to a more deformable structure, hence reducing the effective loading capacity of the film. On the other hand, in FEM simulation the basic mechanical property parameters of the film are assumed to remain unchanged. Moreover, as will be discussed later, the pile-up event is observed around the indentation, which may further reduce the film's loading capacity. Secondly, when the indentation depth reaches the critical depth of 20 nm, the slope of the load segment of the P-h curve for the FEM simulation appears to become steeper than the experimental one. This is attributed to the effect of local nanoindentation-induced dislocation nucleation behavior of film occurring beneath the Berkovich indenter tip [27]. In Figure 3, both FEM and nanoindentation curves display apparent multiple "pop-in" events, which were considered to be the signature corresponding to a clear transition from reversible elastic deformation to irreversible plastic deformation during nanoindentation and a process intimately related to dislocation activities [28,29]. In Figure 3, the first pop-in occurs at indentation depths of 17 nm and 20 nm for the experimental and FEM curves, respectively. Both are in reasonable agreement in indicating the first dislocation nucleation event. Nevertheless, as mentioned above, due to the fact that the material's parameters are assumed to remain unchanged in FEM, which may not reflect the actual film situations, the loading capacity and slope started to deviate. To gain some insight on this scenario, Figure 5 shows the contours of shear stress calculated from FEM. It gives the maximum shear stress of about 0.3 GPa, which is smaller than the nanoindentation experimental result (about 0.7 GPa). These differences can be attributed to the elastic non-linearity of the highly deformed film near the indenter tip, and to the boundary effects because of the finite size of the FEM simulated system, and can at least partially lend some support to the abovementioned arguments.

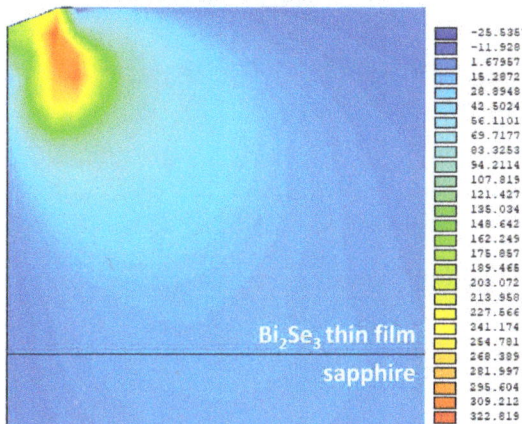

Figure 5. The shear stress of contours plot at first pop-in (the color bar unit: MPa).

Next, we move to discuss the nanoindentation-induced stress distribution within Bi_2Se_3 thin film and at the interface of film–substrate. In order to gain more comprehensive insights on this issue, the geometrical effects of various indenter tips are investigated (fixed at the same indentation depth of 56 nm). Three indenters (Berkovich, flat punch and spherical tips) are used to perform simulations of the nanoindentation process. When the diamond tip draws back, the plastic deformed region undergoes a partial elastic recovery, indicating the irreversibility of the plastic behavior during nanoindentation and indicating that the indentation-induced stress remains within the Bi_2Se_3 thin films. From Figure 6, it can be found that the maximum equivalent stress is about 633 MPa in all three cases (the red part in Figure 6). Moreover, substantial nanoindentation-induced pile-up phenomena are observed with all three indenters.

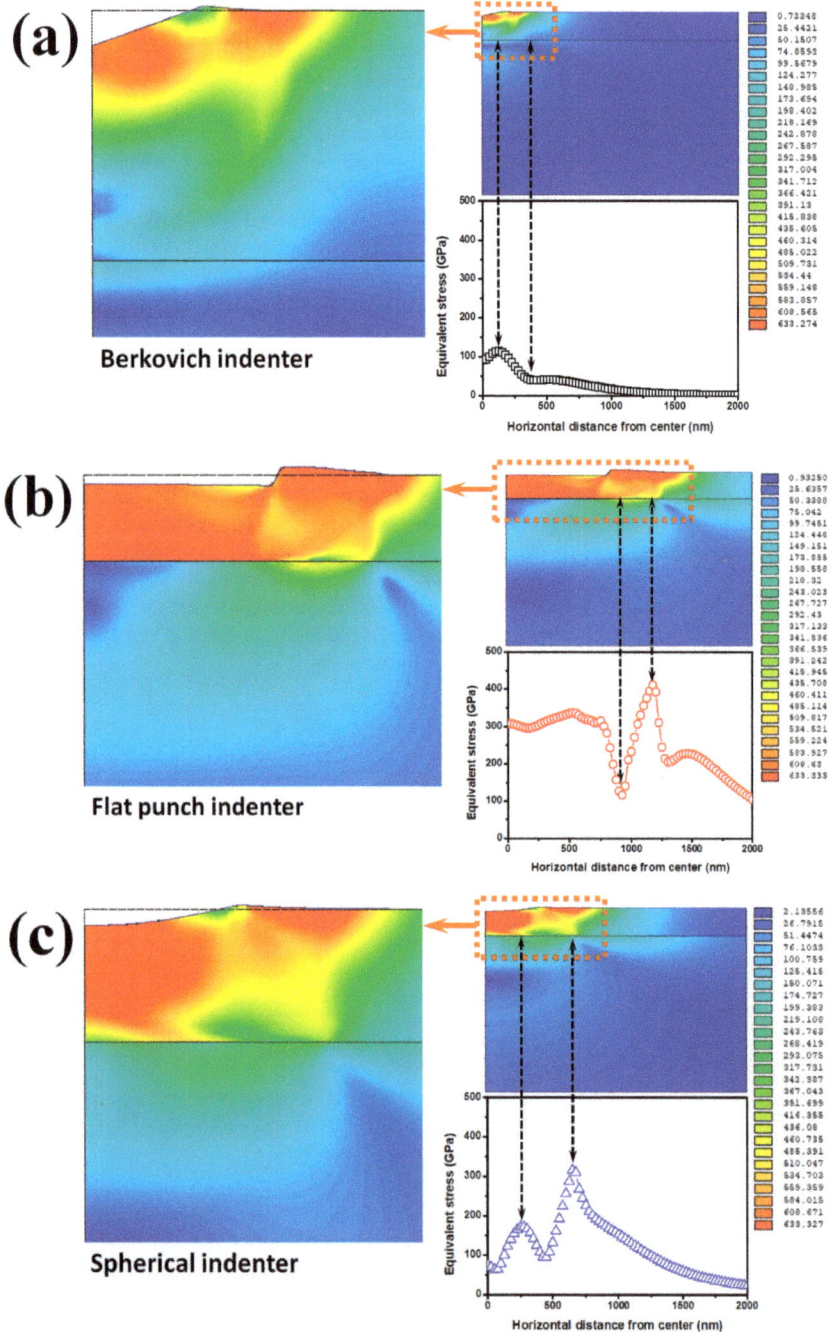

Figure 6. The equivalent stress distributions after unloading of (**a**) Berkovich, (**b**) flat punch and (**c**) spherical indenters (the color bar unit: MPa).

The pile-up event is often observed in systems consisting of soft film on hard substrate. For example, Tsui and Pharr [30] proposed that if the ratio of actual contact area and corner-to-corner area is greater than one, then the pile-up event can be observed, as illustrated in their experiment on an Al/glass system using a Berkovich indenter tip. A similar phenomenon was also observed in our experimental Bi_2Se_3 film/sapphire substrate [7]. It is interesting to note that the pile-up phenomena of modeled Bi_2Se_3 film/sapphire substrate can also be successfully replicated using FEM. On the other hand, it is evident that the residual stress within the Bi_2Se_3 film and/or the interface of Bi_2Se_3 film/sapphire substrate varies significantly with the geometry of the indenters (Figure 6).

To gain some quantitative perspective on the residual stress distribution caused by indenters with various geometrical symmetries, the equivalent stress curves are extracted from the interface of Bi_2Se_3 thin film and sapphire substrate, as displayed by color code in Figure 6. It can be seen that, for the Berkovich indenter, the residual stress is concentrated near the center of the indenter tip and surrounding the indentation (the red part in Figure 6a), and the equivalent stress is about 141 MPa exhibited at the interface between the Bi_2Se_3 thin film and substrate (see the corresponding plot panels shown in Figure 6a). In contrast, as shown in Figure 6b, the residual stresses are distributed quite evenly within the entire Bi_2Se_3 thin film when the flat punch indenter is used. Nevertheless, it is evident that the equivalent stress underneath the flat punch indenter becomes larger close to the edge of indenter (please see the zoom in the small block). Interestingly, in this case the equivalent stress of 410 MPa exhibited at the interface of the Bi_2Se_3 thin film/sapphire substrate is largest among the three types of indenters studied here. Finally, by comparing the results displayed in Figure 6a,c, similar deformation behaviors are observed. However, for spherical indenter tip, the zone of residual stress appears to distribute deeper and wider over the same indentation depth with an equivalent stress of 315 MPa at the interface between the Bi_2Se_3 thin film and substrate.

From the above discussions, it is encouraging to observe that the FEM simulations not only have evidently replicated the pop-in and pile-up events commonly observed in real indentation experiments, but also have provided means for analyzing the equivalent stress distribution within the Bi_2Se_3 thin film/sapphire substrate system during nanoindentation. The simulated results indicated that the deformation of Bi_2Se_3 thin film and the value of equivalent stress are strongly dependent on the geometry of the indenters. The present investigation indicates that the flat punch indenter tip results in the largest maximum equivalent stress (410 MPa) at the interface of the Bi_2Se_3 thin film/sapphire substrate system. Moreover, the simulation also confirmed that the substrate was not deformed during indentation loading and, thus, did not affect the evaluated magnitudes of hardness and Young's modulus for the Bi_2Se_3 thin film used in this study.

4. Conclusions

In summary, we report the nanoindentation responses of Bi_2Se_3 thin film deposited on c-plane sapphire substrate by combining nanoindentation measurements and FEM simulations. Results indicated that the pop-in event is displayed clearly on the loading segment of P-h curves in experiments and the FEM model with the Berkovich indenter tip. Moreover, the FEM simulated value of the critical indentation depth (~20 nm) is in good agreement with that of the experimentally observed pop-in depth (~17 nm). Such calculations indicate the feasibility of using FEM simulations to replicate the main features of actual nanoindentation experiments. In particular, the present study evidently demonstrated that by comparing the nanoindentation data with FEM predictions, a model for the mechanical deformation of Bi_2Se_3 thin films deposited on c-plane sapphire substrate can be reliably developed. Lastly, the comparisons performed on indenters with various geometries may also provide an efficient means to evaluate the contact-induced deformation encountered in practical device fabrication processes, wherein the shape of the contact tip may be different from case to case.

Author Contributions: S.-W.C., B.-S.C., P.H.L. and L.T.C.T. contributed to the experiments and FEM analyses. B.-S.C., S.-R.J., Y.-M.H., P.H.L., L.T.C.T., J.-W.L. and J.-Y.J. contributed to the discussion on materials characterizations. S.-R.J. designed the project of experiments/FEM model and drafted the manuscript. All authors have read and agreed to the published version of the manuscript.

Funding: The authors are thankful for financial support from the Ministry of Science and Technology, Taiwan, under Contract No. MOST 111-2221-E-214-015.

Institutional Review Board Statement: Not applicable.

Informed Consent Statement: Not applicable.

Data Availability Statement: Not applicable.

Conflicts of Interest: The authors declare no conflict of interest.

References

1. Hsieh, D.; Qian, D.; Wray, L.; Xia, Y.; Hor, Y.S.; Cava, R.J.; Hasan, M.Z. A topological Dirac insulator in a quantum spin Hall phase. *Nature* **2008**, *452*, 970–974. [CrossRef] [PubMed]
2. Xia, Y.; Qian, D.; Hsieh, D.; Wray, L.; Pal, A.; Lin, K.; Bansil, A.; Grauer, D.; Hor, Y.S.; Cava, R.J.; et al. Observation of a larger-gap topological-insulator class with a single Dirac cone on the surface. *Nat. Phys.* **2009**, *5*, 398. [CrossRef]
3. Qi, X.L.; Zhang, S.C. Topological insulators and superconductors. *Rev. Mod. Phys.* **2011**, *83*, 1057. [CrossRef]
4. Zhu, H.; Richter, C.A.; Zhao, E.; Bonevich, J.E.; Kimes, W.A.; Jang, H.-J.; Yuan, H.; Li, H.; Arab, A.; Kirillov, O.; et al. Topological Insulator Bi_2Se_3 Nanowire High Performance Field-Effect Transistors. *Sci. Rep.* **2013**, *3*, 1757. [CrossRef]
5. Andzane, J.; Buks, K.; Strakova, M.N.; Zubkins, M.; Bechelany, M.; Marnauza, M.; Baitimirova, M.; Erts, D. Structure and doping determined thermoelectric properties of Bi_2Se_3 thin films deposited by vapour-solid technique. *IEEE Trans. Nanotechnol.* **2019**, *18*, 948. [CrossRef]
6. Yazyev, O.V.; Moore, J.E.; Louie, S.G. Spin polarization and transport of surface states in the topological insulators Bi_2Se_3 and Bi_2Te_3 from first principles. *Phys. Rev. Lett.* **2010**, *105*, 266806. [CrossRef]
7. Lai, H.-D.; Jian, S.-R.; Tuyen, L.T.C.; Le, P.H.; Luo, C.-W.; Juang, J.-Y. Nanoindentation of Bi_2Se_3 Thin Films. *Micromachines* **2018**, *9*, 518. [CrossRef]
8. Wiatrowski, A.; Obstarczyk, A.; Mazur, M.; Kaczmarek, D.; Wojcieszak, D. Characterization of HfO_2 Optical Coatings Deposited by MF Magnetron Sputtering. *Coatings* **2019**, *9*, 106. [CrossRef]
9. Wang, L.; Liu, Y.; Chen, H.; Wang, M. Nanoindentation-induced deformation behaviors of tetrahedral amorphous carbon film deposited by cathodic vacuum arc with different substrate bias voltages. *Appl. Surf. Sci.* **2022**, *576*, 151741. [CrossRef]
10. Cheng, H.-P.; Le, P.H.; Tuyen, L.T.C.; Jian, S.-R.; Chung, Y.-C.; Teng, I.-J.; Lin, C.-M.; Juang, J.-Y. Effects of Substrate Temperature on Nanomechanical Properties of Pulsed Laser Deposited Bi_2Te_3 Films. *Coatings* **2022**, *12*, 871. [CrossRef]
11. Dong, J.; Jiang, H.; Sun, L.; Long, Z.J. Measuring plastic yield stress of magnetron sputtered aluminum thin film by nanoindentation. *Surf. Coat. Technol.* **2015**, *261*, 208. [CrossRef]
12. Zhang, W. Mechanical characterization of YBCO thin films using nanoindentation and finite element method. *Int. J. Mater. Res.* **2017**, *108*, 732–740. [CrossRef]
13. Moćko, W.; Szymanska, M.; Smietana, M.; Kalisz, M. Simulation of nanoindentation experiments of single-layer and double-layer thin films using finite element method. *Surf. Interface Anal.* **2014**, *46*, 1071–1076. [CrossRef]
14. Gupta, A.K.; Porwal, D.; Dey, A.; Mukhopadhyay, A.K.; Sharma, A.K. Evaluation of critical depth ratio for soft V_2O_5 film on hard Si substrate by finite element modeling of experimentally measured nanoindentation response. *J. Phys. D Appl. Phys.* **2016**, *49*, 155302. [CrossRef]
15. Shu, K.; Zhang, C.; Zhang, D.; Cui, S.; Hou, P.; Gu, L. Analysis on the cracking of thin hard films considering the effects of in-terfacial delamination. *Surf. Coat. Technol.* **2020**, *402*, 126284. [CrossRef]
16. Jiang, Z.; Li, Y.; Lei, M. An inverse problem in estimating fracture toughness of TiAlN thin films by finite element method based on nanoindentation morphology. *Vacuum* **2021**, *192*, 110458. [CrossRef]
17. Lichinchi, M.; Lenardi, C.; Haupt, J.; Vitali, R. Simulation of Berkovich nanoindentation experiments on thin films using finite element method. *Thin Solid Films* **1998**, *312*, 240–248. [CrossRef]
18. Li, X.; Bhushan, B. A review of nanoindentation continuous stiffness measurement technique and its applications. *Mater. Charact.* **2002**, *48*, 11–36. [CrossRef]
19. Oliver, W.C.; Pharr, G.M. An improved technique for determining hardness and elastic modulus using load and dis-placement sensing indentation experiments. *J. Mater. Res.* **1992**, *7*, 1564. [CrossRef]
20. Jayaraman, S.; Hahn, G.T.; Oliver, W.C.; Rubin, C.A.; Bastias, P.C. Determination of monotonic stress-strain curve of hard ma-terials from ultra-low-load indentation tests. *Int. J. Solid Struct.* **1998**, *35*, 365. [CrossRef]
21. Rowland, H.D.; King, W.P.; Cross, G.L.W.; Pethica, J.B. Measuring Glassy and Viscoelastic Polymer Flow in Molecular-Scale Gaps Using a Flat Punch Mechanical Probe. *ACS Nano* **2008**, *2*, 419. [CrossRef]

22. Bei, H.; Lu, Z.P.; George, E.P. Theoretical Strength and the Onset of Plasticity in Bulk Metallic Glasses Investigated by Nanoindentation with a Spherical Indenter. *Phys. Rev. Lett.* **2004**, *93*, 125504. [CrossRef]
23. Lim, Y.Y.; Chaudhri, M.M.; Enomoto, Y. Accurate determination of the mechanical properties of thin aluminum films deposited on sapphire flats using nanoindentations. *J. Mater. Res.* **1999**, *14*, 2314–2327. [CrossRef]
24. Kang, J.; Becker, A.; Sun, W. Determining elastic–plastic properties from indentation data obtained from finite element simulations and experimental results. *Int. J. Mech. Sci.* **2012**, *62*, 34–46. [CrossRef]
25. Gupta, A.K.; Porwal, D.; Dey, A.; Sridhara, N.; Mukhopadhyay, A.K.; Sharma, A.K.; Barshilia, H.C. Evaluation of elasto-plastic properties of ITO film using combined nanoindentation and finite element approach. *Ceram. Int.* **2016**, *42*, 1225–1233. [CrossRef]
26. Li, X.D.; Gao, H.; Murphy, C.J.; Caswell, K.K. Nanoindentation of sliver nanowires. *Nano Lett.* **2003**, *3*, 1495. [CrossRef]
27. Vliet, K.J.V.; Li, J.; Zhu, T.; Yip, S.; Suresh, S. Quantifying the early stages of plasticity through nanoscale experiments and sim-ulations. *Phys. Rev. B* **2003**, *67*, 104105. [CrossRef]
28. Tsai, C.-H.; Jian, S.-R.; Juang, J.-Y. Berkovich nanoindentation and deformation mechanisms in GaN thin films. *Appl. Surf. Sci.* **2008**, *254*, 1997–2002. [CrossRef]
29. Mao, W.; Shen, Y.; Lu, C. Nanoscale elastic–plastic deformation and stress distributions of the C plane of sapphire single crystal during nanoindentation. *J. Eur. Ceram. Soc.* **2011**, *31*, 1865–1871. [CrossRef]
30. Tsui, T.Y.; Pharr, G.M. Substrate effects on nanoindentation mechanical property measurement of soft films on hard substrates. *J. Mater. Res.* **1999**, *14*, 292–301. [CrossRef]

Article

Large Negative Photoresistivity in Amorphous NdNiO$_3$ Film

Alexandr Stupakov [1,*], Tomas Kocourek [1], Natalia Nepomniashchaia [1,2], Marina Tyunina [1,3] and Alexandr Dejneka [1,*]

1. Institute of Physics of the Czech Academy of Sciences, Na Slovance 2, 18221 Prague, Czech Republic; kocourek@fzu.cz (T.K.); nepomni@fzu.cz (N.N.); tjunina@fzu.cz (M.T.)
2. Faculty of Nuclear Sciences and Physical Engineering, Czech Technical University in Prague, Technická 2, 16627 Prague, Czech Republic
3. Microelectronics Research Unit, Faculty of Information Technology and Electrical Engineering, University of Oulu, P.O. Box 4500, FI-90014 Oulu, Finland
* Correspondence: stupak@fzu.cz (A.S.); dejneka@fzu.cz (A.D.)

Abstract: A significant decrease in resistivity by 55% under blue lighting with ~0.4 J·mm^{-2} energy density is demonstrated in amorphous film of metal-insulator NdNiO$_3$ at room temperature. This large negative photoresistivity contrasts with a small positive photoresistivity of 8% in epitaxial NdNiO$_3$ film under the same illumination conditions. The magnitude of the photoresistivity rises with the increasing power density or decreasing wavelength of light. By combining the analysis of the observed photoresistive effect with optical absorption and the resistivity of the films as a function of temperature, it is shown that photo-stimulated heating determines the photoresistivity in both types of films. Because amorphous films can be easily grown on a wide range of substrates, the demonstrated large photo(thermo)resistivity in such films is attractive for potential applications, e.g., thermal photodetectors and thermistors.

Keywords: rare-earth nickelates; epitaxial perovskite films; amorphous thin films; photoconductivity

1. Introduction

Rare-earth nickelates is a separate class of perovskite-structure metal oxides, whose main feature is specific resistive behavior. The end-member of the series LaNiO$_3$ (LNO) with a rhombohedral crystallographic structure exhibits the metallic conductivity at all temperatures. Low conductivity and superior catalytic properties make LNO attractive for use as functional sublayers in perovskite heterostructures [1]. Other nickelates with the orthorhombic structure display a sharp metal-to-insulator transition (MIT) from a high-temperature paramagnetic metal state to a low-temperature antiferromagnetic insulator. At MIT, the electrical resistivity changes by several orders of magnitude, which holds promise for many device applications. The most attractive members are NdNiO$_3$ (NNO) and SmNiO$_3$ (SNO), whose MIT temperatures $T_{MI} \approx 200$ and 400 K for bulk ceramics are the closest to room temperature [2,3].

Nowadays, a focus in the research has moved to the investigation of epitaxial thin films and heterostructures, which follow an in-plane structure of similar perovskite substrate. A film–substrate mismatch in the lattice parameters introduces a strain influencing the resistive behavior of the nickelate thin films as compared with their bulk analogues [4,5]. For instance, the most studied NNO films subjected to a tensile strain of ~2% have $T_{MI} \approx 170$ K, whereas a small compressive strain of $<-0.3\%$ fully suppresses MIT in these films [6–9]. Recently, T_{MI} at room temperature range is observed in SNO/NNO repetitive heterostructures with a few unit-cell thicknesses of separate layers [10]. However, the epitaxial growth is restricted by the selection of appropriate expensive substrates.

A hot trend of modern research is to extend the films functionality by an external physical stimulus. One of the attractive possibilities is to stimulate MIT in nickelates by irradiation with light [11]. However, the first attempts to switch MIT optically surmise the dominating influence of local sample heating due to light absorption [10,12]. Another promising direction of modern research is to study the polycrystalline perovskite films, which can be deposited on cheap industrial substrates [1,13–16].

We pursued two main objectives in this work. First, we comprehensively studied the photoresistivity at room temperature in epitaxial nickelate films to clarify a physical origin of the effect. Second, we demonstrated a strong photoresistive response in a novel material: amorphous film of NNO. Whereas a small resistivity increase of a few percent was observed in the epitaxial NNO films, a large drop of resistivity under irradiation was obtained for the amorphous film of the same elemental composition. We believe that this finding can stimulate broader research of non-epitaxial (oriented, polycrystalline and amorphous) nickelate films, which are currently practically unexplored [13,14,17].

2. Materials and Methods

Thin nickelate films were grown by pulsed laser deposition using a KrF excimer laser of 2 J·cm^{-2} energy density and ambient oxygen pressure of 20 Pa. The substrates purchased from the MTI Corporation (Richmond, VA, USA) were kept at temperature of 700 °C during the deposition [9]. In the main part of this paper, we confront the photoresistive response in two NNO films of 80 nm thickness: one film on (001)(La$_{0.3}$Sr$_{0.7}$)(Al$_{0.65}$Ta$_{0.35}$)O$_3$ (LSAT) and another film on SiO$_2$ (quartz) substrates. These films were deposited simultaneously within the same process, which ensured their identical elemental composition. Supporting data for other epitaxial films (LNO, NNO and SNO) of different thicknesses 20–120 nm deposited on LSAT or (001)LaAlO$_3$ (LAO) substrates are shown in Appendix A.

X-ray diffraction patterns were recorded by a multipurpose intelligent Empyrean diffractometer of 3rd generation (Malvern Panalytical Ltd., Malvern, UK) using Cu-Kα radiation with a wavelength 1.540598 Å. Two main films shown in Figure 1 were investigated in the grazing incidence geometry in a 2θ range from 0 to 90°. The epitaxial NNO film on LSAT substrate was also measured using the Bragg–Brentano geometry in a 2θ range from 10 to 120°. For accurate peak separation, the bare LSAT substrate was additionally investigated in the Bragg–Brentano geometry. The data were analyzed using Match! software (version 2.4.7, Crystal Impact, Bonn, Germany).

For electrical characterization, four stripe-type golden contacts of 1 mm width separated by 2 mm each were deposited on the top of the films (10 × 10 mm^2 size) by pulsed laser deposition at 200 °C through a shadow mask. Then the samples were cut into three identical pieces of 3.2 × 10 mm^2 size (photo of the samples is shown in Figure 1e). DC resistance was measured by a digital multimeter DMM6500 (Keithley Instruments, Cleveland, OH, USA) through USB remote interface. A precise 4-wire method was used for the epitaxial films; the amorphous film was measured by a 2-wire method because of its high resistance. The sample temperature was controlled by aT95 system controller supplemented with a liquid nitrogen cooling pump LNP 95. Samples were placed inside a HFS600E-PB4 temperature-controlled probe system (all devices are manufactured by Linkam Scientific Instruments Ltd., Tadworth, UK) with a transparent quartz window and 4 tungsten gold-plated tip probes ensuring ohmic contacts.

Commercial laser diodes in a TO-56 CAN package were mounted inside the standard focusing optical modules of ⌀12 mm, which were integrated with a fan cooler. The emitted optical power was controlled by a constantly applied current and calibrated by an optical power meter PM100D equipped with a photodiode sensor S121C (Thorlabs Inc, Newton, MA, USA). Four different high-power laser diodes were used: a blue laser (OSRAM PLPT5 447KA) with a 447 nm wavelength and a maximal optical power 1.2 W; a green laser (OSRAM PLT5 520B) with 520 nm, up to 85 mW; a red laser (Panasonic LNCQ28PS01WW) with 660 nm, up to 0.1 W; and an infra-red laser (ADL-80V01NL) with 808 nm, up to 0.4 W. A central uncovered part of the films 3.2 × 2 mm^2 between two inner golden contacts was irradiated. Optical absorption of the film-substrate stacks was also estimated using the same power meter putting the central part of the films between the laser beam and the photodiode sensor S121C. For each laser and film, 3–4 different levels of the optical power were adjusted to estimate the mean absorption factor with a maximal standard error of the averaging of 2%.

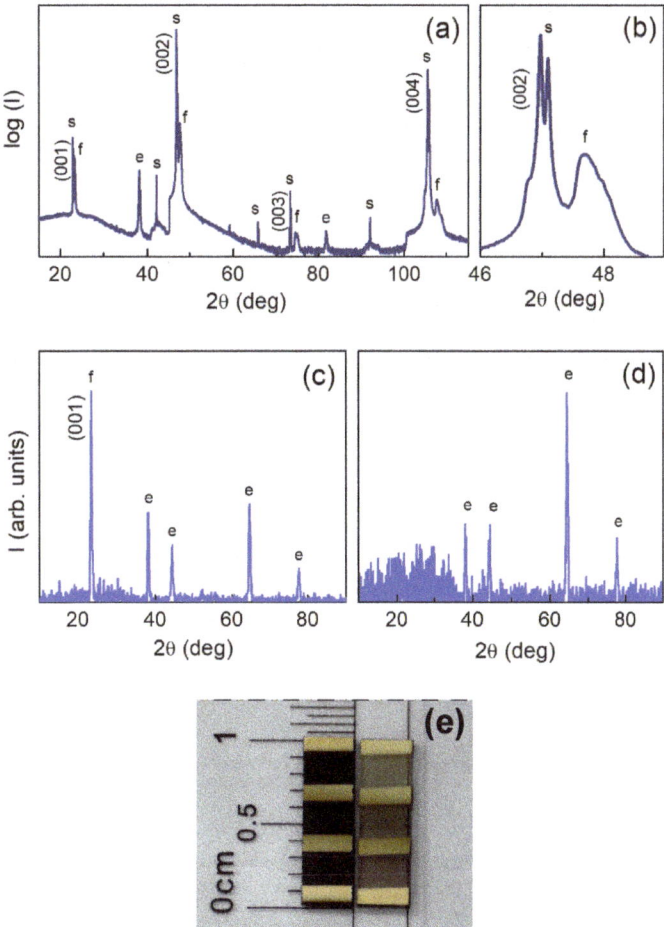

Figure 1. X-ray diffraction analysis of the studied films: (**a,b**) Bragg–Brentano θ–2θ scans and (**c,d**) grazing incidence scans of the NNO films on (**a–c**) LSAT and (**d**) SiO$_2$ substrates. Diffraction peaks from the films, the LSAT substrate and the Au electrodes are marked by f, s, and e, correspondingly. Details of the (002) peaks are shown in (**b**). (**e**) Photo of the films: the darker epitaxial film on LSAT is on the left side.

3. Results

Different substrates enabled either epitaxial film growth or the creation of amorphous structure. Epitaxial NNO films that we created on LSAT substrate (clear X-ray diffraction peaks from the film and the substrate are shown in Figure 1a–c), and amorphous NNO films were grown on SiO$_2$ (the absence of the corresponding peaks is shown in Figure 1d).

Figure 2 presents the resistivity as a function of temperature for both films. As illustrated by Figure 2a, the epitaxial NNO film demonstrates a typical behavior with a pronounced MIT at $T_{MI} \approx -105\ °C$ (170 K), a narrow low-temperature hysteresis ~5 °C and relevant resistivity values $\rho \approx 0.6\ m\Omega \cdot cm$ at room temperature [3–9]. Contrary to that, the amorphous film shows a strong insulating behavior with a fivefold exponential drop of resistivity with a temperature increase from 26 to 70 °C (see Figure 2b). This dependence can be fitted with a good accuracy (Pearson correlation coefficient is 0.9998) using an adiabatic polaron hopping model $\sigma T \propto \exp(-E_h/kT)$ [18]. The corresponding linear fit shown in Figure 2c gives a hopping activation energy $E_h \approx 350$ meV.

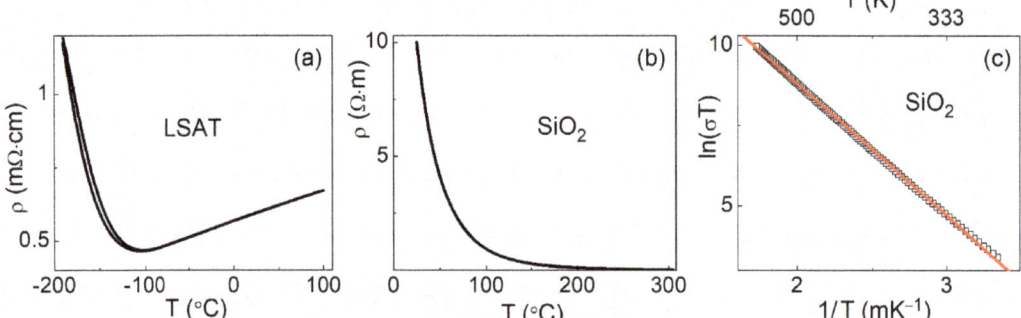

Figure 2. Resistivity temperature dependence at cooling–heating loop for NNO films on (**a**) LSAT and (**b**) SiO$_2$ substrates. (**c**) Adiabatic polaron hopping fit for the latter dependence.

Figure 3 illustrates the impact of the light irradiation on the resistive properties of both NNO films. Few-second illumination results in a stepwise change of the films resistivity. During 3–5 s of irradiation, the resistivity becomes stable with time if the opposite side of the substrate is hold at 26 °C by the temperature controller (see Figures 4a and A1a). For quantitative assessment of the observed photoresistive effect, illumination by the blue laser with 0.8-W optical power is taken as a reference. Such illumination of the central films region for three seconds providing ~0.4 J·mm^{-2} energy density leads to a 7.7% increase and 55% decrease in the electrical resistivity for the epitaxial and the amorphous film, respectively. With active temperature controller, the repeatable laser switching leads to reversible changes in the resistivity. If the controller does not operate, additional continuous change in the resistivity and rise in the inbuilt thermocouple reading are observed (see Figures 4b and A1b). Five 15-s pulses of 1.2 W power increase the temperature of a massive silver-coated stage (ceramic disk of 2 cm diameter and 1 cm thickness) by almost 10 °C. The stage temperature increases continuously during each light pulse and decreases much slower between the pulses. At continuous lighting, the resistivity change saturates after ~1000 s together with stabilization of the stage temperature on a level >40 °C.

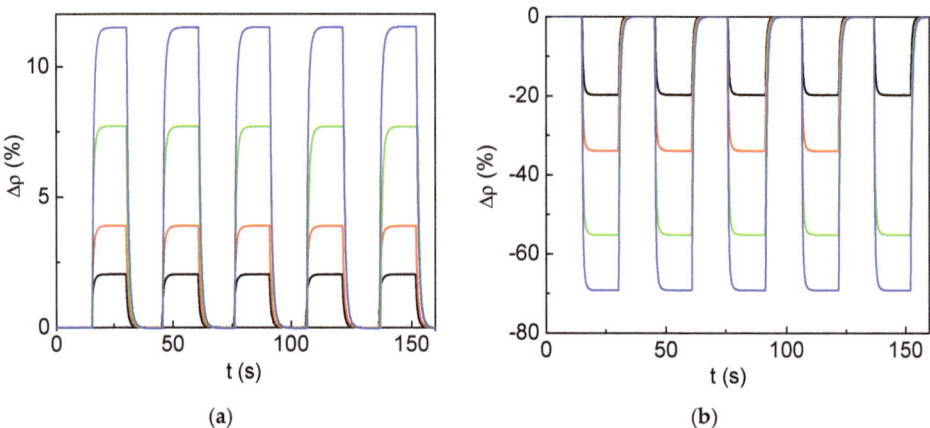

Figure 3. Percentage change of the resistivity for NNO films on (**a**) LSAT and (**b**) SiO$_2$ substrates at fixed temperature $T = 26$ °C and periodical 15-s illumination by the blue laser at different optical power: 0.2, 0.4, 0.8 and 1.2 W.

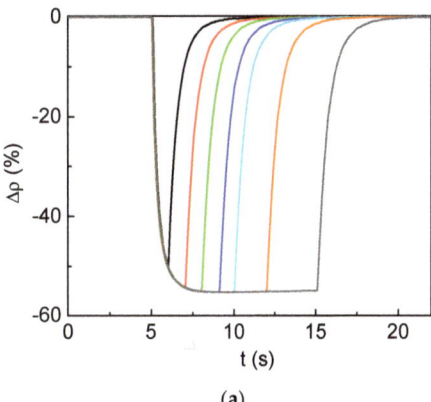 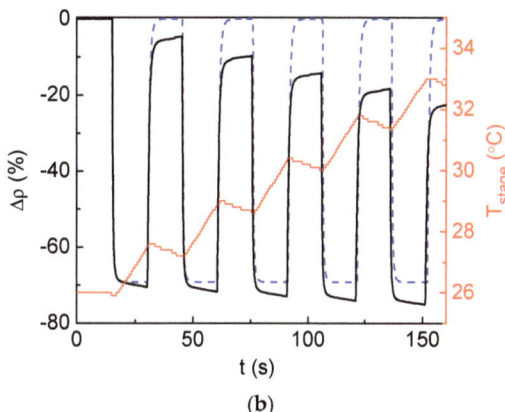

Figure 4. Percentage change of the resistivity at lighting by the blue laser on the amorphous NNO film. (**a**) Optical power is set to 0.8 W, whereas the durations of the light pulse are varied as 1, 2, 3, 4, 5, 7 and 10 s. (**b**) Optical power is set to 1.2 W, whereas the temperature controller holds the temperature at 26 °C or does not operate. The reversible resistivity changes at the active controller are shown by blue dashed line. In the latter case, an additional decrease in the resistivity shown by black line is accompanied by heating of the sample stage (reading of the inbuild thermo couple is shown by red line and referred to the right ordinate axis).

A small positive photoresistive response, similar to that presented in Figure 3a, was obtained for other epitaxial NNO and LNO films in the room-temperature metallic phase (see Figures A2 and A3). Epitaxial films of SNO with similar decaying dependences $\rho(T)$ but smaller resistivity values at room temperature than the amorphous NNO film display a negative photoresistive drop, whose amplitude under the reference blue illumination of ~0.4 J·mm^{-2} is ~20% (see Figure A4 and Table A1).

For the epitaxial NNO film, a nearly linear increase in the resistivity with optical power of the incident light is well correlated with its temperature dependence (see Figure 5a). For the amorphous film, a decrease in the photoresistivity with optical power also follows the corresponding $\rho(T)$ dependence, as illustrated by Figure 5b (values of the film resistance R are indicated on the right ordinate axis). Therefore, the magnitude of the resistivity drop is not simply proportional to the incident optical power as for the epitaxial film but is an exponential function of the power. Evidently, different resistivity temperature dependences shown in Figure 2 are responsible for significant distinctions of the photoresistive response in the studied films. The photoresistive effect in the case of illumination with other light wavelengths is qualitatively the same to that obtained with the blue laser (compare Figures 3 and A5). Quantitative difference is discussed in the next section.

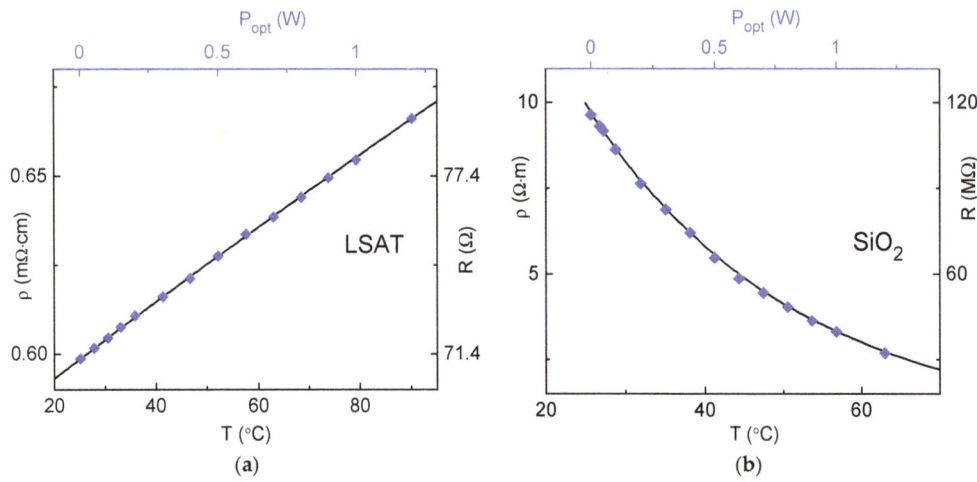

Figure 5. Change of the resistivity with optical power of the blue laser superimposed on the corresponding temperature dependence for NNO films on (**a**) LSAT and (**b**) SiO$_2$ substrates. The right ordinate axes indicate the corresponding resistance values. The temperature controller holds T = 26 °C.

4. Discussion

Assuming the local film heating is the main reason of the observed photoresistive effect, the appropriate temperature increments can be estimated from the corresponding $\rho(T)$ dependences shown in Figure 5. As illustrated by Figures 6 and A6, such estimated increments ΔT are proportional to the optical power P_{opt}, which gives a constant effective heat capacity of the film-substrate stack. However, the slopes k of the presented linear fits $\Delta T(P_{opt})$ with the Pearson correlation coefficients > 0.9996 are varied with the films and the light wavelengths. These distinctions in the heating ability are mostly ascribed to the different optical absorptions of the films because both substrates are transparent and no significant reflection from the film surface is observed (see Table 1). Even visually, the epitaxial film is much darker than the amorphous one, proving a twice as high absorption factor (see Figure 1e and Ref. [14]). Both film-substrate stacks absorb the light of lower wavelengths better; however, this dependence is not so pronounced for the epitaxial films. Normalization of the fit slopes k by the absorption factor α reduces the parameter variation for the 8 considered cases: total mean values <k> = 37.3 ± 15.9 (±43%) but <k/α> = 58 ± 3.7 (±6.4%). The latter standard deviation ~ 6% can be assigned to the light reflection at the film surface and the film-substrate interface (estimated on a level of ~7%) as well as to wavelength-dependent light absorption of the LSAT substrate [19,20]. Absorption factor of the commercial crystal quartz is known to be about 10% within a broad wavelength range from ultraviolet to far infrared light. The obtained value <k/α> = 58 °C/W is physically reasonable: the energy of 3 J fully absorbed during the 3-s illumination with 1-W optical power could heat the thermally isolated substrates of LSAT and quartz by 49 and 96 °C, respectively.

The magnitude of the negative photoresistivity in the amorphous film can be increased by preparing thicker films for higher optical absorption, as well as tuned by a proper manufacturing of the sensitive film structure to optimize its resistance value. Thermal isolation of the film can also tune the photoresistive changes. Improving the substrate-stage contact by thermally conductive grease decreases the time of the resistivity switch up to one second (compare Figures 4a and 7a). However, the magnitude of the photoresistive change also decreases twice. Contrariwise, putting a sheet of paper between the substrate and the temperature-controlled stage increases the magnitude of the photoresistive change as well as the switching time twice (see Figure 7b).

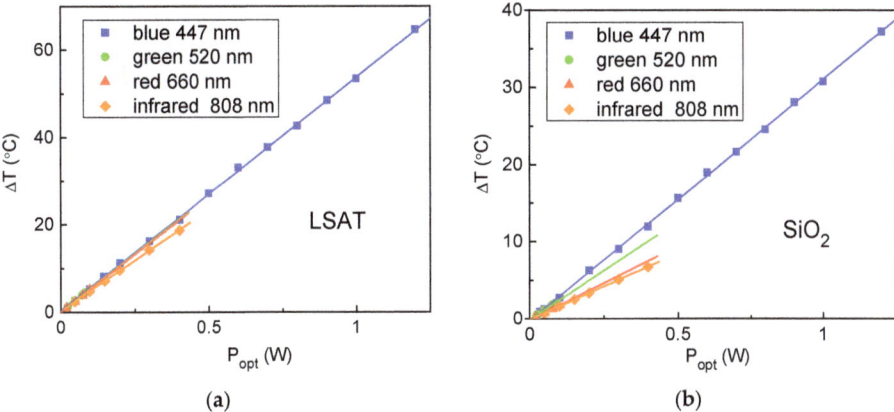

Figure 6. Estimated temperature increments as functions of the optical power for all light wavelengths with the corresponding linear fits for NNO films on (**a**) LSAT and (**b**) SiO$_2$ substrates.

Table 1. The slope k of the linear fits $\Delta T(P_{opt})$ shown in Figure 6, the measured absorption factor α of the film-substrate stack and the ratio k/α for both NNO films and all light wavelengths λ. The absorption factor of the bare LSAT substrate α_s is shown in the fifth column.

λ, nm	LSAT				SiO$_2$		
	k, °C/W	α, %	k/α, °C/W	α_s, %	k, °C/W	α, %	k/α, °C/W
447	53.5	91.8	58.3	33.6	31.2	52.8	59.1
520	53.1	88.2	60.3	26.9	25.3	41.7	60.8
660	52.8	83.8	63	24.5	19.1	36.9	51.7
808	46.6	82.8	56.3	8.8	17.1	31.2	54.6

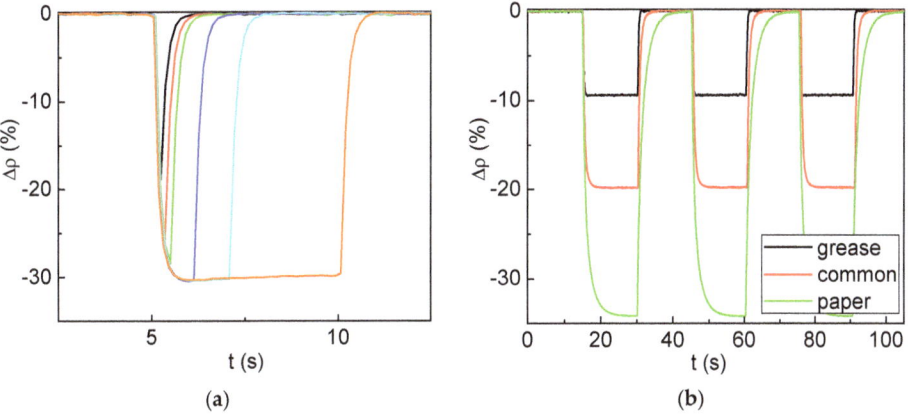

Figure 7. The percentage change of the resistivity at lighting by the blue laser on the amorphous NNO film, whose thermal contact to the temperature-controlled stage (T = 26 °C) is varied. (**a**) Optical power is set to 0.8 W and the substrate-stage contact is improved by thermally conductive grease. Durations of the light pulse are varied as 0.1, 0.25, 0.5, 1, 2 and 5 s. (**b**) Optical power is set to 0.2 W, whereas the thermal contact between the substrate back-side and the temperature-controlled stage is varied as follows: (i) improved by thermally conductive grease; (ii) leave in the common condition of a freely lying sample; and (iii) a paper sheet of 0.1 mm thickness is placed between the substrate and the cooling stage.

The achievable magnitudes of the photoresistive effect in the amorphous NNO film can be attractive for using in thermal photodetectors [21]. This type of the light detector utilizes the considered principle: it measures temperature rises resulting from the light absorption. The thermal detectors are applied when photon semiconducting detectors are ineffective, in particular, for measuring the long-wavelength infrared light and the high-power lasers. Usually, a black coating covers the temperature sensor for homogeneous broadband absorption. The thermal detectors expectedly suffer from their temperature sensitivity; therefore, an air/water cooling is frequently used. Another strategy is to modulate the incident light with a chopper and to estimate a dark-light relative difference. Typical response time of the thermal detectors is a few seconds. Sensitivity of such industrial sensors is not declared by the producers. However, the observed huge photoresistance change ~527 MΩ/W·mm^{-2} with the similar time response at the reference 0.8 W blue light illumination should be competitive (see Figure 5b).

The competitive analysis can be performed estimating an amorphous film potential for industrial temperature sensing since the same physical principle is in play; the exponential decay of the resistivity with heating is illustrated by Figures 2b and 5b. Such a dependence is typical for thermistors with negative temperature coefficient, i.e., thermally sensitive resistors with negative dR/dT, which are generally made of ceramics or polymers. An important figure of merit for the thermometer layer material is the temperature coefficient of resistance $\alpha = (dR/dT)/R$, whose typical values range from −3 to −6% °C^{-1} [22]. Another important characteristic of the thermistors is a so-called B value defined by Equation (1) (both temperatures are in Kelvin):

$$B_{T1/T2} = \frac{T_2 \cdot T_1}{T_2 - T_1} \cdot \ln\left(\frac{R_1}{R_2}\right) \qquad (1)$$

Typical $B_{25/100}$ values for the thermistors with negative temperature coefficient are between ~3000 and 5000 K. As illustrated by Figure 8, values of these parameters estimated for the presented amorphous film are industrially competitive. This is especially true for the range of rapid resistivity decay T = 25–75 °C, which is an operation range of semiconductor electronics. Therefore, a thin film element of the amorphous NNO can be effective as a microchip-inbuilt temperature sensor. The range of the highest R changes can be optimized varying the NNO film thickness/length. Moreover, amorphous SNO films are expected to provide similar resistivity decay at much higher temperatures, which could be also industrially attractive. The thermistors are typically suitable for use within a temperature range between −55 and 200 °C. For the higher temperatures in the order of 600 °C, thermocouples are usually used instead of the thermistors.

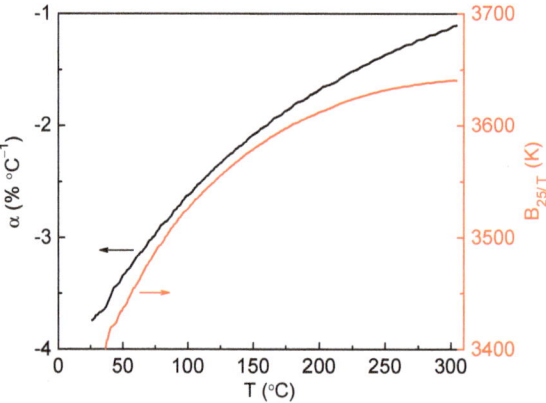

Figure 8. Temperature coefficient of resistance α and B_{25} value (black/red lines referred to the left/right axes, respectively) as functions of temperature for the amorphous NNO film.

5. Conclusions

The novel material, amorphous film of NdNiO$_3$, demonstrates useful resistive properties at room temperature: measurable resistance ~110 MΩ, which exponentially drops with increasing temperature (temperature coefficient of resistance is about -3% $°C^{-1}$). This opens a promising opportunity for developing sensitive microelectronic temperature sensors (thermistors). A large negative photoresistivity (~55% under blue lighting with ~0.4 J·mm^{-2} energy density) is additionally observed in the amorphous NdNiO$_3$ film, which make it an attractive material for thermal photodetectors. Contrary to the epitaxial films deposited on expensive high-quality perovskite substrates, the amorphous films can be deposited on any industrially attractive substrates such as glass, quartz or silicon, which provides a serious technical advantage. The presented results evidence that a local film heating plays an important role in the observed photoresistive response. Any noticeable excitation of the electronic states by visible and infra-red light was not observed for the studied epitaxial and amorphous NdNiO$_3$ films in the room temperature range. This important outcome should prevent further speculations in the field of nickelate research and indicate challenging directions for further development of these perovskite materials.

Author Contributions: Conceptualization, M.T. and A.S.; methodology, T.K., N.N. and A.S.; software, A.S.; validation, A.S., M.T. and A.D.; formal analysis, A.S., N.N. and M.T.; investigation, A.S.; resources, A.D.; writing—original draft preparation, A.S.; writing—review and editing, M.T. and A.D.; visualization, A.S.; supervision, M.T. and A.D.; project administration, A.S. All authors have read and agreed to the published version of the manuscript.

Funding: This research was funded by the Czech Science Foundation (GA ČR), grant number 20-21864S.

Institutional Review Board Statement: Not applicable.

Informed Consent Statement: Not applicable.

Data Availability Statement: The data that support the findings of this study are available from the corresponding author upon reasonable request.

Conflicts of Interest: The authors declare no conflict of interest.

Appendix A

Figures A2a and A3a present the resistivity as a function of temperature for other tested epitaxial films of NNO and LNO deposited on LSAT or LAO. All these films have metallic resistive behavior at room temperature. Only another NNO film on LSAT of 20 nm thickness demonstrates a similar MIT behavior as the reference epitaxial film considered in the main text (compare Figures 2a and A2a). The rest of the tested films do not display MIT above -190 °C, which is common for these materials and substrates [6–9]. Blue light illumination with the reference optical power 0.8 W (~0.4 J·mm^{-2} energy density) results in a comparable increase in the film resistivity by 5–8% (compare Figures 3a, A2b and A3b). Figure A4a presents the resistivity as a function of temperature for tested epitaxial films of SNO deposited on LSAT or LAO. All SNO films demonstrate insulating behavior at room temperature with a lower resistivity drop and increasing temperature as compared to the amorphous NNO film (compare with Figure 2b). This results in a lower photoresistivity drop ~20%, as illustrated by Figure A4b. Table A1 compares the magnitude of the observed photoresistive effect for all tested films at room temperature under the reference blue light illumination with ~0.4 J·mm^{-2} energy density.

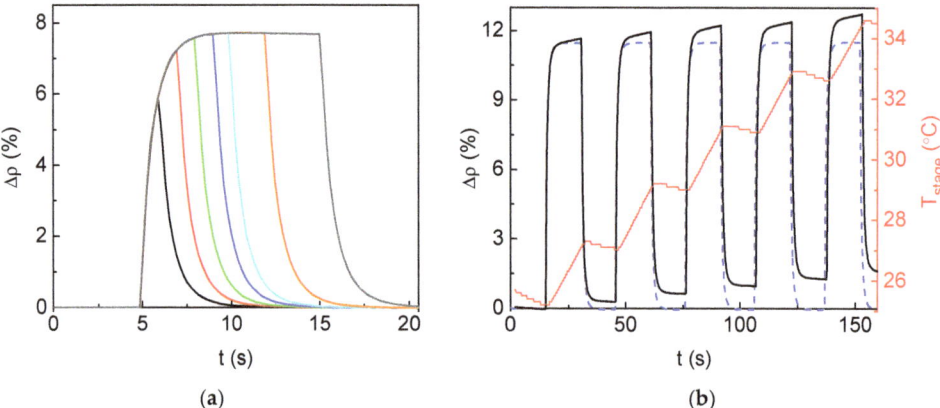

Figure A1. Percentage change of the resistivity at lighting by the blue laser on the epitaxial NNO film. (**a**) Optical power is set to 0.8 W, whereas the durations of the light pulse are varied as 1, 2, 3, 4, 5, 7 and 10 s. (**b**) Optical power is set to 1.2 W, whereas the temperature controller holds the temperature at 26 °C or does not operate. The reversible resistivity changes at the active controller are shown by blue dashed line. In the latter case, an additional increase in the resistivity shown by black line is accompanied by heating of the sample stage (reading of the inbuild thermocouple is shown by red line and referred to the right ordinate axis).

It is worth noting that the data presented in Figures A2b and A3b were obtained in a simpler mode than those in Figure 3. Here the illumination was applied by supplying a constant current to the laser diode. The disadvantage of this simple method is that the diode is heated by ~5–7 °C during the first few minutes after the current supply, which is accompanied by a loss in the emitted optical power. This causes small initial leaps in the photoresistive profiles. During ~100 s, the laser temperature stabilizes together with the emitted power, which was checked by the optical power meter. Additional experiments were performed where the stabilized laser beam is chopped mechanically (dashed profiles in Figures A2b and A3b). The photoresistive changes obtained at periodical 15-s illumination, including all data presented in the main text, were measured using the stabilized laser, whose beam was chopped by a half-disk shutter attached to a stepping motor.

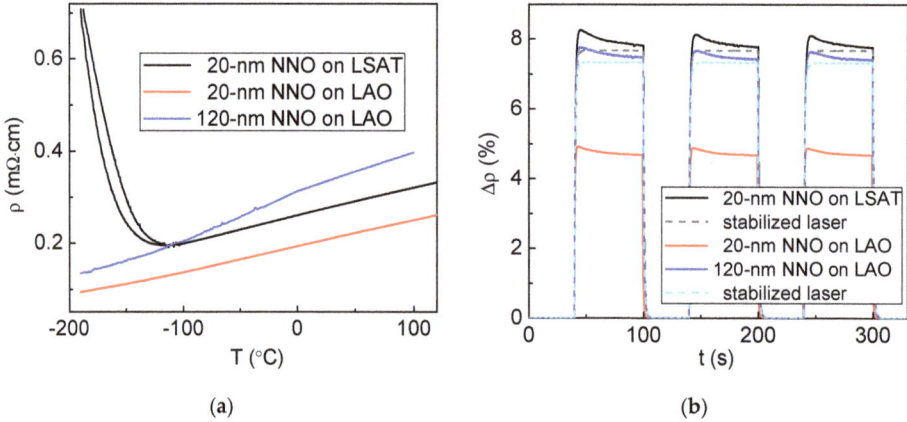

Figure A2. (**a**) Resistivity temperature dependence for other tested NNO films: two 20-nm thick films deposited on LSAT and LAO and a 120-nm thick film on LAO. (**b**) Percentage change of their room-temperature resistivity at periodical 60-s illumination by the blue laser at optical power 0.8 W.

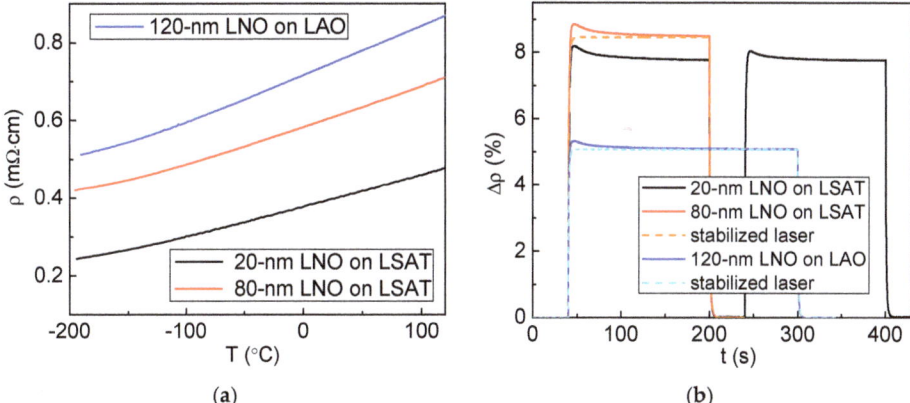

Figure A3. (**a**) Resistivity temperature dependence for other tested LNO films: two films deposited on LSAT with thicknesses 20 and 80 nm as well as a 120-nm thick film on LAO. (**b**) Percentage change of their room-temperature resistivity at long-term illuminations by the blue laser at optical power 0.8 W.

By analyzing Table A1 in general, it is evident that the magnitude of the photoresistive change can be tuned by a proper manufacturing of the sensitive film structure with defined values of the resistivity and its derivative $d\rho/dT$. However, the thermal factors, namely optical absorption and heat removal, prevail. The absorption factor of the dark perovskite films is high: ~80% for the thin 20-nm films and >90% for the thicker films (see Figure 1e). This determines a higher heating of the thicker films. A lower heating of the films on LAO is caused by its twice bigger thermal conductivity 10 W/m·K as compared to that of LSAT: 5.1 W/m·K. It is also proved by smaller slopes k of the linear fits $\Delta T(P_{opt})$ in Figure A6: 21.7 and 27.6 °C/W, respectively, for the blue light (20.1 and 24.1 °C/W for the infra-red illumination). The corresponding ratios k/α are 28.2 and 32.1 °C/W, which is very close to a half of that value obtained for the NNO film on LSAT (see Table 1).

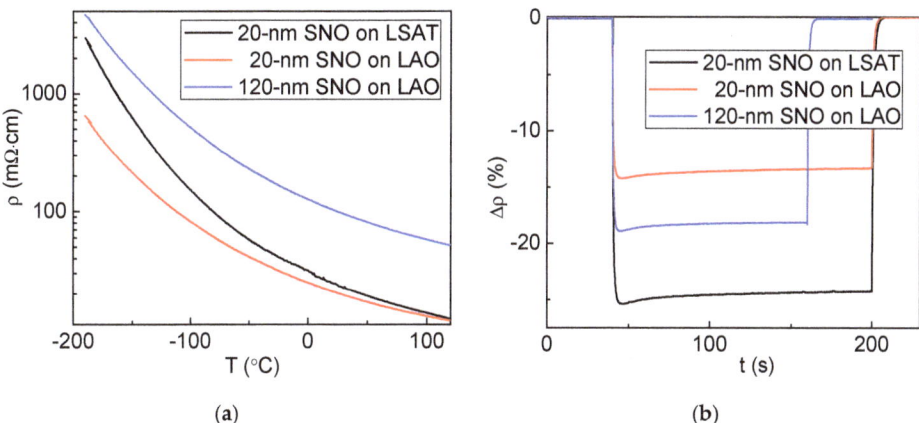

Figure A4. (**a**) Resistivity temperature dependence for other tested SNO films: two 20-nm thick films deposited on LSAT and LAO and a 120-nm thick film on LAO. (**b**) Percentage change of their room-temperature resistivity at long-term illuminations by the blue laser at optical power 0.8 W.

Table A1. Photoresistivity parameters for all tested films at $T = 26\ °C$ and the reference blue light illumination of 0.8 W power: percentage changes of the resistivity $\Delta\rho$, the measured absorption factors α of the film-substrate stacks, resistivity values ρ, derivatives $d\rho/dT$ and estimated rises of the film temperature ΔT.

Film	NNO					LNO			SNO		
Substrate	LSAT		LAO		SiO$_2$	LSAT		LAO	LSAT	LAO	
Thickness, nm	20	80	20	120	80	20	80	120	20	20	120
$\Delta\rho$, %	7.7	7.7	4.7	7.4	−55	7.8	8.5	5.1	−24	−13.4	−18
α, %	83.7	91.8	77.2	96.5	52.8	79.8	92	96.2	80.3	71.9	86
ρ, mΩ·cm	0.277	0.6	0.21	0.34	10^6	0.4	0.61	0.76	25.2	22.3	99.5
$d\rho/dT$, μΩ·cm/K	0.61	1.1	0.57	0.88	$-36\cdot 10^6$	0.88	1	1.3	−231	−172	−731
ΔT, °C	36	42.7	17	29	25	35	50	30	35	20	25

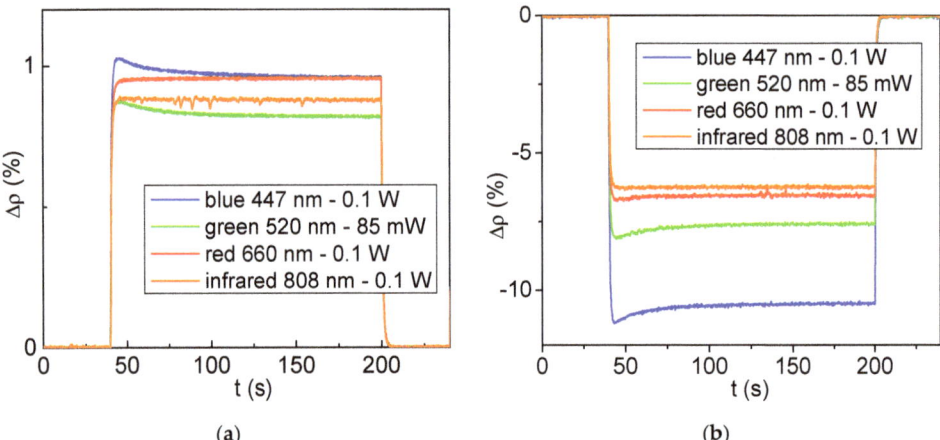

Figure A5. Percentage change of the resistivity for NNO films on (**a**) LSAT and (**b**) SiO$_2$ substrates at fixed temperature $T = 26\ °C$ and long-term illumination by different lasers at optical power 0.1 W (except the green laser, which maximal power is 85 mW).

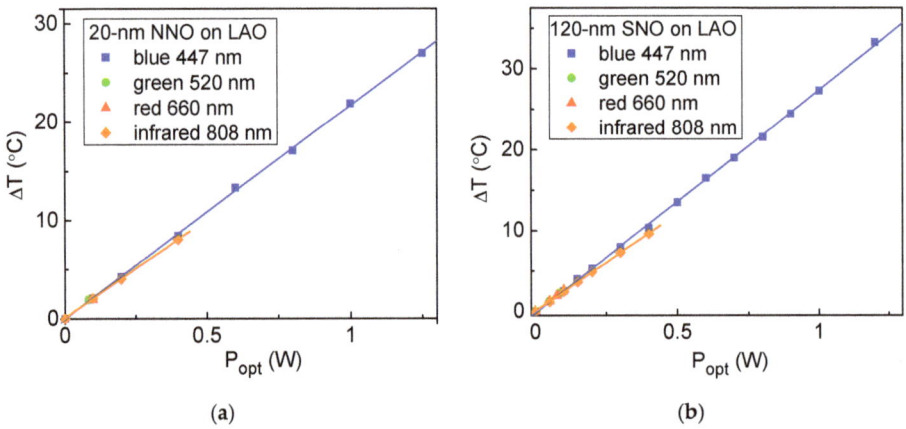

Figure A6. Estimated temperature increments as functions of the optical power for all light wavelengths with the corresponding linear fits for (**a**) the 20-nm thick NNO and (**b**) the 120-nm thick SNO films on LAO.

Table values of the heat capacity and the density (0.57 and 0.733 J/g·K as well as 6.74 and 2.65 g/cm^3 for LSAT and quartz, respectively) are taken to estimate a heating of the isolated substrates of 3.2 × 10 mm size in the main text.

References

1. Park, J.; Lim, Y.; Kong, S.; Lee, H.; Kim, Y.-B. Rapid fabrication of chemical solution-deposited lanthanum nickelate thin films via intense pulsed-light process. *Coatings* **2019**, *9*, 372. [CrossRef]
2. Medarde, M.L. Structural, magnetic and electronic properties of RNiO$_3$ perovskites (R = rare earth). *J. Phys. Condens. Matter.* **1997**, *9*, 1679–1707. [CrossRef]
3. Catalan, G. Progress in perovskite nickelate research. *Phase Transit.* **2008**, *81*, 729–749. [CrossRef]
4. Middey, S.; Chakhalian, J.; Mahadevan, P.; Freeland, J.W.; Millis, A.J.; Sarma, D.D. Physics of ultrathin films and heterostructures of rare-earth nickelates. *Annu. Rev. Mater. Res.* **2016**, *46*, 305–334. [CrossRef]
5. Catalano, S.; Gibert, M.; Fowlie, J.; Íñiguez, J.; Triscone, J.-M.; Kreisel, J. Rare-earth nickelates RNiO$_3$: Thin films and heterostructures. *Rep. Prog. Phys.* **2018**, *81*, 046501. [CrossRef] [PubMed]
6. Liu, J.; Kargarian, M.; Kareev, M.; Gray, B.; Ryan, P.J.; Cruz, A.; Tahir, N.; Chuang, Y.D.; Guo, J.; Rondinelli, J.M.; et al. Heterointerface engineered electronic and magnetic phases of NdNiO$_3$ thin films. *Nat. Commun.* **2013**, *4*, 2714. [CrossRef] [PubMed]
7. Wang, L.; Ju, S.; You, L.; Qi, Y.; Guo, Y.; Ren, P.; Zhou, Y.; Wang, J. Competition between strain and dimensionality effects on the electronic phase transitions in NdNiO$_3$ films. *Sci. Rep.* **2015**, *5*, 18707. [CrossRef] [PubMed]
8. Mikheev, E.; Hauser, A.J.; Himmetoglu, B.; Moreno, N.E.; Janotti, A.; Van de Walle, C.G.; Stemmer, S. Tuning bad metal and non-Fermi liquid behavior in a Mott material: Rare-earth nickelate thin films. *Sci. Adv.* **2015**, *1*, e1500797. [CrossRef] [PubMed]
9. Stupakov, A.; Pacherova, O.; Kocourek, T.; Jelinek, M.; Dejneka, A.; Tyunina, M. Negative magnetoresistance in epitaxial films of neodymium nickelate. *Phys. Rev. B* **2019**, *99*, 085111. [CrossRef]
10. Liao, Z.; Gauquelin, N.; Green, R.J.; Müller-Caspary, K.; Lobato, I.; Li, L.; Van Aert, S.; Verbeeck, J.; Huijben, M.; Grisolia, M.N.; et al. Metal–insulator-transition engineering by modulation tilt-control in perovskite nickelates for room temperature optical switching. *Proc. Natl. Acad. Sci. USA* **2018**, *115*, 9515–9520. [CrossRef]
11. Ahn, C.; Cavalleri, A.; Georges, A.; Ismail-Beigi, C.; Millis, A.J.; Triscone, J.-M. Designing and controlling the properties of transition metal oxide quantum materials. *Nat. Mater.* **2021**, *20*, 1462–1468. [CrossRef]
12. Mattoni, G.; Manca, N.; Hadjimichael, M.; Zubko, P.; van der Torren, A.J.; Yin, C.; Catalano, S.; Gibert, M.; Maccherozzi, F.; Liu, Y.; et al. Light control of the nanoscale phase separation in heteroepitaxial nickelates. *Phys. Rev. Mater.* **2018**, *2*, 085002. [CrossRef]
13. Chen, B.; Gauquelin, N.; Jannis, D.; Cunha, D.M.; Halisdemir, U.; Piamonteze, C.; Lee, J.H.; Belhadi, J.; Eltes, F.; Abel, S.; et al. Strain-engineered metal-to-insulator transition and orbital polarization in nickelate superlattices integrated on silicon. *Adv. Mater.* **2020**, *32*, 2004995. [CrossRef]
14. Sun, Y.; Wang, Q.; Park, T.J.; Gage, T.E.; Zhang, Z.; Wang, X.; Zhang, D.; Sun, X.; He, J.; Zhou, H.; et al. Electrochromic properties of perovskite NdNiO$_3$ thin films for smart windows. *ACS Appl. Electron. Mater.* **2021**, *3*, 1719–1731. [CrossRef]
15. Lin, T.-K.; Chang, H.-W.; Chou, W.-C.; Wang, C.-R.; Wei, D.-H.; Tu, C.-S.; Chen, P.-Y. Multiferroic and nanomechanical properties of Bi$_{1-x}$Gd$_x$FeO$_3$ polycrystalline films (x = 0.00–0.15). *Coatings* **2021**, *11*, 900. [CrossRef]
16. Thiruchelvan, P.S.; Lai, C.-C.; Tsai, C.-H. Combustion processed nickel oxide and zinc doped nickel oxide thin films as a hole transport layer for perovskite solar cells. *Coatings* **2021**, *11*, 627. [CrossRef]
17. Simandan, I.-D.; Sava, F.; Buruiana, A.-T.; Burducea, I.; Becherescu, N.; Mihai, C.; Velea, A.; Galca, A.-C. The effect of the deposition method on the structural and optical properties of ZnS thin films. *Coatings* **2021**, *11*, 1064. [CrossRef]
18. Thiessen, A.; Beyreuther, E.; Werner, R.; Koelle, D.; Kleiner, R.; Eng, L.M. Quantifying the electrical transport characteristics of electron-doped La$_{0.7}$Ce$_{0.3}$MnO$_3$ thin films through hopping energies, Mn valence, and carrier localization length. *J. Phys. Chem. Solids* **2015**, *80*, 26–33. [CrossRef]
19. Nunley, T.N.; Willett-Gies, T.I.; Cooke, J.A.; Manciu, F.S.; Marsik, P.; Bernhard, C.; Zollner, S. Optical constants, band gap, and infrared-active phonons of (LaAlO$_3$)$_{0.3}$(Sr$_2$AlTaO$_6$)$_{0.35}$ (LSAT) from spectroscopic ellipsometry. *J. Vac. Sci. Technol. A* **2016**, *34*, 051507. [CrossRef]
20. Müllerová, J.; Šutta, P.; Holá, M. Optical absorption in Si:H thin films: Revisiting the role of the refractive index and the absorption coefficient. *Coatings* **2021**, *11*, 1081. [CrossRef]
21. Voshell, A.; Terrones, M.; Rana, M. Thermal and photo sensing capabilities of mono- and few-layer thick transition metal dichalcogenides. *Micromachines* **2020**, *11*, 693. [CrossRef] [PubMed]
22. Le, D.T.; Ju, H. Solution synthesis of cubic spinel Mn–Ni–Cu–O thermistor powder. *Materials* **2021**, *14*, 1389. [CrossRef] [PubMed]

Article

Tunable Electrical Properties of Ti-B-N Thin Films Sputter-Deposited by the Reactive Gas Pulsing Process

Charalampos Sakkas [1], Jean-Marc Cote [1], Joseph Gavoille [1], Jean-Yves Rauch [1], Pierre-Henri Cornuault [1], Anna Krystianiak [2], Olivier Heintz [2] and Nicolas Martin [1,*]

[1] Institut FEMTO-ST, UMR 6174, CNRS, ENSMM, Univ. Bourgogne Franche-Comté, 15B, Avenue des Montboucons, 25030 Besancon, France
[2] Laboratoire Interdisciplinaire Carnot de Bourgogne (ICB), UMR 6303, CNRS, Univ. Bourgogne Franche-Comté, 9, Avenue Alain Savary, BP 47 870, 21078 Dijon, France
* Correspondence: nicolas.martin@femto-st.fr; Tel.: +33-363-08-2431

Citation: Sakkas, C.; Cote, J.-M.; Gavoille, J.; Rauch, J.-Y.; Cornuault, P.-H.; Krystianiak, A.; Heintz, O.; Martin, N. Tunable Electrical Properties of Ti-B-N Thin Films Sputter-Deposited by the Reactive Gas Pulsing Process. *Coatings* **2022**, *12*, 1711. https://doi.org/10.3390/coatings12111711

Academic Editors: Sheng-Rui Jian and Phuoc Huu Le

Received: 15 October 2022
Accepted: 8 November 2022
Published: 9 November 2022

Publisher's Note: MDPI stays neutral with regard to jurisdictional claims in published maps and institutional affiliations.

Copyright: © 2022 by the authors. Licensee MDPI, Basel, Switzerland. This article is an open access article distributed under the terms and conditions of the Creative Commons Attribution (CC BY) license (https://creativecommons.org/licenses/by/4.0/).

Abstract: Titanium-boron-nitrogen (Ti-B-N) thin films were deposited by RF reactive magnetron sputtering using a titanium diboride (TiB_2) target in an argon + nitrogen mixture. The argon mass flow rate was kept constant, whereas that of nitrogen was pulsed during the deposition. A constant pulsing period of P = 10 s was used, and the introduction time of the nitrogen gas (duty cycle (dc)) was systematically varied from dc = 0 to 100% of the pulsing period. This reactive gas pulsing process allowed the deposition of Ti-B-N thin films with various boron and nitrogen concentrations. Such adjustable concentrations in the films also led to changes in their electronic transport properties. Boron and nitrogen contents exhibited a reverse evolution as a function of the nitrogen duty cycle, which was correlated with the transition from a metallic to semiconducting-like behavior. A percolation model was applied to the electrical conductivity as a function of the nitrogen pulsing parameters, assuming some correlations with the evolution of the Ti-B-N thin film nanostructure.

Keywords: Ti-B-N; reactive sputtering; gas pulsing; electrical conductivity; percolation

1. Introduction

The development of nanostructured thin films by reactive sputtering for a wide range of applications is among the challenging tasks associated with creating innovative functional materials. Binary compounds combining two metallic elements or a single metal with a light element, such as carbon, boron, nitrogen or oxygen, have been extensively investigated in recent decades. The addition of a third element to form ternary materials remains relevant today, as the association of three elements may lead to the formation of one-, two- or even multiphase systems, as well as nanocomposite materials [1–3]. As a result, one of the most interesting features of these multiphasic materials is that their properties can be tailored by adjusting the size, volume fraction, distribution, composition, etc., of the appropriate phases. Among the large family of ternary compounds, many investigations have been focused on combinations of two metals with a light and reactive element [4–8]. Two metals are commonly associated with nitrogen or carbon for hard coatings [9,10] or with oxygen for optical applications [11,12]. Other studies report metallic oxynitrides (MO_xN_y), oxycarbides (MO_xC_y) or carbonitrides (MN_xC_y) from one metal with the two corresponding elements (i.e., oxygen + nitrogen, oxygen + carbon or nitrogen + carbon, respectively) [13–17]. For these thin film materials, one of the challenging tasks is the high reactivity of light elements towards the sputtered metal, which may restrain some reachable compositions, thus limiting the range of final properties.

Among ternary compounds, the ternary metal boron nitrides, namely M-B-N, were compiled by Rogl [18] for several metals. Such experimental data show that the phase diagram feature strongly depends on the metal affinity towards boron and nitrogen and that these ternary systems can be classified with respect to the chemical ability of the metal element to form binary borides, binary nitrides and/or ternary boron nitride compounds. Furthermore, consistent with this compilation, most relevant works devoted to metal boron nitride thin films make systematic claims about the key role played by nitrogen incorporation during the processing stage. This is particularly true when metal boron nitride films are prepared by reactive sputtering, whereby control of the process (and the resulting film properties) strongly depends on the nitrogen partial pressure [19–22]. Titanium boron nitrides (Ti-B-N) are no exception; a phase mixture consisting of materials with differing properties can be produced [23–25]. Because hard phases. such as TiN, TiB$_2$, and cubic BN, can coexist with soft phases, such as hexagonal BN or amorphous BN, the majority of studies on Ti-B-N films have been focused on their mechanical and tribological properties as hard and wear-resistant coatings [26–29]. It is also worth noting that enhanced hardness and wear resistance of sputter-deposited Ti-B-N thin films have been accordingly connected to the produced nanocomposite structure, the latter being closely related to the film composition and thus dependent on the nitrogen partial pressure implemented during the deposition process.

Although many investigations have been conducted with the aim of understanding the mechanical performances of Ti-B-N films, little work has been performed on their electrical behaviors as a function of B and N contents. It is of particular interest to determine how electronic transport properties change as a function of B and N concentrations, as pure TiB$_2$ and TiN materials both exhibit metallic-like behaviors. Combining these two materials often results in optimized mechanical and tribological properties with given amounts of TiB$_2$ and TiN in an amorphous matrix of BN. Because BN is an insulating material and assuming a nanocomposite structure of Ti-B-N films, the motivation of the present study is to understand how the electrical conductivity of Ti-B-N films can be tuned vs. light element concentration and to identify correlations with the most relevant sputtering process parameters.

In this work, we prepared Ti-B-N thin films by reactive sputtering using the reactive gas pulsing process (RGPP) [30]. Nitrogen mass flow was periodically supplied during the deposition stage with an increasing time of injection. The chemical composition was first determined so as to prove that B and N contents are adjustable by means of nitrogen gas pulsing. Electrical properties (resistivity, charge carrier mobility and concentration) were systematically measured, illustrating significant variations in the conductivity of Ti-B-N films under given nitrogen pulsing parameters with increasing nitrogen concentration in the films. Assuming a nanocomposite structure of Ti-B-N films, a percolation model is suggested to explain their electronic transport properties.

2. Materials and Methods

Ti-B-N thin films were prepared by RF reactive magnetron sputtering. The sputtering machine was a 110 L vacuum chamber evacuated by a cryogenic pump backed with a dry primary pump, achieving an ultimate pressure of 5×10^{-8} mbar. A TiB$_2$ rectangular target (381 × 127 × 6.35 mm^3; purity, 99.9%) was fixed 100 mm from the center of the substrate holder. The argon flow rate was kept constant at q_{Ar} = 29 sccm, and a constant pumping speed S_{Ar} = 95 L s^{-1} was used, leading to an argon partial pressure of 0.6 Pa. Before any deposition, an etching period of 15 min and 250 V of bias was applied to the substrates. Then, the TiB$_2$ target was RF-sputtered using a constant electric power of 1 kW. Furthermore, a presputtering time was applied for 5 min to remove the contamination layer on the target surface and stabilize the process, leading to a target self-bias potential of V_{SB} = 210 V. A nitrogen flow rate of q_{N2} was pulsed during Ti-B-N deposition by means of RGPP [30]. A rectangular signal was used to pulse nitrogen gas with a constant pulsing period of P = 10 s. This period was selected because such operating conditions correspond to a freedom of alternating between the boride- and nitride-poisoned state and vice-versa. For all depositions, the nitrogen injection time (t_{ON}) was varied from 0 to 10 s, corresponding to duty cycles (dc ($dc = t_{ON}/P$)) from 0 to 100% of P. The maximum nitrogen rate was q_{N2Max} = 5 sccm. This value corresponds to the amount of nitrogen required to completely avalanche the reactive sputtering process in the nitride sputtering mode [31]. During the t_{OFF} time, the nitrogen mass flow rate was completely stopped (q_{N2min} = 0 sccm). The total sputtering pressure and TiB$_2$ target voltage were alternated in the range of 6.0–6.3 × 10^{-3} mbar and 210–220 V for t_{OFF} and t_{ON} times, respectively. During deposition, the substrate holder (191 × 142 mm^2) was biased with a negative voltage of 20 V, and no external heating was added. Ti-B-N films were deposited on glass and (100) Si substrates. The deposition rate was measured based on the film thickness and the corresponding deposition time. The latter was adjusted in order to obtain a constant film thickness of 300 nm.

Electrical resistivity measurements from room temperature (25 °C) to 200 °C were conducted with a homemade system equipped with an annealing hot plate using the four-probe van der Pauw method. Carrier concentration and carrier mobility were assessed from setup when applying a magnetic field of B = 0.8 T in a device operating according to the van der Pauw method. Infrared analysis was performed with a Perkin Elmer Spectrum Two Lita FT-IR spectrometer in the wavenumber range of 400 to 4000 cm^{-1}. The film composition was obtained by X-ray photoelectron spectroscopy (XPS) with a PHI VersaProbe 1 system using a monochromatized and focalized Al Kα X-ray source (hυ = 1486.7 eV, spot diameter = 200 µm). The base pressure during analysis was more than 5 × 10^{-7} Pa. CasaXPS software (2.325) was used for data treatment [32].

3. Results

The deposition rate (R) of Ti-B-N films was first measured as function of the nitrogen duty cycle (dc) (Figure 1) and obtained as the thickness measurements of the films (average of 10 measurements, giving rise to error bars of the deposition rate) by mechanical profilometry (step height method) and the time of deposition.

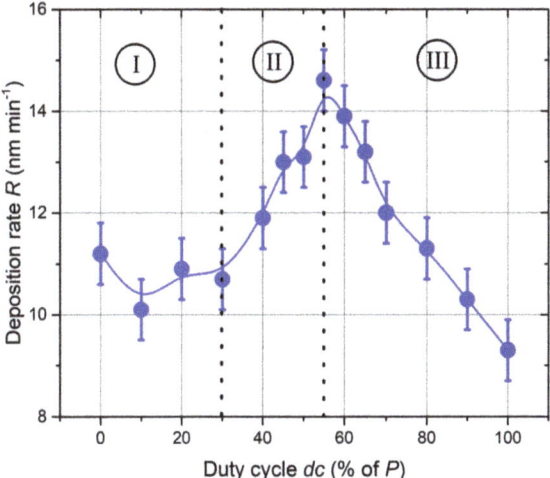

Figure 1. Ti-B-N deposition rate (R) as a function of the duty cycle (dc) when a TiB$_2$ target is RF-sputtered and nitrogen gas is injected with a pulsing period of P = 10 s. The deposition rate gradually changes, exhibiting a maximum dc close to 55% of P.

The evolution of deposition rate vs. duty cycle can be divided into three regions (I, II and III in Figure 1). For duty cycles lower than 30% of P (region I), a nearly constant rate can be assumed. The nitrogen injection time (less than 30% of P = 10 s) is too short for full nitriding of the TiB$_2$ target surface. The boride sputtering mode prevails, and Ti-B-N films can be expected to exhibit characteristics similar to those of the TiB$_2$ compound, i.e., metallic-like behavior. An increase in duty cycle from 30% to 55% of the pulsing period (P) (region II) gives rise to enhanced deposition rates from 10.7 to 14.6 nm min^{-1}, respectively. The deposition rate exhibits a maximum for dc = 55% of P, and a further increase in duty cycle (region III) produces a continuous and linear drop in the deposition rate to 9.3 nm min^{-1} when the nitrogen gas is constantly supplied (i.e., for dc = 100% of P). For duty cycles lower than 55% of P, the nitrogen injection time (t_{ON} time) is too short to completely set the reactive sputtering process in the nitride mode. Alternations between nitride and boride modes occur during t_{ON} and t_{OFF} times, respectively. Nitride mode prevails for duty cycles higher than 55% of P, and the TiB$_2$ target surface becomes increasingly covered by a nitride compound, which reduces the sputtering yield. Similarly, nitrogen atoms are progressively incorporated into the Ti-B-N films as the nitrogen injection t_{ON} time is increased, corresponding to the formation of a poisoning nitride layer on the target surface and a decrease in the deposition rate. This optimized deposition rate vs. nitrogen injection corroborates with results previously reported by Chaleix and Machet [33] and later by Pierson et al. [34], who also obtained a maximum rate for a given range of nitrogen flow rates. Similarly, we assigned this maximum rate of nitrogen incorporation to the growing films and a progressive nitriding of the target leading to a decrease in the deposition rate (typical of minimally reactive systems).

3.1. Composition and Structure

The films composition as a function of the duty cycle was first determined from XPS analyses. XPS spectra show clear signals corresponding to Ti 2p, N 1s and B 1s peaks. (Figure 2).

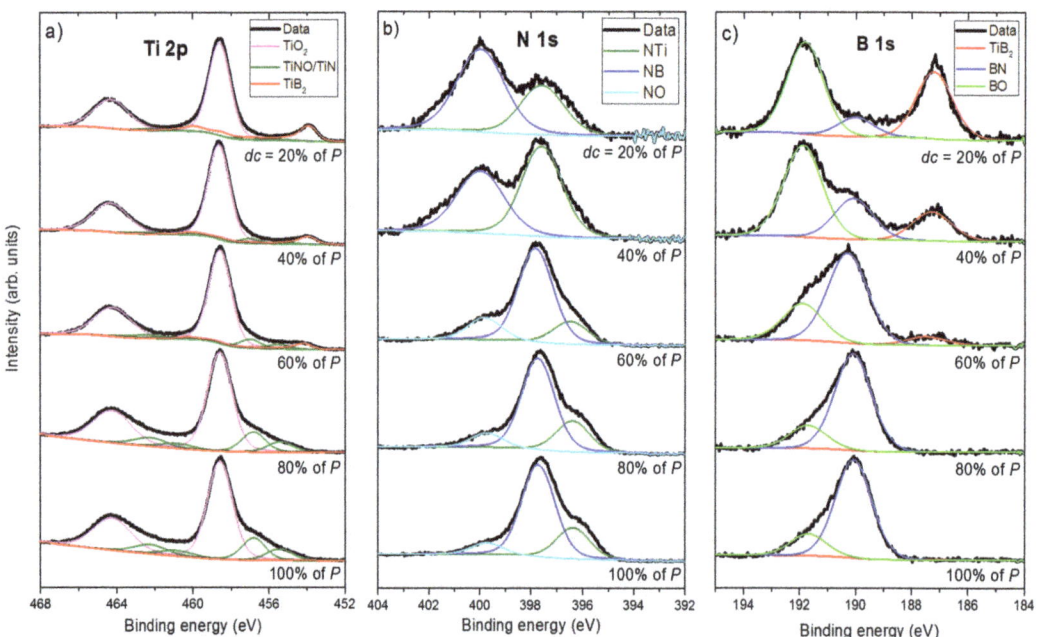

Figure 2. XPS spectra of (**a**) Ti 2p, (**b**) N 1s and (**c**) B 1s of Ti-B-N thin films prepared with various duty cycles (dc).

The position of these peaks is in agreement with the results recorded by other investigators [35,36]. A curve fitting-procedure was systematically performed for all recorded signals and for various duty cycles. The Ti 2p peak consists of three components corresponding to the contributions from TiO$_2$ (458.2 and 464.1 eV due to Ti 2p$^{1/2}$ and Ti 2p$^{3/2}$, respectively), TiNO/TiN (456.7 and 462.3 eV) and TiB$_2$ (454.3 eV), as illustrated in Figure 2a. For the N 1s spectra (Figure 2b), decomposition leads to three contributions: NTi (396.9 eV),

NB (398.3 eV) and NO (402.8 eV). The B 1s spectra also display three contributions assigned to TiB$_2$ (187.8 eV), BN (190.0 eV) and BO (192.1 eV).

For the 2p levels of titanium (Figure 2a), the decomposition is more complicated. The contributions of TiO$_2$ are still visible at 458.2 and 464.1 eV due to Ti 2p$^{1/2}$ and Ti 2p$^{3/2}$, respectively. The main difficulty is associated with determining the contributions of TiN/TiNO on the one hand and those of TiB$_2$ on the other hand. For each of these compounds, the forms are complex and cannot be reduced to a simple component. Moreover, because pure standards of TiN and TiB$_2$ compounds are not accurately produced, the best prepared samples were the TiN sample with 54% nitrogen and 46% titanium with no boron detected and a sample as close as possible to TiB$_2$ with 58% boron, 34% titanium and 8% nitrogen (atomic concentrations). The maximum of intensity was at 456.7 eV for TiNO/TiN and at 454.3 eV for TiB$_2$.

For the lowest duty cycles and up to dc = 40% of P, curve fitting of Ti 2p and B 1s peaks gives rise to signals corresponding to Ti–B bonds with TiB$_2$ peak at 454 eV and 187 eV, as shown in Figure 2a,c. As expected, a significant contribution of film oxidation is measured from Ti–O and B–O bonds with TiO$_2$ peaks at 458 and 464 eV (2p$^{1/2}$ and 2p$^{3/2}$ signals in Figure 2a), as well as with BO peak at 192 eV (Figure 2c). As the duty cycle increases (dc = 60% of P and higher), the TiB$_2$ contribution recorded from the Ti 2p and B 1s peaks reduces and vanishes for a constant supply of nitrogen. Similarly, the BN peak from N 1s (398 eV) and B 1s (190 eV) becomes increasingly intense, as well as the TiN peak at 396 eV (Figure 2b). The film oxidation remains for any duty cycle, but it mainly originates from Ti signals (high reactivity of this metal towards oxygen), with a significant reduction in the influence of B–O and N–O bonds, as reflected by NO and BO peaks at 400 eV (Figure 2b) and 192 eV (Figure 2c), respectively. As expected, increasing the duty cycle from 0 to 100% of P favors the amount of Ti-N and B-N bonds, whereas that of Ti–B bonds decreases without completely preventing film oxidation due to the high reactivity between titanium and oxygen.

[Ti], [B] and [N] atomic concentrations in the films as a function of duty cycle (neglecting the amount of oxygen due to surface oxidation) were determined based on Ti 2p, N 1s and B 1s signals (Figure 3).

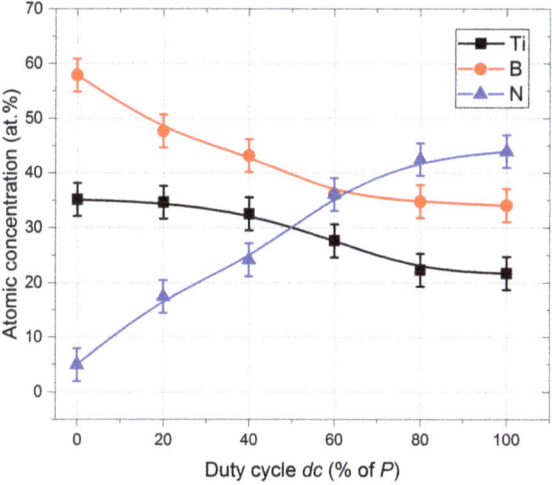

Figure 3. [Ti], [B] and [N] atomic concentrations vs. duty cycle (dc) obtained from XPS measurements.

These three elemental concentrations exhibit a continuous and smooth variation with increased nitrogen injection time. B and N concentrations show a reverse evolution, whereas the Ti concentration slightly reduces (note that concentrations were determined assuming homogeneous compositions through the film thickness, which is not completely relevant and induces an assumed inaccuracy of ±3 at.% for the chemical concentration). Without nitrogen pulsing (dc = 0% of P), the [B]/[Ti] atomic concentration ratio is close to 1.8, which is lower than that of the stoichiometric TiB$_2$ compound due to the presence of oxygen (more than 10 at.%) and a few at.% of nitrogen. When nitrogen is constantly injected (dc = 100% of P), nitrogen-rich Ti-B-N films are prepared with [N] content of more than 44 at.% and [B] content of approximately 35 at.%. For duty cycles higher than 60% of P, the three element concentrations tend to stabilize. The reverse evolution of nitrogen and boron contents vs. nitrogen supply is in agreement with results previously reported by Pierson et al. [37] and Han et al. [38], who also sputter-deposited Ti-B-N films from a TiB$_2$ target in an Ar + N$_2$ atmosphere. WE also observed the same kind of saturation for Ti, B and N atomic concentrations as the nitrogen flow rate increased (stabilization of the concentrations for nitrogen injections depending on the operating conditions).

The composition of the films is plotted within the Ti-B-N ternary phase diagram (Figure 4).

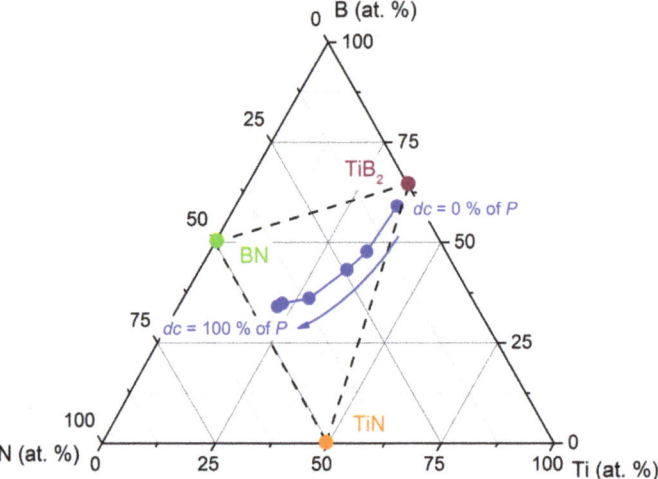

Figure 4. The composition of Ti-B-N films (blue symbols) sputter-deposited by pulsing the nitrogen gas from dc = 0 to 100% of the pulsing period of P = 10 s within a ternary phase diagram. The positions of TiB$_2$, TiN and BN stoichiometric compounds are indicated, as well as quasi-binary tie lines between the individual compounds (dashed lines).

TiB$_2$, TiN and BN stoichiometric compounds are indicated, as well as quasi-binary tie lines between the individual compounds (dashed lines). Nitrogen pulsing does not produce a chemical composition evolution along the quasi-binary TiN-TiB$_2$ line, as reported for Ti-B-N films prepared by PACVD [39]. Instead, it moves away from such a line as the duty cycle increases with a BN enrichment of the films, as the nitrogen supply tends to be constant. This trend was previously reported by others [40,41], giving rise to a comparable saturation of nitrogen content in the Ti-B-N films, even by sputtering a TiB$_2$ target in a pure nitrogen atmosphere. At first, as proposed by Mayrhofer et al. [42], we suggest the formation of a composite structure composed of TiN and TiB$_2$ nanocrystals randomly distributed in the film and embedded into a disordered and amorphous a-BN$_x$ compound. Assuming this nanocrystalline Ti-B-N structure, the corresponding phase proportions (mole fractions) are 42% TiN, 10% TiB$_2$ and 48% BN in the equilibrium phase diagram for Ti-B-N films prepared with dc = 100% of P. As expected, reducing the duty cycle leads to reduced contents of TiN and BN phases (e.g., for dc = 40% of P, phase proportions in mole fractions are 28% TiN, 53% TiB$_2$ and 19% BN). Furthermore, the use of a short nitrogen injection time (dc = 20% of P) leads to a significant incorporation of N in Ti-B-N films due to the high reactivity of nitrogen towards titanium and boron elements.

The occurrence of chemical bonds can be highlighted by FTIR analyses, especially for light elements, such as B and N. Figure 5 shows the FTIR spectra recorded for Ti-B-N films deposited on (100) silicon and for various duty cycles. No clear signals are recorded for duty cycles lower than 45% of P. Some peaks appear as the nitrogen injection time increases and become more significant when the duty cycle exceeds 55% of P. A broad asymmetric band appears close to 1390 cm^{-1} and clearly develops as dc reaches 90% of P, which is mainly assigned to the characteristic absorption of hexagonal BN (in-plane B-N bond stretching) in a Ti-B-N single layer [43,44].

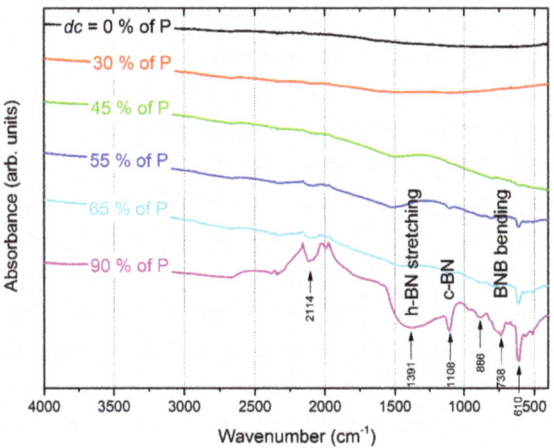

Figure 5. FTIR spectra measured for Ti-B-N films prepared by pulsing nitrogen gas with an increasing duty cycle from 0 to 90% of the pulsing period of P = 10 s.

Similarly, a broad and significant signal is measured around 738 cm^{-1} related to the B-N-B bending mode occurring in the amorphous boron nitride [37], and a second signal is associated with the transverse optical mode of the cubic BN phase at 1108 cm^{-1} [45,46]. A narrow and intense peak can also be noticed at 610 cm^{-1}, with a few weak peaks between 500 and 600 cm^{-1}. This group of signals corresponds to Ti-N bonds (typical stretching mode of TiN around 600 cm^{-1} [47]). All these FTIR signals support the formation and increasing amount of B-N and Ti-N bonds for high duty cycles, maintaining an amorphous film structure. These results are in agreement with X-ray diffraction analyses results in which no diffraction peaks were detected due to boron nitride or other phases (not shown here).

3.2. Electrical Properties

The DC electrical resistivity of Ti-B-N thin films deposited on glass at room temperature (ρ_{300K}) continuously rises as a function of the duty cycle, spanning more than six orders of magnitude as dc changes from 0 to 100% of P (Figure 6). Because the nitrogen pulsing period is very short (P = 10 s) and as a result of variation in deposition rates from 9.3 to 14.6 nm min^{-1} (Figure 1), a random N and B distribution is first produced through the film thickness rather than a multilayered structure, which could influence electrical conductivity, as previously reported for metal/metal oxide coatings [48]. For a dc lower than 40% of P, the resistivity is in the range of typical metals, i.e., below 10^{-5} Ω m. Afterwards, Ti-B-N films abruptly become more resistive for duty cycles between 40 and 80% of P and tend to stabilize to a few Ω m as nitrogen gas is constantly supplied. Carrier mobility (μ_{300K}) and carrier concentration (n_{300K}) at 300 K are also both influenced by the nitrogen injection for a similar range of duty cycles. The carrier concentration is more than 10^{28} m^{-3} for films prepared with duty cycles of less than 40% of P, typically corresponding to metallic-like behaviors. The carrier concentration loses several orders of magnitude as the duty cycle changes from 60 to 100% of P, where n_{300K} = 2.9 × 10^{19} m^{-3}.

Similarly, the carrier mobility exhibits a reverse evolution with μ_{300K} lower than a few 10^{-4} m^2 V^{-1} s^{-1} for the lowest duty cycles, whereas it reaches 2.0×10^{-2} m^2 V^{-1} s^{-1} for dc = 100% of P. As a result, this drop in conductivity is mainly assigned to the strong decrease in the free carrier concentration, which prevails with enhanced mobility. These significant variations in electronic transport properties of Ti-B-N films correlate with their nitrogen enrichment and thus with an increasing amount of amorphous BN phase. These results support conclusions previously reported by Rogl [18], who claimed that metallic-like behavior can be expected for boron-rich compounds but that with increasing nitrogen content, semiconducting insulating properties can be expected to develop.

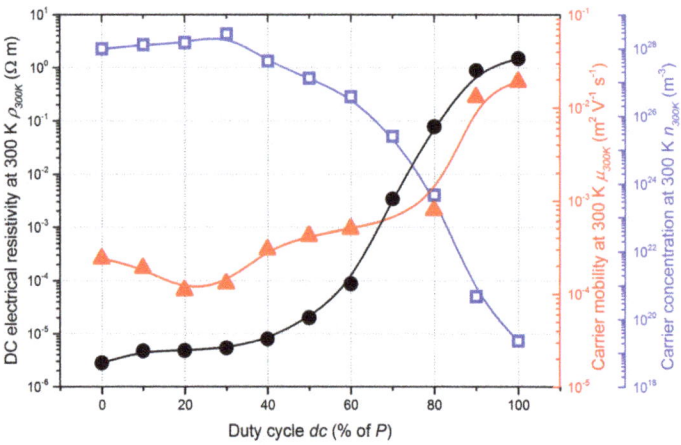

Figure 6. DC electrical resistivity (ρ_{300K}), carrier mobility (μ_{300K}) and carrier concentration (n_{300K}) measured at room temperature as a function of the duty cycle (dc) of Ti-B-N thin films.

Moreover, this abrupt increase in resistivity obtained for duty cycles between 40 and 80% of P also corresponds to the transition from absorbent (metallic-like) to transparent (semiconducting-like) Ti-B-N films in the visible region (optical transmittance spectra of Ti-B-N films deposited on glass not shown here).

The relationship of resistivity vs. temperature also supports this gradual metal-to-semiconducting transition as the duty cycle and thus the nitrogen content in Ti-B-N films increases (Figure 7). A nearly temperature-independent resistivity can be noticed for the lowest duty cycles, where ρ is maintained nearly constant from 30 to 200 °C, with, for example, a negative temperature coefficient of resistance at 300 K (TCR$_{300K}$) of -8.6×10^{-5} K^{-1} for films prepared without nitrogen pulsing. This is consistent with the results reported by Pierson et al. [49], who also reported negative TCR values and the loss of metallic character, even for Ti-B-N films prepared with low nitrogen flow rates. Their results also showed a significant increase in TiBN electrical resistance for a given range of nitrogen flow rates corresponding to negative TCRs, which is, in some ways similar, to the sudden increase in resistivity measured for duty cycles higher than 40% of P (Figure 6). In our TiBN films, TCR$_{300K}$ becomes even more negative for dc = 50% of P, with TCR$_{300K}$ = -3.9×10^{-4} K^{-1}, and a further increase in the nitrogen injection time leads to a semiconducting-like behavior, with an exponential decrease in the film resistivity as the temperature rises. An Arrhenius plot (electrical conductivity vs. reciprocal temperature) gives rise to an activation energy of E_a = 12 meV for dc = 60% of P, reaching 124 meV for a constant supply of nitrogen flow rate (dc = 100% of P). This increasing activation energy is connected to the nitrogen enrichment of Ti-B-N films, as well as to the formation of an amorphous and insulating BN phase. However, the highest value of activation energy is quite low compared to typical semiconducting materials, although it remains in the order of magnitude of some nitride or boride semiconductors, for which the temperature-dependent resistivity strongly depends on their composition [50].

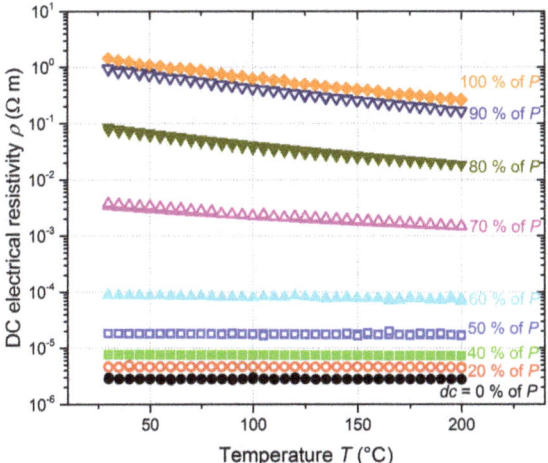

Figure 7. DC electrical resistivity (ρ) vs. temperature (T) measured on Ti-B-N thin films deposited on glass for various duty cycles ($dc = 0$ to 100% of the constant pulsing period of $P = 10$ s).

Our results clearly show that electronic transport characteristics of Ti-B-N films progressively but rapidly change from metallic to semiconducting behaviors as a function of the duty cycle, in direct association with variations in composition and structure. At first and as frequently modeled for Ti-B-N mechanical properties, we can suppose that films exhibit a nanocomposite structure [39,42,51]. The latter is made of nanometric TiN and TiB$_2$ crystallites (conductive phase) embedded in a disordered BN matrix (insulating medium).

Based on the sudden change in electrical properties of Ti-B-N films for a given range of duty cycles (Figure 6) and assuming a nanocomposite structure, a percolation model was used to describe the evolution of thin film conductivity at 300 K (σ_{300K}) as a function of 1-ndc, where ndc is defined as the normalized duty cycle (Figure 8). Sputtering conditions with 1-ndc = 0 correspond to the most N-rich Ti-B-N films (i.e., dc = 100% of P) and therefore the most resistive films. Conductivity abruptly increases when 1-ndc reaches 0.4, and then tends to saturate. This σ_{300K} vs. 1-ndc evolution typically behaves like a percolation phenomenon. Two electrically distinct media (insulating amorphous BN matrix and conducting TiN and TiB$_2$ phases) are mixed in Ti-B-N films. The proportion of the conducting phase increases with increasing 1-ndc. As a result, the insulating–conducting transition corresponds to the percolation threshold, which can be determined assuming an effective medium theory (EMT) [52]. The percolation theory allows for description of the electrical conductivity of a composite medium before and after the percolation threshold (ϕ_c) through the following equations [53]:

$$\sigma = \sigma_i \left(\frac{\phi_c - \phi}{\phi_c} \right)^{-s}, \quad (1)$$

before the percolation threshold (i.e., $\phi < \phi_c$), where $\sigma = \sigma_{300K}$ is the DC electrical conductivity at 300 K (S m^{-1}), σ_i is the DC electrical conductivity of the insulating medium (S m^{-1}), $\phi = 1\text{-}ndc$ is the conducting phase ratio (arb. units) and s is the critical exponent in the insulating region.

$$\sigma = \sigma_c \left(\frac{\phi - \phi_c}{1 - \phi_c} \right)^t, \quad (2)$$

after the percolation threshold (i.e., $\phi > \phi_c$), where σ_c is the DC electrical conductivity of the conducting phase (S m^{-1}), and t is the critical exponent in the conducting region.

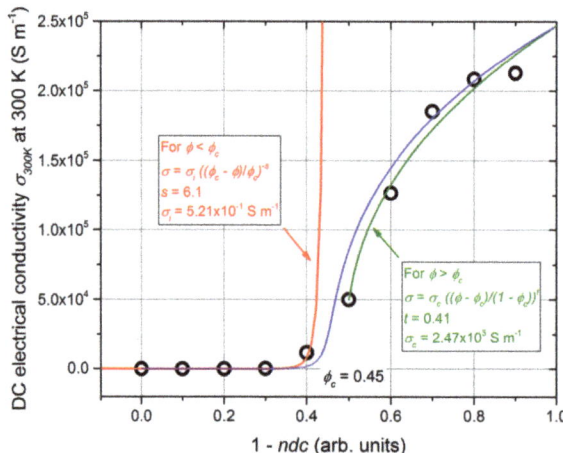

Figure 8. DC electrical conductivity at room temperature (σ_{300K}) vs. $\phi = 1\text{-}ndc$ for Ti-B-N thin films (ndc is the normalized duty cycle). Red and green lines are related to the conductivity calculated from the fitting before (Equation (1)) and after (Equation (2)) the percolation threshold. The blue line is the best fit obtained from Equation (3).

For the full range of conducting phase ratios (ϕ), a phenomenological relationship, can be used, as suggested by McLachlan et al. [54]:

$$(1-\phi)\left(\frac{\sigma_i^{\frac{1}{s}} - \sigma^{\frac{1}{s}}}{\sigma_i^{\frac{1}{s}} - A\sigma^{\frac{1}{s}}}\right) + \phi\left(\frac{\sigma_c^{\frac{1}{t}} - \sigma^{\frac{1}{t}}}{\sigma_c^{\frac{1}{t}} - A\sigma^{\frac{1}{t}}}\right) = 0, \qquad (3)$$

with $A = \frac{1-\phi_c}{\phi_c}$.

The best fittings were performed from these equations, as shown in Figure 8 (red and green lines for Equations (1) and (2), respectively). A percolation threshold of $\phi_c = 0.45$ was obtained, which corresponds to dc = 55% of P. Based on the chemical compositions determined from XPS analyses (Figure 3), this duty cycle produced comparable B and N concentrations in Ti-B-N films. This result supports Mayrhofer et al., who claimed that this range of compositions matches with a film composed of two nanoscale phases (2–3 nm) encapsulated by a high-volume fraction (about 50%) of a disordered phase [38,41].

According to the experimental data at $\phi = 1$, the conductivity of the conducting phase is $\sigma_c = 2.47 \times 10^5$ S m^{-1}. This value is lower than that of single TiN and TiB$_2$ films ($\sigma_{TiN} = 1.85 \times 10^6$ S m^{-1} [55] and $\sigma_{TiB2} = 3.52 \times 10^6$ S m^{-1} [56]), as the conducting phase certainly contains defects and is composed of small grains (a few nm), although it remains in the order of magnitude of metallic materials. With respect to the insulating medium, results of the fits yield $\sigma_i = 5.21 \times 10^{-1}$ S m^{-1}. This value is higher than conductivities usually reported for BN compounds (σ is below 10^{-6} S m^{-1} for pure c-BN at room temperature [57]). This difference is understandable, as the electronic transport properties of BN strongly depend on crystallinity, preparation method, and the nature and type of doping [57]. In our Ti-B-N films, the insulating matrix certainly contains a significant amount of defects and Ti, favoring mobility and the concentration of free carriers, thus enhancing the overall conductivity.

Critical exponents s and t are also of interest, and universal values between $s_{un} = 0.7$–1 and $t_{un} = 1.6$–2 are commonly reported for three-dimensional systems with an inverse Swiss cheese model [52]. For our Ti-B-N films, fittings from experimental data lead to s = 6.1 and t = 0.41, which different significantly from universal values. These discrepancies can be assigned to an inhomogeneous structure of the films (percolation theory is based on perfect spherical particles for both insulating and conducting media). Sputtering deposition at room temperature often leads to columnar architectures [58], and RGPP may produce periodic multilayers [59]. As a result, our Ti-B-N films can be considered an anisotropic system affecting the percolation law, as the assumed homogeneity and sphericity of media are not fully established. However, fitting with Equation (3) and using parameters obtained from each region allows for a reliable description of the electrical conductivity of Ti-B-N films for the full range of duty cycles (blue line in Figure 8). This two-exponent phenomenological percolation equation can be used to assess the complex DC electrical conductivity of Ti-B-N films exhibiting reverse B and N concentrations.

4. Conclusions

In this study, 300 nm thick Ti-B-N thin films were sputter-deposited by RF reactive sputtering. A TiB_2 target was sputtered in a reactive atmosphere composed of Ar and N_2 gases. A reactive gas pulsing process (RGPP) was implemented to inject nitrogen gas with a constant pulsing period of $P = 10$ s. The duty cycle (dc) was systematically varied (t_{ON} injection time) in order to tune the chemical composition of the films. Nitrogen enrichment was achieved as the dc gradually increased from 0 to 100% of P. In the same way, the boron concentration was inversely reduced, maintaining a nearly constant Ti content. Furthermore, increasing the nitrogen injection time favored the occurrence of B-N bonds in the films, with a phase proportion close to 48% of BN (mole fraction) when nitrogen was constantly supplied ($dc = 100\%$ of P).

For duty cycles between 40% and 80% of P, electronic transport properties measured at room temperature significantly changed, with a resistivity enhancement of four orders of magnitude, whereas carrier mobility and concentration correspondingly changed for the same range of dc. A regular but rapid transition from metallic to semiconducting-like behavior was highlighted as a function of the duty cycle, as supported by the resistivity vs. temperature measurements. Assuming a nanocomposite structure in Ti-B-N films made of nanometric TiN and TiB_2 crystallites (conductive phase) embedded in a disordered BN matrix (insulating medium), a percolation model adequately describes the conductivity of the tunable films as a function of nitrogen injection.

These results demonstrate that the proposed nanocomposite structure for mechanical properties and wear resistance of Ti-B-N films can also be applied to describe the electrical properties of such films as a function of B and N concentrations. Furthermore, our results indicate that the RGPP technique is a suitable approach to adjust electronic transport properties by means of a simple and easy change in pulsing parameters. A further incorporation of nitrogen in films would be relevant to further extend multiphase nanostructuring growth.

Author Contributions: Conceptualization and data curation, C.S.; methodology and software, J.-M.C.; formal analysis and validation, J.G.; formal analysis and validation, J.-Y.R.; formal analysis and validation, P.-H.C.; data curation and investigation, A.K.; formal analysis and validation, O.H.; writing—review and editing and funding acquisition, N.M. All authors have read and agreed to the published version of the manuscript.

Funding: This research was partially funded by Fonds Européen de Développement Régional—FEDER (grant number CTE 6059) in the framework of the TOOLEXPERT Interreg V project.

Institutional Review Board Statement: Not applicable.

Informed Consent Statement: Not applicable.

Data Availability Statement: All data are presented in the current manuscript.

Acknowledgments: This work was supported by the Région Bourgogne Franche-Comté and by EIPHI Graduate School (contract 'ANR-17-EURE-0002').

Conflicts of Interest: The authors declare no conflict of interest. The funders had no role in the design of the study; in the collection, analyses or interpretation of data; in the writing of the manuscript; or in the decision to publish the results.

References

1. McManus-Driscoll, J.L.; Zerrer, P.; Wang, H.; Yang, H.; Yoon, J.; Fouchet, A.; Yu, R.; Blamire, M.G.; Jia, Q. Strain and spontaneous phase ordering in vertical nanocomposite heteroepitaxial thin films. *Nature Mater.* **2008**, *7*, 314–320. [CrossRef]
2. Chen, A.; Bi, Z.; Jia, Q.; MacManus-Driscoll, J.L.; Wang, H. Microstructure, vertical strain control and tunable functionality in self-assembled vertically aligned nanocomposite thin films. *Acta Mater.* **2013**, *61*, 2783–2792. [CrossRef]
3. Wu, J.F.; Yu, L.H.; Hu, H.B.; Asempah, I.; Xu, J.H. Structural, mechanical and tribological properties of NbCN-Ag nanocomposite films deposited by reactive magnetron sputtering. *Coatings* **2018**, *8*, 50. [CrossRef]
4. Lopez, D.A.S.; Chagas, L.G.; Batista, A.D.; Guaita, M.G.D.; Amorin, L.H.C.; Da Silva, P.R.C.; Yamashini, G.; Zaia, D.A.M.; De Santana, H.; Urbano, A. Effect of RF magnetron sputtering parameters on the optimization of the discharge capacity of ternary lithium oxide thin films. *J. Mater. Sci.* **2021**, *32*, 17462–17472. [CrossRef]
5. Txintxurreta, J.; G-Berasategui, E.; Ortiz, R.; Hernandez, O.; Mendizabal, L.; Barriga, J. Indium oxide thin film deposition by magnetron sputtering at room temperature for the manufacturing of efficient transparent heaters. *Coatings* **2021**, *11*, 92. [CrossRef]
6. Saikumar, A.K.; Sundaresh, S.; Nehate, S.D.; Sundaram, K.B. Properties of RF magnetron sputtered copper gallium oxide ($CuGa_2O_4$) thin films. *Coatings* **2021**, *11*, 921. [CrossRef]
7. Rebholz, C.; Leyland, A.; Schneider, J.M.; Voevodin, A.A.; Matthews, A. Structure, hardness and mechanical properties of magnetron-sputtered titanium-aluminium boride thin films. *Surf. Coat. Technol.* **1999**, *120*, 412–417. [CrossRef]
8. Minemoto, T.; Negami, T.; Nishiwaki, S.; Takakura, H.; Hamakawa, Y. Preparation of $Zn_{1-x}Mg_xO$ films by radio frequency magnetron sputtering. *Thin Solid Films* **2000**, *372*, 173–176. [CrossRef]
9. Greczynski, G.; Lu, J.; Jensen, J.; Bolz, S.; Kolker, W.; Schiffers, C.; Lemmer, O.; Greene, J.E.; Hultman, L. A review of metal-ion-flux-controlled growth of metastable TiAlN by HIPIMS/DCMS co-sputtering. *Surf. Coat. Technol.* **2014**, *257*, 15–25. [CrossRef]
10. Palmquist, J.P.; Birch, J.; Jansson, U. Deposition of epitaxial ternary metal carbide films. *Thin Solid Films* **2002**, *405*, 122–128. [CrossRef]
11. Glynn, C.; Aureau, D.; Collins, G.; O'Hanlon, S.; Etcheberry, A.; O'Dwyer, C. Solution processable broadband transparent mixed metal oxide nanofilm optical coatings via substrate diffusion doping. *Nanoscale* **2015**, *7*, 20227–20237. [CrossRef] [PubMed]
12. Steinecke, M.; Kiedrowski, K.; Jupé, M.; Ristau, D. Very thick mixture oxide ion beam sputtering films for investigation of nonlinear material properties. *Eur. Phys. J. Appl. Phys.* **2017**, *80*, 30301–30305. [CrossRef]
13. Martin, N.; Banakh, O.; Santo, A.M.E.; Springer, S.; Sanjinès, R.; Takadoum, J.; Lévy, F. Correlation between processing and properties of TiO_xN_y thin films sputter deposited by the reactive gas pulsing technique. *Appl. Surf. Sci.* **2001**, *185*, 123–133. [CrossRef]
14. Vaz, F.; Cerqueira, P.; Rebouta, L.; Nascimento, S.M.C.; Alves, E.; Goudeau, P.; Rivière, J.P.; Pischow, K.; de Rijk, J. Structural, optical and mechanical properties of coloured TiN_xO_y thin films. *Thin Solid Films* **2004**, *447*, 449–454. [CrossRef]
15. Fernandes, A.C.; Carvalho, P.; Vaz, F.; Lancero-Mendez, S.; Machado, A.V.; Parreira, N.M.G.; Pierson, J.F.; Martin, N. Property change in multifunctional TiC_xO_y thin films: Effect of the O/Ti ratio. *Thin Solid Films* **2006**, *515*, 866–871. [CrossRef]
16. Sanchez-Lopez, J.C.; Abad, M.D.; Carvalho, I.; Galindo, R.E.; Benito, N.; Ribeiro, S.; Henriques, M.; Cavaleiro, A.; Carvalho, S. Influence of silver content on the tribomechanical behavior on Ag-TiCN bioactive coatings. *Surf. Coat. Technol.* **2012**, *206*, 2192–2198. [CrossRef]
17. Ji, A.L.; Ma, L.B.; Liu, C.; Li, C.R.; Cao, Z.X. Synthesis and characterization of superhard aluminum carbonitride thin films. *Diam. Relat. Mater.* **2005**, *14*, 1348–1352. [CrossRef]
18. Rogl, P. Materials science of ternary metal boron nitrides. *Int. J. Inorg. Mater.* **2001**, *3*, 201–209. [CrossRef]
19. Lu, Y.H.; Zhou, Z.F.; Sit, P.; Shen, Y.G.; Li, K.Y.; Chen, H. X-ray photoelectron spectroscopy characterization of reactively sputtered Ti-B-N thin films. *Surf. Coat. Technol.* **2004**, *187*, 98–105. [CrossRef]
20. Tian, C.X.; Wang, Z.S.; Zou, C.W.; Tang, X.S.; Xie, X.; Lie, S.Q.; Liang, F.; Li, Z.J.; Liu, Y.F.; Su, F.H. Ternary and quaternary TiBN and TiBCN nanocomposite coatings deposited by arc ion plating. *Surf. Coat. Technol.* **2019**, *359*, 445–450. [CrossRef]
21. Lu, Y.H.; Shen, Y.G.; Li, K.Y. Nanostructured two-phase nc-TiN/a-(TiB_2, BN) nanocomposite thin films. *Appl. Surf. Sci.* **2006**, *253*, 1631–1638. [CrossRef]
22. Mayrhofer, P.H.; Willmann, H.; Mitterer, C. Recrystallization and grain growth of nanocomposite Ti-B-N coatings. *Thin Solid Films* **2003**, *440*, 174–179. [CrossRef]
23. Aoudi, S.M.; Namavar, F.; Gorishnyy, T.Z.; Rohde, S.L. Characterization of TiBN films grown by ion beam assisted deposition. *Surf. Coat. Technol.* **2002**, *160*, 145–151. [CrossRef]
24. Andrievski, R.A. Structure and properties of nanostructured boride/nitride materials. *Int. J. Refract. Met. Hard Mater.* **1999**, *17*, 153–155. [CrossRef]
25. Mayrhofer, H.; Mitterer, C.; Clemens, H. Self-organized nanostructures in hard ceramic coatings. *Adv. Eng. Mater.* **2005**, *7*, 1071–1082. [CrossRef]
26. Kainz, C.; Schalk, N.; Tkadletz, M.; Mitterer, C.; Czettl, C. The effect of B and C addition on microstructure and mechanical properties of TiN hard coatings grown by chemical vapor deposition. *Thin Solid Films* **2019**, *688*, 137283–8. [CrossRef]
27. Zhou, S.Y.; Pelenovich, V.O.; Han, B.; Yousaf, M.I.; Yan, S.J.; Tian, C.X.; Fu, D.J. Effects of modulation period on microstructure, mechanical properties of TiBN/TiN nanomultilayered films deposited by multi arc ion plating. *Vacuum* **2016**, *126*, 34–40. [CrossRef]

28. Asempah, I.; Xu, J.H.; Yu, L.H.; Luo, H.; Liu, J.L.; Yu, D.; Ding, N. The role of copper incorporation on the microstructure, mechanical and tribological properties of TiBN-Cu films by reactive magnetron sputtering. *J. Alloy. Compd.* **2019**, *801*, 112–122. [CrossRef]
29. Cicek, H.; Baran, O.; Demirci, E.E.; Tahmasebian, M.; Totik, Y.; Efeoglu, I. The effect of nitrogen flow rate on TiBN coatings deposited on cold work tool steel. *J. Adhes. Sci. Technol.* **2014**, *28*, 1140–1148. [CrossRef]
30. Martin, N.; Lintymer, J.; Gavoille, J.; Chappé, J.M.; Sthal, F.; Takadoum, J.; Vaz, F.; Rebouta, L. Reactive sputtering of TiO_xN_y coatings by the reactive gas pulsing process—Part I: Pattern and period of pulses. *Surf. Coat. Technol.* **2007**, *201*, 7720–7726. [CrossRef]
31. El Mouatassim, A.; Pac, M.J.; Pailloux, F.; Amiard, G.; Henry, P.; Rousselot, C.; Eydi, D.; Tuilier, M.H.; Cabioc'h, T. On the possibility of synthesizing multilayered coatings in the (Ti, Al)N system by RGPP: A microstructural study, *Surf. Coat. Technol.* **2019**, *374*, 845–851. [CrossRef]
32. Fairley, N.; Fernandez, V.; Richard-Plouet, M.; Guillot-Deudon, C.; Walton, J.; Smith, E.; Flahaut, D.; Greiner, M.; Biesinger, M.; Tougaard, S.; et al. Systematic and collaborative approach to problem solving using X-ray photoelectron spectroscopy. *Applied Surface Sci. Adv.* **2021**, *5*, 100112. [CrossRef]
33. Chaleix, L.; Machet, J. Study of the composition and of the mechanical properties of TiBN films obtained by d. c. magnetron sputtering. *Surf. Coat. Technol.* **1997**, *91*, 74–82. [CrossRef]
34. Pierson, J.F.; Chapusot, V.; Billard, A.; Alnot, M.; Bauer, P. Characterisation of reactively sputtered of Ti-B-N and Ti-B-O coatings. *Surf. Coat. Technol.* **2002**, *151–152*, 526–530. [CrossRef]
35. Holzschuh, H. Deposition of Ti-B-N (single and multilayer) and Zr-B-N coatings by chemical vapor deposition techniques on cutting tools. *Thin Solid Films* **2004**, *469–470*, 92–98. [CrossRef]
36. Hahn, R.; Tymoszuk, A.; Wojcik, T.; Kirnbauer, A.; Kozak, T.; Capek, J.; Sauer, M.; Foelske, A.; Hunold, O.; Polcik, P.; et al. Phase formation and mechanical properties of reactively and non-reactively sputtered Ti-B-N hard coatings. *Surf. Coat. Technol.* **2021**, *420*, 127327. [CrossRef]
37. Pierson, J.F.; Tomasella, E.; Bauer, P. Reactively sputtered Ti-B-N nanocomposite films: Correlation between structure and optical properties. *Thin Solid Films* **2002**, *408*, 26–32. [CrossRef]
38. Han, B.; Neena, D.; Wang, Z.; Kondamareddy, K.K.; Li, N.; Zuo, W.; Yan, S.; Liu, C.; Fu, D. Investigation of structure and mechanical properties of plasma vapor deposited nanocomposite TiBN films. *Plasma Sci. Technol.* **2017**, *19*, 045503. [CrossRef]
39. Mayrhofer, P.H.; Stoiber, M. Thermal stability of superhard Ti-B-N coatings. *Surf. Coat. Technol.* **2007**, *201*, 6148–6153. [CrossRef]
40. Lin, J.; Moore, J.J.; Mishra, B.; Pinkas, M.; Sproul, W.D. The structure and mechanical and tribological properties of TiBCN nanocomposite coatings. *Acta Mater.* **2010**, *58*, 1554–1564. [CrossRef]
41. Karuna Purnapu Rupa, P.; Chakraborti, P.C.; Mishra, S.K. Structure and indentation behavior of nanocomposite Ti-B-N films. *Thin Solid Films* **2014**, *564*, 160–169. [CrossRef]
42. Mayrhofer, P.H.; Mitterer, C.; Wen, J.G.; Petrov, I.; Greene, J.E. Thermally induced self-hardening of nanocrystalline Ti-B-N thin films. *J. Appl. Phys.* **2006**, *100*, 044301. [CrossRef]
43. Chu, K.; Shen, Y.G. Mechanical and tribological properties of nanostructured TiN/TiBN multilayer films. *Wear* **2008**, *265*, 516–524. [CrossRef]
44. Tsai, P.C. The deposition of characterization of BCN films by cathodic arc plasma evaporation. *Surf. Coat. Technol.* **2007**, *201*, 5108–5113. [CrossRef]
45. Kurooka, S.; Ikeda, T.; Kohama, K.; Tanaka, T.; Tanaka, A. Formation and characterization of BN films with Ti added. *Surf. Coat. Technol.* **2003**, *166*, 111–116. [CrossRef]
46. Moreno, H.; Caicedo, J.C.; Amaya, C.; Munoz-Saldana, J.; Yate, L.; Esteve, J.; Prieto, P. Enhancement of surface mechanical properties by using TiN/[BCN/BN]$_n$/c-BN multilayer system. *Appl. Surf. Sci.* **2010**, *257*, 1098–11104. [CrossRef]
47. Sedira, S.; Achour, S.; Avci, A.; Eskizeybek, V. Physical deposition of carbon doped titanium nitride film by DC magnetron sputtering for metallic implant coating use. *Appl. Surf. Sci.* **2014**, *295*, 81–85. [CrossRef]
48. Cacucci, A.; Tsiaoussi, I.; Potin, V.; Imhoff, L.; Martin, N.; Nyberg, T. The interdependence of structural and electrical properties in TiO_2/TiO/Ti periodic multilayers. *Acta Mater.* **2013**, *61*, 4215–4225. [CrossRef]
49. Pierson, J.F.; Bertran, F.; Bauer, J.P.; Jolly, J. Structural and electrical properties of sputtered titanium boronitride films. *Surf. Coat. Technol.* **2001**, *142–144*, 906–910. [CrossRef]
50. Wang, C.C.; Akbar, S.A.; Chen, W.; Patton, V.D. Electrical properties of high-temperature oxides, borides, carbides and nitrides. *J. Mater. Sci.* **1995**, *30*, 1627–1641. [CrossRef]
51. Mitterer, C.; Mayrhofer, P.H.; Beschliesser, M.; Losbichler, P.; Warbichler, P.; Hover, F.; Gibson, P.N.; Gissler, W.; Hruby, H.; Musil, J.; et al. Microstructure and properties of nancomposite Ti-B-N and Ti-B-C coatings. *Surf. Coat. Technol.* **1999**, *120–121*, 405–411. [CrossRef]
52. Nan, C.W. Physics of inhomogeneous inorganic materials. *Progr. Mater. Sci.* **1993**, *37*, 1–116. [CrossRef]
53. Fabreguette, F.; Maglione, M.; Imhoff, L.; Domenichini, B.; Marco de Lucas, M.C.; Sibillot, P.; Bourgeois, S.; Sacilotti, M. Conductimetry and impedance spectroscopy study of low pressure metal organic chemical vapor deposition TiN_xO_y films as a function of the growth temperature: A percolation approach. *Appl. Surf. Sci.* **2001**, *175–176*, 574–578. [CrossRef]
54. McLachlan, D.S.; Cai, K.; Sauti, G. AC and DC conductivity-based microstructural characterization. *Int. J. Refract. Met. Hard Mater.* **2001**, *19*, 437–445. [CrossRef]

55. Liang, H.; Xu, J.; Zhou, D.; Sun, X.; Chu, S.; Bai, Y. Thickness dependent microstructural and electrical properties of TiN thin films prepared by DC reactive magnetron sputtering. *Ceram. Int.* **2016**, *42*, 2642–2647. [CrossRef]
56. Raman, M.; Wang, C.C.; Chen, W.; Akbar, S.A. Electrical resistivity of titanium diboride and zirconium diboride. *J. Am. Ceram. Soc.* **1995**, *78*, 1380–1382. [CrossRef]
57. Zhang, X.W. Doping and electrical properties of cubic boron nitride thin films: A critical review. *Thin Solid Films* **2013**, *544*, 2–12. [CrossRef]
58. Anders, A. A structure zone diagram including plasma-based deposition and ion etching. *Thin Solid Films* **2010**, *518*, 4087–4090. [CrossRef]
59. Zaoui, M.; Bourceret, A.; Gaillard, Y.; Giljean, S.; Rousselot, C.; Pac, M.J.; Richard, F. Relation between hardness of (Ti, Al)N based multilayered coatings and periods of their stacking. *Acta Polytech.* **2020**, *27*, 79–83. [CrossRef]

Article

Bi Layer Properties in the Bi–FeNi GMR-Type Structures Probed by Spectroscopic Ellipsometry

Natalia Kovaleva [1,*], Dagmar Chvostova [2], Ladislav Fekete [2] and Alexandr Dejneka [2]

1 Lebedev Physical Institute, Russian Academy of Sciences, Leninsky Prospect 53, 119991 Moscow, Russia
2 Institute of Physics, Academy of Sciences of the Czech Republic, Na Slovance 2, 18221 Prague, Czech Republic; chvostov@fzu.cz (D.C.); fekete@fzu.cz (L.F.)
* Correspondence: kovalevann@lebedev.ru (N.K.); dejneka@fzu.cz (A.D.)

Abstract: Bismuth (Bi) having a large atomic number is characterized by a strong spin–orbit coupling (SOC) and is a parent compound of many 3D topological insulators (TIs). The ultrathin Bi films are supposed to be 2D TIs possessing a nontrivial topology, which opens the possibility of developing new efficient technologies in the field of spintronics. Here we aimed at studying the dielectric function properties of ultrathin Bi/FeNi periodic structures using spectroscopic ellipsometry. The $[Bi(d)–FeNi(1.8 \text{ nm})]_N$ GMR-type structures were grown by rf sputtering deposition on Sitall-glass (TiO_2) substrates. The ellipsometric angles $\Psi(\omega)$ and $\Delta(\omega)$ were measured for the grown series ($d = 0.6, 1.4, 2.0,$ and 2.5 nm, N = 16) of the multilayered film samples at room temperature for four angles of incidence of 60°, 65°, 70°, and 75° in a wide photon energy range of 0.5–6.5 eV. The measured ellipsometric angles, $\Psi(\omega)$ and $\Delta(\omega)$, were simulated in the framework of the corresponding multilayer model. The complex (pseudo)dielectric function spectra of the Bi layer were extracted. The GMR effects relevant for the studied Bi–FeNi MLF systems were estimated from the optical conductivity zero-limit (optical GMR effect). The obtained results demonstrated that the Bi layer possessed the surface metallic conductivity induced by the SOC effects, which was strongly enhanced on vanishing the semimetallic-like phase contribution on decreasing the layer thickness, indicating its nontrivial 2D topology properties.

Citation: Kovaleva, N.; Chvostova, D.; Fekete, L.; Dejneka, A. Bi Layer Properties in the Bi–FeNi GMR-Type Structures Probed by Spectroscopic Ellipsometry. *Coatings* **2022**, *12*, 872. https://doi.org/10.3390/coatings12060872

Academic Editor: Sheng-Rui Jian

Received: 18 May 2022
Accepted: 17 June 2022
Published: 20 June 2022

Publisher's Note: MDPI stays neutral with regard to jurisdictional claims in published maps and institutional affiliations.

Copyright: © 2022 by the authors. Licensee MDPI, Basel, Switzerland. This article is an open access article distributed under the terms and conditions of the Creative Commons Attribution (CC BY) license (https:// creativecommons.org/licenses/by/ 4.0/).

Keywords: optical GMR effect; bismuth–permalloy multilayers; spectroscopic ellipsometry

1. Introduction

The relativistic effect of spin–orbit (SOC) coupling is involved in the so-called Rashba effect [1]. This phenomenon arises from the apparent loss of crystalline inversion symmetry near the surface or heterojunction leading to the lifting of the spin degeneracy and generating spin-polarized surface metallic states. In this respect, 3D (2D) topological insulators (TIs) also exhibit spin-polarized surface metallic states due to SOC. However, contrary to the Rashba effect, the surface metallic bands of a TI are determined by its bulk characteristics. The TIs host metallic surface states in a bulk energy gap, which are topologically protected. The surface (or interface) states of TIs can be topologically trivial or nontrivial. In the latter case, for example, electrons cannot be backscattered by impurities. Bismuth (Bi), having a large atomic number, is characterized by a strong SOC and is a parent compound of many 3D TIs, such as $Bi_{1-x}Sb_x$ or Bi_2Se_3, even though 3D bulk Bi itself is topologically trivial. The specific feature of the electronic band structure of bulk Bi having $R\bar{3}m$ rhombohedral symmetry [2–4] is its inverted band gaps at both the Γ and M points of the Brillouin zone due to the strong SOC. The uniqueness of Bi films associated with surface metallic states [5,6] and the semiconductor-to-metal transition [7,8] are well documented in the literature.

Theoretical analyses predict a 1-bilayer (BL) Bi(111) film to be a 2D TI [9,10]. If there is no or weak inter-BL coupling, a stack of the odd–even 1-BL films will exhibit nontrivial to trivial oscillations of topology (where the topological number ν [11] is equal to 1 or 0, respectively). However, for the nontrivial topology in a stack of 1-BL films, the intermediate inter-BL coupling strength, which is, for example, higher than the van der Waals strengths, is a mandatory condition. The direct (Γ point) and indirect band gap values were calculated by Liu et al. as a function of the Bi film thickness [12]. It was established that below four BLs the film is a semiconductor with the direct gap open at the Γ point and the positive indirect band gap leading to nontrivial topology characteristic of an intrinsic 2D TI. Above four BLs the indirect band gap becomes negative resulting in a semiconductor–semimetal transition due to the overlapping of two bands at the Fermi level around the Γ and M points. This suggests that the Bi films from five to eight BLs represent a 2D TI situated between two trivial metallic surfaces [12].

A comprehensive study of the associated SOC effects in ultrathin Bi layers opens the possibility of developing new efficient technologies in the field of spintronics. For this purpose, here we aimed at studying the dielectric function properties of ultrathin periodic structures $Bi/Ni_{79}Fe_{21}$, prepared by rf sputter deposition, which is one of the most common technologies used to grow coatings and multilayered films (MLFs) exhibiting a giant magnetoresistance (GMR) effect for various existing and modern nanotechnological applications. In earlier work, we demonstrated that the electronic band structure and surface electronic properties of ultrathin Bi layers in real GMR-type $(Bi-FeNi)_N$ MLF structures incorporating nanoisland FeNi layers could be successfully studied by spectroscopic ellipsometry (SE) [13]. Here, by applying the elaborated SE approach, we investigated (Bi–FeNi) MLFs, where the thickness of the FeNi layer was 1.8 nm, corresponding to the FeNi layer structural percolation threshold [14,15], and the Bi spacer layer was 0.6, 1.4, 2.0, and 2.5 nm thick, incorporating about two, four, six, and eight Bi(012)-type planes, respectively. We found that the Bi spacer layers have a metallic surface conductivity, which demonstrates strongly enhanced metallicity properties on vanishing the Bi semimetallic-like phase contribution on decreasing the layer thickness, which can be constructive in finding new nontrivial 2D topology properties of the (Bi–FeNi) GMR-type structures for their different nanotechnological applications.

2. Materials and Methods

The $(Bi-FeNi)_N$ MLFs were prepared in a sputter deposition system by cathode sputtering from 99.95% pure Bi and $Fe_{21}Ni_{79}$ targets in an alternative way. The base pressure in a sputter deposition chamber was 2×10^{-6} Torr. The multilayers were deposited at approximately 80 °C in an argon atmosphere of 6×10^{-4} Torr on insulating glassy Sitall (TiO_2) substrates. We utilized the substrates having typical dimensions $15 \times 5 \times 0.6$ mm^3. The nominal thicknesses of the FeNi and Bi layers were controlled by the layer deposition times in accordance with the material deposition rates. A series consisting of four MLF samples was prepared. In the series of the grown $(Bi-FeNi)_N$ samples, the nominal thickness of the FeNi layer was 1.8 nm, the Bi layer thickness was 0.6, 1.4, 2.0, and 2.5 nm, and the number N of periodically repeated Bi/FeNi layers was 16. The thickness of the FeNi layer was chosen to be 1.8 nm, matching the structural percolation threshold [14,15]. The Bi layer thicknesses were chosen in such a way that the conditions for ferromagnetic (FM) or antiFM coupling in the GMR-type structures would be optimized. To prevent degradation of the MLFs, the deposited $(Bi-FeNi)_{16}$–FeNi/Sitall samples were covered in situ with a 2.1 nm thick Al_2O_3 capping layer.

The related $[Bi-FeNi(0.8,1.2 \text{ nm})]_N$ samples prepared by rf sputtering deposition onto the Sitall substrates under similar conditions were investigated by X-ray diffraction (XRD) as well as by the X-ray reflectivity (XRR) experimental techniques from our previous study (see Supplementary online information for the article [13]). The XRR spectra proved to have a good periodicity and consistency with the corresponding nominal thicknesses of the FeNi and Bi slices in the Bi/FeNi MLF structures, as well as a relatively low interface roughness between the constituent layers. The XRD characterization suggested a (012)-type Bi plane orientation, where the interlayer distance was 3.28 Å. It followed from this that in the studied MLF structures, the Bi layers with a thickness corresponding to 0.6, 1.4, 2.0, and 2.5 nm incorporated two, four, six, and eight Bi(012)-type planes, respectively.

In the present study, the surface morphology of the Bi–FeNi(1.8 nm) MLF samples, prepared by rf sputtering deposition on the Sitall (TiO$_2$) substrates, was studied at room temperature using an ambient AFM (Bruker, Dimension Icon) in the PeakForce Tapping mode with ScanAsyst-Air tips (Bruker, k = 0.4 N/m, nominal tip radius 2 nm). The SE measurements for the investigated Al$_2$O$_3$/(Bi–FeNi)$_{16}$/Sitall samples were performed at room temperature in a wide photon energy range of 0.5–6.5 eV using a J.A. Woollam VUV–VASE ellipsometer (see the scheme illustrating the SE study of the (Bi–FeNi)$_N$ MLFs in Ref. [13], Figure 1a). The measured ellipsometry spectra are represented by real values of the angles $\Psi(\omega)$ and $\Delta(\omega)$, which are defined through the complex Fresnel reflection coefficients for light-polarized parallel r_p and perpendicular r_s to the plane of incidence, $\tan \Psi \, e^{i\Delta} = \frac{r_p}{r_s}$. The ellipsometric angles, $\Psi(\omega)$ and $\Delta(\omega)$, measured for the Bi–FeNi MLF samples were simulated using the multilayer model simulation available in J.A. Woollam VASE software [16]. From the multilayer model simulations, the (pseudo)dielectric function spectra of the ultrathin 0.6, 1.4, 2.0, and 2.5 nm Bi layers and 1.8 nm FeNi layer inside the Bi–FeNi MLF structures were extracted. The corresponding calculated optical conductivity spectra were analyzed.

3. Results
3.1. Atomic Force Microscopy Study

The retrieved 5×5 µm^2 and 1×1 µm^2 AFM images of the Al$_2$O$_3$(2.1 nm)/[Bi(0.6, 1.4, 2.0, 2.5 nm)–FeNi(1.8 nm)]$_N$/Sitall multilayered films (where the given layer thicknesses correspond to their nominal values) presented in Figure 1a–h show a discernible contrast because of the available surface height deviations. The surface roughness of the Sitall glass (TiO$_2$) substrates was investigated by AFM in our earlier publication [17]. The height profile of the Sitall substrates (see Ref. [17], Figure 2a) demonstrated a height deviation within the range 1–3 nm characteristic of the relatively large 0.3–1 µm lateral scale, which characterizes the Sitall substrate surface roughness. From the AFM measurements on the areas 5×5 µm^2 and 1×1 µm^2 the root-mean square (RMS) surface roughness values were evaluated, which are presented in the caption to Figure 1. The corresponding RMS roughness values are notably higher for the Al$_2$O$_3$(2.1 nm)/[Bi(2.5 nm)–FeNi(1.8 nm)]$_{16}$/Sitall MLF sample. The smaller-scale (1×1 µm^2) images clearly exhibit a fine grainy-like structure of the surface morphology, which seems to be characteristic for all studied film samples (see Figure 1e–h). The typical grain size, of about 50 nm, is notably larger for the FeNi(1.8 nm)-Bi MLF sample incorporating the 2.5 nm thick Bi layers, and, following the estimated RMS roughness values, the average grain size decreases to about 20 nm when decreasing the Bi layer thickness to 1.4 nm. As one can see from the typical height profiles presented in Figure 1i,j, when decreasing the Bi layer thickness from 2.5 to about 0.6 nm, the surface morphology becomes highly irregular due to the formation of conglomerates of nanoislands separated by rather flat (relatively small roughness) areas of about 20 nm.

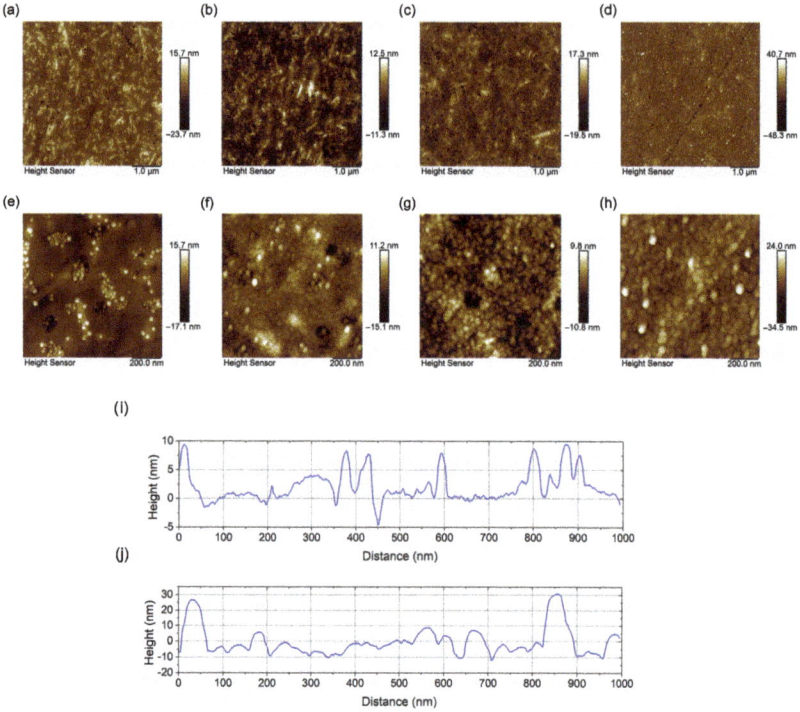

Figure 1. AFM images (**a–d**) 5 × 5 μm^2 and (**e–h**) 1 × 1 μm^2 of the Al$_2$O$_3$/(Bi–FeNi)$_{16}$/Sitall MLF samples, where the nominal Al$_2$O$_3$ and FeNi layer thicknesses are 2.1 and 1.8 nm and the nominal Bi layer thicknesses are 0.6, 1.4, 2.0, and 2.5 nm, respectively. The estimated surface RMS roughness values are in (**a–d**), 3.6, 3.0, 3.1, and 5.2 nm, and in (**e–h**) 3.2, 2.6, 2.7, and 5.3 nm, respectively. (**i,j**) The typical height profiles for the MLF samples with the nominal Bi layer thicknesses of 0.6 and 2.5 nm, respectively.

3.2. Spectroscopic Ellipsometry Study of the Ultrathin Bi–FeNi Multilayer Film Samples

The ellipsometric angles $\Psi(\omega)$ and $\Delta(\omega)$ were measured for the prepared Al$_2$O$_3$/(Bi–FeNi)$_{16}$/Sitall MLF samples at the angles of incidence of 60°, 65°, 70°, and 75°. Figure 2 demonstrates the ellipsometric angles $\Psi(\omega)$ and $\Delta(\omega)$ recorded at 65° and 70°. To model the contributions from free charge carriers and interband optical transitions, the complex dielectric function $\tilde{\varepsilon}(\omega) = \varepsilon_1(\omega) + i\varepsilon_2(\omega)$ of the Bi and FeNi layers was interpreted in terms of the Drude and Lorentz parts, respectively,

$$\tilde{\varepsilon}(E \equiv \hbar\omega) = \epsilon_\infty - \frac{A_D}{E^2 + iE\gamma_D} + \sum_j \frac{A_j \gamma_j E_j}{E_j^2 - E^2 - iE\gamma_j}, \qquad (1)$$

where ε_∞ is the high-frequency dielectric constant, which takes into account the contribution from the higher-energy interband transitions. The fitted Drude parameters were A_D and the free charge carrier's scattering rate γ_D. The fitted parameters of Lorentz bands were E_j, γ_j, and A_j of the band maximum energy, the full width at half-maximum, and the ε_2 band height, respectively. The obtained ellipsometric angles $\Psi(\omega)$ and $\Delta(\omega)$ measured at different angles of incidence of 60°, 65°, 70°, and 75° were fitted for each sample simultaneously using J.A. Woollam VASE software [16] in the framework of the designed multilayer model. The multilayer model for the studied Al$_2$O$_3$/(Bi–FeNi)/Sitall multilayers was constructed as it is schematically presented in Figure 3, exactly so, as the layers were deposited. In addition, we attempted to take into account the roughness properties of the surface by using the conventional approach of effective medium approximation (EMA) based on the (50% Al_2O_3–50% vacuum) Bruggeman model. The dispersion model for the Bi layers included three or four Lorentz terms as well as the Drude part. The dispersion model for the 1.8 nm permalloy layers incorporated in the studied MLF structures included the Drude term responsible for the free charge carrier contribution and one Lorentz oscillator to account for the most pronounced interband optical transition. In addition, the dielectric function spectra of the bare Sitall substrate derived from our earlier SE studies [18,19] were introduced to the elaborated multilayer model. The dielectric response of the Al_2O_3 capping layer was represented by the tabular complex dielectric function spectra [20]. The thicknesses of the Bi and FeNi layers, as well as of the surface layers, were fitted. The unknown parameters were allowed to vary until the minimum of the mean squared error (MSE) was reached. The best simulation result for the studied [Bi(0.6, 1.4, 2.0, 2.5 nm)–FeNi(1.8 nm)]$_{16}$ MLF samples corresponded to the lowest obtained MSE values of 0.3843, 0.297, 0.2934, and 0.4508, respectively. The good quality of the fit allowed us to estimate the actual Bi and FeNi layer thicknesses in the MLFs under study. The quality of the fit is demonstrated by Figure 2, where we plotted the measured ellipsometric angles along with the simulation results. The Drude and Lorentz parameters resulting from the simulation of the Al_2O_3/[Bi(d)–FeNi(1.8 nm)]$_{16}$/Sitall MLF samples are given in Tables 1 and 2, and the resulting $\varepsilon_1(\omega)$ and $\varepsilon_2(\omega)$ parts of the Bi and FeNi (pseudo)dielectric function spectra are presented in Figure 4.

Table 1. Drude–Lorentz parameters for the Bi spacer layer in the [Bi(0.6, 1.4, 2.0, 2.5 nm)–NiFe(1.8 nm)]$_{16}$-multilayered films obtained from the model simulations of the dielectric functions by using Equation (1). The values of E_j, γ_j, and γ_D are given in eV and optical conductivity limit $\sigma_{1(\omega \to 0)}$ in $\Omega^{-1} \cdot cm^{-1}$.

	Parameters	0.6 nm	1.4 nm	2.0 nm	2.5 nm
Drude	A_D	46.(9) ± 4	66.(7) ± 4	24.(5) ± 4	25.(1) ± 2
	γ_D	1.2(5) ± 0.09	1.51(0) ± 0.06	2.7(2) ± 0.4	3.1(3) ± 0.2
	$\sigma_{1(\omega \to 0)}$	6300 ± 540	8970 ± 540	3290 ± 540	3370 ± 270
Lorentz oscillator	E_1	–	0.45(8) ± 0.05	0.35(9) ± 0.01	0.38(6) ± 0.004
	A_1	–	15.(0) ± 6	96.(0) ± 10	70.(8) ± 2
	γ_1	–	0.52(6) ± 0.09	0.79(1) ± 0.02	0.67(6)
Lorentz oscillator	E_2	4.67	5.31(5) ± 0.03	5.08(7) ± 0.04	4.77(5) ± 0.04
	A_2	10.2(7) ± 0.6	2.53(2) ± 0.05	1.2(5) ± 0.1	0.67(6) ± 0.08
	γ_2	4.2(1) ± 0.07	3.99(3) ± 0.07	3.4(7) ± 0.2	2.5(5) ± 0.2
Lorentz oscillator	E_3	11.1	7.8	7.7	7.7
	A_3	7.2	4.1	4.1	4.1
	γ_3	8.9	2.8	2.8	2.8

Table 2. Drude–Lorentz parameters for the 1.8 nm thick NiFe layer in the [Bi(0.6, 1.4, 2.0, 2.5 nm)–NiFe]$_{16}$-multilayered films obtained from the simulations of the model dielectric function described by Equation (1). The values of E_1, γ_1, and γ_D are given in eV and optical conductivity limit $\sigma_{1(\omega \to 0)}$ in $\Omega^{-1} \cdot cm^{-1}$.

	Parameters	0.6 nm	1.4 nm	2.0 nm	2.5 nm
Drude	A_D	33.(8) ± 2	15.(0) ± 1	21.(7) ± 2	13.(1) ± 2
	γ_D	0.876(5) ± 0.04	2.8(2) ± 0.3	3.4(2) ± 0.4	3.1(3) ± 0.2
	$\sigma_{1(\omega \to 0)}$	4540 ± 270	2020 ± 130	2920 ± 270	1760 ± 270
Lorentz oscillator	E_1	1.87	3.32	3.32	3.32
	A_1	14.76	14.28	15.23	14.74
	γ_1	3.62	5.88	5.65	5.95

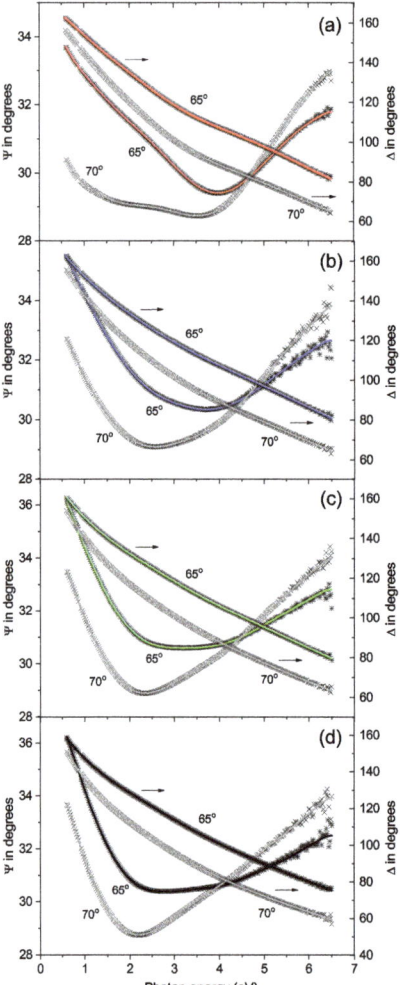

Figure 2. (a–d) Ellipsometric angles, $\Psi(\omega)$ and $\Delta(\omega)$ (symbols), measured at the angles of incidence of 65° and 70° for the Al_2O_3/[Bi(d)–NiFe(1.8 nm)]$_{16}$/Sitall multilayered films where the Bi spacer layer thicknesses d = 0.6, 1.4, 2.0, and 2.5 nm, respectively. The solid red, blue, green, and black curves show the corresponding simulation results for a 65° angle by the dielectric function model using Equation (1).

From Figure 4a,b one can see that the complex (pseudo)dielectric functions of the 0.6, 1.4, 2.0, and 2.5 nm thick Bi spacers inside the investigated Bi–FeNi MLFs demonstrate a metallic character. Moreover, the $\varepsilon_1(\omega)$ function progressively decreases while the Bi thickness decreases from 2.5 to 2.0 to 1.4 nm and the $\varepsilon_2(\omega)$ increases at low photon energies, respectively. According to our simulation results, we expect that the best metallicity properties are demonstrated by the Bi layer in the [Bi(1.4 nm)–NiFe(1.8 nm)]$_{16}$ structure. At the same time, the complex (pseudo)dielectric functions of the thinnest 0.6 nm thick Bi layer look somewhat different. Here, in addition to the low-energy metallic Drude response identified by the characteristic behavior of $\varepsilon_1(\omega)$ and $\varepsilon_2(\omega)$, the Lorentz band around 4–5 eV makes an essential contribution to the dielectric function response (the corresponding Drude (A_D and γ_D) and Lorentz (A_j, E_j, and γ_j) parameters are listed in Table 1). Next, being similar, the dielectric functions of the 1.8 nm thick permalloy layers in the [FeNi–Bi(1.4, 2.0, 2.5 nm)] MLFs are dominated by the $\varepsilon_2(\omega)$ resonance and $\varepsilon_1(\omega)$

antiresonance features, indicating the predominant contribution from the Lorentz oscillator peaking at around 3 eV (see Figure 4c,d). An upturn evident in the $\varepsilon_2(\omega)$ at low photon energies indicates an additional Drude contribution, which is relatively less pronounced. Following our simulation results, we expect the advanced metallicity properties of the FeNi layer in the [Bi(0.6 nm)–NiFe(1.8 nm)]$_{16}$ structure (see the corresponding Drude (A_D and γ_D) and Lorentz (A_j, E_j, and γ_j) parameters listed in Table 2).

Figure 5a–d present the evolution of the Bi intralayer optical conductivity, $\sigma_1(\omega) = \varepsilon_2(\omega)\omega(\text{cm}^{-1})/60$, upon decreasing the Bi spacer layer thickness in the [FeNi(1.8 nm)–Bi(2.5, 2.0, 1.4, 0.6 nm)]$_{16}$ structures, and Figure 5e–h show the associated optical conductivity spectra of the 1.8 nm FeNi permalloy layer. Here, the contributions from the Drude and Lorentz oscillators following the multilayer model simulations using Equation (1) are evidently demonstrated. The presented optical conductivity spectra of the Bi and FeNi layers follow the main trends identified in their complex dielectric function spectra presented in Figure 4.

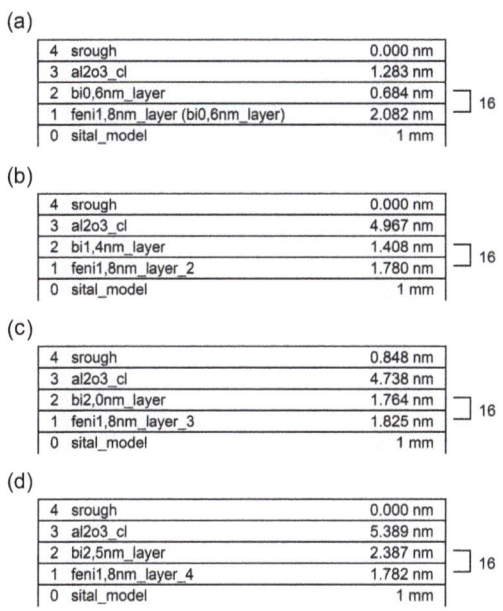

Figure 3. The multilayer model applied for the simulation of the Al$_2$O$_3$/[Bi(0.6, 1.4, 2.0, and 2.5 nm)–FeNi(1.8 nm)]$_{16}$/Sitall samples. The Bi and FeNi thicknesses estimated from the model simulations are (**a**) 0.684 ± 0.037 nm and 2.082 ± 0.116 nm, (**b**) 1.408 ± 0.574 nm and 1.780 ± 0.65 nm, (**c**) 1.764 ± 0.194 nm and 1.825 ± 0.358 nm, and (**d**) 2.387 ± 0.128 nm and 1.782 ± 0.171 nm. Note the good agreement between the thicknesses of the FeNi and Bi layers estimated from the model simulations and their respective nominal thickness values. The roughness and Al$_2$O$_3$ capping layer thicknesses estimated from the model simulations are (**a**) 0.00 ± 3.85 nm and 1.283 ± 2.37 nm, (**b**) 0.000 ± 4.97 nm and 4.967 ± 2.17 nm, (**c**) 0.848 ± 5.86 nm and 4.738 ± 2.92 nm, and (**d**) 0.000 ± 2.95 nm and 5.389 ± 1.23 nm.

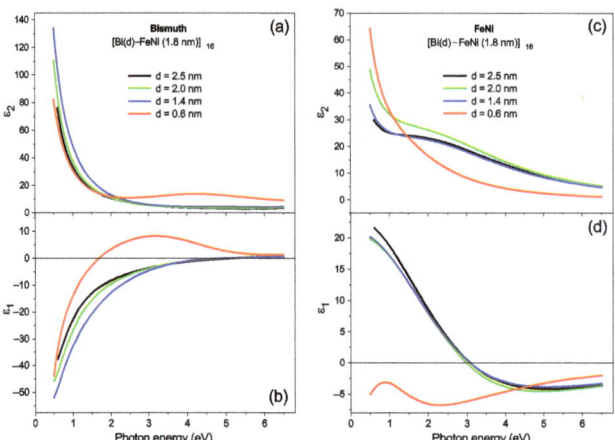

Figure 4. The complex (pseudo)dielectric function spectra, $\varepsilon_2(\omega)$ and $\varepsilon_1(\omega)$, of the (**a,b**) Bi layers and (**c,d**) FeNi layers in the $[\text{Bi}(d)\text{–FeNi}(1.8\,\text{nm})]_{16}$ structures shown for the Bi layer nominal thickness values $d = 0.6, 1.4, 2.0,$ and 2.5 nm by solid red, blue, green, and black curves, respectively.

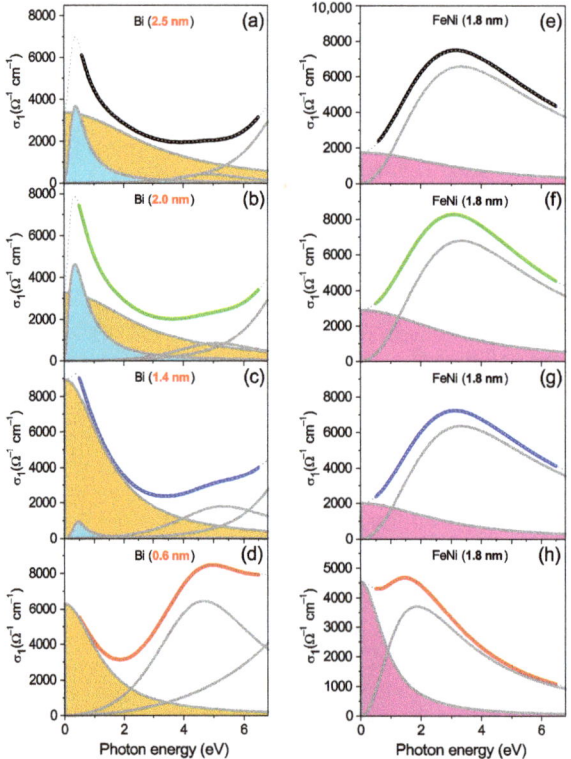

Figure 5. The intralayer optical conductivity, $\sigma_1(\omega) = \varepsilon_2(\omega)\omega[\text{cm}^{-1}]/60$, for the (**a–d**) Bi layers and (**e–h**) FeNi layers in the $[\text{Bi}(d)\text{–FeNi}(1.8\,\text{nm})]_{16}$ structures shown for the Bi layer nominal thickness values $d = 2.5, 2.0, 1.4,$ and 0.6 nm by solid curves (**a,e**) black, (**b,f**) green, (**c,g**) blue, and (**d,h**) red, respectively. The contributions from the Drude term and the Lorentz oscillator in (**a–d**) are displayed by the yellow- and cyan-shaded areas. In (**e–h**), the Drude term for the FeNi layers is displayed by the magenta-shaded area. Shown by the dotted curves are the summary of the Drude and Lorentz contributions.

4. Discussion

Initially, we would like to discuss the GMR effects relevant to the studied MLF systems. Our simulations of the dielectric functions for the 1.8 nm thick NiFe layer inside the [Bi(0.6,1.4,2.0,2.5 nm)–NiFe(1.8 nm)] MLFs showed the presence of the Drude term complemented with the pronounced Lorentz band located at around 2–3 eV (see Table 2). From the corresponding optical conductivity spectra presented in Figure 5e–h one can notice that the associated Drude dc limit, $\sigma_{1\omega \to 0}$, displays an oscillating character (in agreement with the results deduced for the corresponding Drude parameter A_D, see Table 2 and Figure 6). We can expect that the Bi spacer thicknesses for which the FeNi layers are preferentially antiFM-coupled in the studied MLFs are around 1.4 and 2.5 nm implying that the [Bi(1.4,2.5 nm)–NiFe(1.8 nm)]$_{16}$ film structures will exhibit a drop in resistance (being negative magnetoresistance) when exposed to an external magnetic field. It is well known from the literature that the first antiFM maximum exhibits a negative magnetoresistance of about 20%, while the second antiFM maximum decreases to about 10%, and the presence of the third antiFM maximum cannot confidently be retrieved (see, for example, [21] and references therein). Using a simple model of a two-current series resistor [22], the magnetoresistance $\frac{\Delta R}{R}$ can be estimated as

$$\frac{\Delta R}{R} = 100\% \frac{(\alpha - \beta)^2}{4\left(\alpha + \frac{d_{Bi}}{d_{FeNi}}\right)\left(\beta + \frac{d_{Bi}}{d_{FeNi}}\right)}, \qquad (2)$$

where d_{Bi} and d_{FeNi} are the thicknesses of the Bi and FeNi layers, and $\alpha = \frac{\downarrow \rho_{FeNi}}{\rho_{Bi}}$ and $\beta = \frac{\uparrow \rho_{FeNi}}{\rho_{Bi}}$ are the ratios of the resistivity in the FeNi layer to that in the Bi layer in the spin down and spin up current channel, respectively. Exploiting values for $\rho = \sigma_{1\omega \to 0}^{-1}$ estimated for the 1.4 nm Bi and 1.8 nm FeNi layers from the current model simulations (see Tables 1 and 2), namely, $\rho_{Bi} = \frac{1}{8970} \Omega \cdot cm$, $\downarrow \rho_{FeNi} = \frac{1}{2020} \Omega \cdot cm$, and $\uparrow \rho_{FeNi} = \frac{1}{4540} \Omega \cdot cm$ (the latter estimate is given by the FM coupling for the 0.6 nm Bi spacer), we obtain $\alpha = 4.4$, and $\beta = 2.0$. Then, using Equation (2), we have $\frac{\Delta R}{R} = 10\%$. This means that the 1.4 nm Bi spacer corresponds to the second antiFM maximum. Following the same approach for the 2.5 nm Bi spacer, where $\rho_{Bi} = \frac{1}{3370} \Omega \cdot cm$, $\downarrow \rho_{FeNi} = \frac{1}{1760} \Omega \cdot cm$ and $\uparrow \rho_{FeNi} = \frac{1}{2920} \Omega \cdot cm$ (corresponding to the FM coupling for the 2.0 nm Bi spacer), we obtain $\alpha = 1.9$ and $\beta = 1.2$. Using Equation (2), we have $\frac{\Delta R}{R} = 1.4\%$, which may correspond to the very weakly pronounced third antiFM maximum. From the analysis presented above, we may expect that the first antiFM maximum, corresponding to the magnetoresistance of about 20%, occurs for the Bi spacer thickness of about 0.9 nm, which is in agreement with the results presented in [21].

Further, in the XRD patterns of the investigated Al$_2$O$_3$/[Bi(1.4,2.0,2.5 nm)–NiFe(1.8 nm)]$_{16}$/Sitall film samples, the peak of the $R\bar{3}m$ crystalline Bi phase was identified at $2\theta \approx 26.2°$ suggesting a (012) orientation of the Bi layers, which is characterized by the interlayer distance of 3.28 Å. Using STM and reflection high-energy electron diffraction (RHEED) techniques, it was shown that the initial growth of Bi(012)-type films occurs in the form of islands with a height increment of about 6.6 Å, indicating an even-number layer stability leading to the laterally flat morphology of the Bi(012)-type islands [23]. Consequently, we can expect that the 0.6, 1.4, 2.0, and 2.5 nm Bi spacer layers in the investigated MLFs incorporate about two, four, six, and eight (012)-type Bi planes, respectively.

The model simulations for the [Bi(2.5, 2.0 nm)–FeNi(1.8 nm)]$_{16}$ film samples revealed that the low-energy dielectric function of the Bi intralayers had competing contributions from the Drude term and from the intense Lorentz band around 0.36–0.39 eV with a ε_2 maximum height of 70–100 (see Table 1). The Drude and Lorentz contributions were more clearly pronounced in the corresponding optical conductivity spectra (see Figure 5a,b). The obtained Drude and Lorentz parameters were in excellent agreement with those deduced in our previous study [13] for the Bi spacer layer incorporated in the [Bi(2.5, 2.0 nm)–FeNi(1.2 nm)]$_{16}$ structures under study. The pronounced Lorentz band found at low photon energies for Bi single crystals (rhombohedral symmetry, space group $R\bar{3}m$) [24,25] and bulk Bi layers [26,27] is characteristic of the semimetallic-like electronic band structure due to the contributions from the interband transitions near the Γ point, $\Gamma_6^+ - \Gamma_6^-$ and $\Gamma_{45}^+ - \Gamma_6^-$ [2], and near the T point, $T_6^- - T_{45}^-$ [4]. The estimated values (see Table 1) of the Drude dc limit $\sigma_{1\omega \to 0}$ (2750–3830 $\Omega^{-1} \cdot cm^{-1}$) as well as the free charge carrier's γ_D (2.3–3.3 eV) were consistent with those characteristic of the metallic surface states related to the Rashba SOC in Bi(111) films, $\sigma_{1\omega \to 0}$ = 2300 $\Omega^{-1} \cdot cm^{-1}$ and γ_D = 2.0 eV) [6]. Meanwhile, the model simulation for the [Bi(1.4 nm)–FeNi(1.8 nm)]$_{16}$ structure indicated that for the 1.4 nm Bi layer, the Drude dc limit significantly increased to 8970 ± 540 $\Omega^{-1} \cdot cm^{-1}$, while the γ_D essentially decreased to 1.50 ± 0.06 eV. In this case, the Lorentz band was nearly suppressed. The associated found Drude parameters for the ultrathin Bi layer inside the [Bi(0.6 nm)–FeNi(1.8 nm)]$_{16}$ structure were slightly different, namely, $\sigma_{1\omega \to 0}$ = 6300 ± 540 $\Omega^{-1} \cdot cm^{-1}$ and γ_D = 1.2 ± 0.1 eV, and the Lorentz band was clearly not present (see Figure 5c,d and Table 1).

Thus, we discovered that, on the one hand, the optical conductivity spectra of the 2.0 and 2.5 nm thick Bi spacer layers in the (Bi–FeNi) MLFs incorporating eight and six Bi(012)-type monolayers, respectively, had contributions from the pronounced low-energy Lorentz oscillator and from the free charge carrier Drude term (for details, see Figure 5a,b and Table 1). Here, the presence of the low-energy Lorentz band points on the Bi semimetallic phase contribution and the parameters obtained for the Drude conductivity indicate that its origin can be associated with the surface metallic states [6]. Therefore, the 2.0 and 2.5 nm Bi layers can be associated with the semimetallic Bi phase sandwiched between two metallic layers on the top and bottom surfaces. On the other hand, the contribution from the intrinsic Lorentz band was strongly suppressed for the 1.4 and 0.6 nm layers, where the Drude conductivity displayed notably improved metallicity properties, as one can see from the optical conductivity spectra shown in Figure 5c,d (for details, see Table 1).

From the above discussion of the obtained results, we can conclude that the Bi layer consisting of four Bi(012)-type monolayers represents a kind of crossover regarding the contributions from the semimetallic Bi phase and/or surface metallic-like states. Here, we notice some similarity with the theory results presented for the ultrathin Bi(111) layers by Liu et al. [12]. In their paper, it was established that below four Bi(111) BLs the film is a semiconductor with a direct gap open at the Γ point and a positive indirect band gap, leading to a nontrivial Z_2 topology ($\nu = 1$) characteristic of an intrinsic 2D TI. However, above four Bi(111) BLs, the indirect band gap becomes negative resulting in a semiconductor–semimetal transition, due to the overlapping of two bands at the Fermi level around the Γ and M points. It was argued by Liu et al. [12] that the Bi layers consisting of five to eight Bi(111) BLs represented a 2D TI placed between two "trivial" metallic surfaces [12]. This means that for the surface considered as an individual 2D system, its Z_2 number is trivial ($\nu = 0$). The surface bands have no contribution to the nontrivial Z_2 topology and, therefore, these trivial metallic surfaces are not robust and can easily be removed by surface defects or impurities. We found [13] that the Bi layers in the [Bi(2.0, 2.5 nm)–FeNi(0.8 nm)] multilayers, incorporating the nanoisland permalloy layer, exhibited bulk-like semimetallic properties of the electronic band structure, although the surface (Drude) metallic conductivity was absent there (see Ref. [13], Figure 4d). Indeed, a strong magnetic and spatial disorder induced by magnetic FeNi nanoislands, as well as long-range many-body interactions between the magnetic moments of permalloy nanoislands [17], may lead to a specific localization of free charge carriers [28]. However, the surface conductivity (or interface) states for the 1.4 nm layer in the Bi–FeNi(1.8 nm) multilayers may be topologically nontrivial and, in this case, the electrons cannot be backscattered by impurities. Here, the Drude dc limit was 8970 ± 540 $\Omega \cdot cm^{-1}$ and the scattering rate γ_D = 1.5 ± 0.06 eV. We found that the 0.6 nm thick Bi layer exhibited somewhat different Drude dc limit (6300 ± 540 $\Omega \cdot cm^{-1}$) and γ_D (1.2 ± 0.1 eV), see Table 1 and Figure 6, which can be attributed to the discontinuous nanoisland structure of this layer.

Finally, we would like to note that it will be challenging to investigate the dc transport and superconductivity properties of the ultrathin Bi films possessing 2D TI surface states following the approach presented in [29], where the subkelvin superconductivity without any external stimuli was discovered in 3D TI Cd_3As_2 films [30,31].

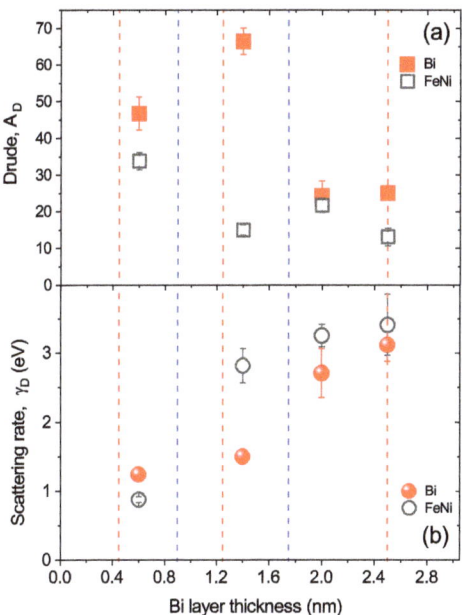

Figure 6. (**a**,**b**) Parameters of the Drude term (A_D and γ_D) for the Bi (filled symbols) and FeNi (empty symbols) layers in the [Bi(0.6, 1.4, 2.0, 2.5 nm)–FeNi(1.8 nm)] MLF structures.

5. Conclusions

In summary, using wide-band (0.5–6.5 eV) spectroscopic ellipsometry, we studied the optical properties of the [Bi(0.6, 1.4, 2.0, 2.5 nm)–FeNi(1.8 nm)]$_{16}$ MLFs prepared by rf sputtering. The XRD analysis suggested that the 0.6, 1.4, 2.0, and 2.5 nm Bi layers in the studied MLFs corresponded to about two, four, six, and eight Bi(012)-type monolayers, respectively. From the multilayer model simulations of the measured ellipsometric data, we extracted the Bi and FeNi layer dielectric functions. The dielectric function for the 2.0 and 2.5 nm Bi spacer layers were represented by the Drude resonance due to the surface states and the low-energy Lorentz band peaking at around 0.3–0.4 eV. The pronounced Lorentz band was characteristic of the semimetallic bulk-like Bi electronic zone structure due to the contributions from the interband transitions near the Γ point. We discovered that the 2.0 and 2.5 nm Bi spacer layers could be associated with the semimetallic Bi phase sandwiched between two trivial (where the topology number $\nu = 0$) metallic layers on the top and bottom surfaces. The contribution from the low-photon-energy Lorentz band was strongly suppressed for the 1.4 and 0.6 nm Bi layers, where the Drude conductivity displayed notably improved metallicity properties. This indicated that the Bi layer consisting of four Bi(012)-type monolayers represented a kind of crossover regarding the contributions from the semimetallic Bi phase and/or surface metallic-like states. Therefore, the properties of Bi layers below four monolayers may be associated with the nontrivial topology (where the topology number $\nu = 1$) characteristic of an intrinsic 2D TI. We expect that the Bi layers having thickness of 0.9 nm will exhibit a maximal GMR effect of about 20% in the (Bi–FeNi) MLFs, where the Drude dc limit is about $8970 \pm 540\,\Omega\cdot cm^{-1}$. These states may be protected from backscattering, which makes them promising in spintronic devices and quantum computing.

Author Contributions: Conceptualization, N.K. and A.D.; methodology, D.C. and L.F.; validation, N.K. and A.D.; investigation, D.C. and N.K.; data curation, D.C. and N.K.; writing—original draft preparation, N.K.; writing—review and editing, N.K. and A.D.; visualization, N.K. and L.F.; supervision, N.K. and A.D.; project administration, A.D.; funding acquisition, A.D. All authors have read and agreed to the published version of the manuscript.

Funding: This work was supported in part by the European Structural and Investment Funds and the Czech Ministry of Education, Youth, and Sports (Project No. SOLID21, Cz 02.1.01/0.0/0.0/16_019/0000760).

Institutional Review Board Statement: Not applicable.

Informed Consent Statement: Not applicable.

Data Availability Statement: Not applicable.

Acknowledgments: We thank F.A. Pudonin for providing us with the Bi/FeNi multilayer film samples and O. Pacherova for their XRD analysis. We thank A. Muratov for participation in the spectroscopic ellipsometry measurements.

Conflicts of Interest: The authors declare no conflict of interest.

Sample Availability: Samples of the Bi/FeNi multilayer films are available from the P.N. Lebedev Physical Institute, Moscow, Russia.

Abbreviations

The following abbreviations are used in this manuscript:

GMR	Giant magnetoresistance
SOC	Spin–orbit coupling
TI	Topological insulator
MLF	Multilayered film
SE	Spectroscopic ellipsometry
AFM	Atomic force microscopy
FM	Ferromagnetic
XRD	X-ray diffraction
XRR	X-ray reflectivity

References

1. Bychkov, Y.A.; Rashba, E.I. Properties of a 2D electron gas with lifted spectral degeneracy. *JETP Lett.* **1984**, *39*, 78–81.
2. Golin, S. Band Structure of Bismuth: Pseudopotential Approach. *Phys. Rev. B* **1968**, *166*, 643–651. [CrossRef]
3. Gonze, X.; Michenaud, J.-P.; Vigneron, J.-P. First-principles study of As, Sb, and Bi electronic properties. *Phys. Rev. B* **1990**, *41*, 11827–11836. [CrossRef] [PubMed]
4. Liu, Y.; Allen, R.E. Electronic structure of the semimetals Bi and Sb. *Phys. Rev. B* **1995**, *52*, 1566–1577. [CrossRef] [PubMed]
5. Hofmann, P. The surfaces of bismuth: Structural and electronic properties. *Prog. Surf. Sci.* **2006**, *81*, 191–245. [CrossRef]
6. Yokota, Y.; Takeda, J.; Dang, C.; Han, G.; McCarthy, D.N.; Nagao, T.; Hishita, S.; Kitajima, K.; Katayama, I. Surface metallic states in ultrathin Bi(001) films studied with terahertz time-domain spectroscopy. *Appl. Phys. Lett.* **2012**, *100*, 251605. [CrossRef]
7. Hoffman, C.A.; Meyer, J.R.; Bartoli, F.J. Semimetal-to-semiconductor transition in Bismuth thin films. *Phys. Rev. B* **1993**, *48*, 11431–11434. [CrossRef]
8. Koroteev, Y.M.; Bihlmayer, G.; Chulkov, E.V.; Blugel, S. First-principles investigation of structural and electronic properties of ultrathin Bi films. *Phys. Rev. B* **2008**, *77*, 045428–. [CrossRef]
9. Wada, M.; Murakami, S.; Freimuth, F.; Bihlmayer, G. Localized edge states in two-dimensional topological insulators: Ultrathin Bi films. *Phys. Rev. B* **2011**, *83*, 121310(R). [CrossRef]
10. Murakami, S. Quantum Spin Hall Effect and Enhanced Magnetic Response by Spin-Orbit Coupling. *Phys. Rev. Lett.* **2006**, *97*, 236805. [CrossRef]
11. Fu, L.; Kane, C.L.; Mele, E.J. Topological Insulators in Three Dimensions. *Phys. Rev. Lett.* **2007**, *98*, 106803. [CrossRef] [PubMed]
12. Liu, Z.; Liu, C.-X.; Wu, Y.-S.; Duan, W.-H.; Liu, F.; Wu, J. Stable nontrivial Z_2 topology in ultrathin Bi(111) films: A first principles study. *Phys. Rev. Lett.* **2011**, *107*, 136805. [CrossRef] [PubMed]
13. Kovaleva, N.N.; Chvostova, D.; Pacherova O.; Muratov A.V.; Fekete L.; Sherstnev I.A.; Kugel K.I.; Pudonin F.A.; Dejneka A. Bismuth layer properties in the ultrathin Bi–FeNi multilayer films probed by spectroscopic ellipsometry. *Appl. Phys. Lett.* **2021**, *119*, 183101. [CrossRef]
14. Sherstnev, I.A. Electronic Transport and Magnetic Structure of Nanoisland Ferromagnetic Systems. Ph.D. Thesis, P.N. Lebedev Physical Institute, Moscow, Russia, 2014.

15. Boltaev, A.P.; Pudonin, F.A.; Shertnev, I.A.; Egorov, D.A. Detection of the metal-insulator transition in disordered systems of magnetic nanoislands. *JETP* **2017**, *125*, 465–468. [CrossRef]
16. Woollam, J.A. *VASE Spectroscopic Ellipsometry Data Analysis Software*; J.A. Woollam, Co.: Lincoln, NE, USA, 2010.
17. Stupakov, A.; Bagdinov, A.V.; Prokhorov, V.V.; Bagdinova, A.N.; Demikhov, E.I.; Dejneka, A.; Kugel, K.I.; Gorbatsevich, A.A.; Pudonin, F.A.; Kovaleva, N.N. Out-of-plane and in-plane magnetization behaviour of dipolar interacting FeNi nanoislands around the percolation threshold. *J. Nanomater.* **2016**, 3190260. http://dx.doi.org/10.1155/2016/3190260. [CrossRef]
18. Kovaleva, N.N.; Chvostova, D.; Bagdinov, A.V.; Petrova M.G.; Demikhov E.I.; Pudonin F.A.; Dejneka A. Interplay of electronic correlation and localization in disordered β-tantalum films: Evidence from dc transport and spectroscopic ellipsometry study. *Appl. Phys. Lett.* **2015**, *106*, 051907. [CrossRef]
19. Kovaleva, N.; Chvostova, D.; Dejneka, A. Localization phenomena in disordered tantalum films. *Metals* **2017**, *7*, 257. [CrossRef]
20. Palik, E.D. *Handbook of Optical Constants of Solids*; Elsevier Science: San Diego, CA, USA, 1991.
21. Hütten, A.; Mrozek, S.; Heitmann, S.; Hempel, T.; Brückl H.; Reiss, G. Evolution of the GMR-Effect Amplitude in Copper-Permalloy-Multilayered Thin Films. *Acta Mater.* **1999**, *47*, 4245–4252. [CrossRef]
22. Mathon, J. Exchange Interactions and Giant Magnetoresistance in Magnetic Multilayers. *Contemp. Phys.* **1991**, *32*, 143–156. [CrossRef]
23. Nagao, T.; Sadowski, J.T.; Saito, M.; Yaginuma, S.; Fujikawa, Y.; Kogure, T.; Ohno, T.; Hasegawa, S.; Sakurai, T. Nanofilm allotrope and phase transformation of ultrathin Bi film on Si(111)-7 × 7. *Phys. Rev. Lett.* **2004**, *93*, 105501. [CrossRef]
24. Wang, P.Y.; Jain, A.L. Modulated Piezoreflectance in Bismuth. *Phys. Rev. B* **1970**, *2*, 2978–2983. [CrossRef]
25. Lenham, A.P.; Treherne, D.M.; Metcalfe, R.J. Optical Properties of Antimony and Bismuth Crystals. *J. Opt. Soc. Am.* **1965**, *55*, 1072–1074. [CrossRef]
26. Hunderi, O. Optical properties of crystalline and amorphous bismuth films. *J. Phys. F* **1975**, *5*, 2214–2225. [CrossRef]
27. Toudert, J.; Serna, R. Interband transitions in semi-metals, semiconductors, and topological insulators: A new driving force for plasmonics and nanophotonics [Invited]. *Opt. Mater. Express* **2017**, *7*, 2299–2325. [CrossRef]
28. Kovaleva, N.N.; Kusmartsev, F.V.; Mekhiya, A.B.; Trunkin, I.N.; Chvostova, D.; Davydov, A.B.; Oveshnikov, L.N.; Pacherova, O.; Sherstnev, I.A.; Kusmartseva, A.; et al. Control of Mooij correlations at the nanoscale in the disordered metallic Ta-nanoisland FeNi multilayers. *Sci. Rep.* **2020**, *10*, 21172. [CrossRef] [PubMed]
29. Suslov, A.V.; Davydov, A.B.; Oveshnikov, L.N.; Morgun, L.A.; Kugel, K.I.; Zakhvalinskii, V.S.; Pilyuk, E.A.; Kochura, A.V.; Kuzmenko, A.P.; Pudalov, V.M.; et al. Observation of subkelvin superconductivity in Cd_3As_2 thin films. *Phys. Rev. B* **2019**, *99*, 094512. [CrossRef]
30. Kochura, A.V.; Zakhvalinskii, V.S.; Htet, A.Z.; Ril', A.I.; Pilyuk, E.A.; Kuz'menko, A.P.; Aronzon, B.A.; Marenkin, S.F. Growth of thin cadmium arsenide films by magnetron sputtering and their structure. *Inorg. Mater.* **2019**, *55*, 879–886. [CrossRef]
31. Kovaleva, N.; Chvostova, D.; Fekete, L.; Muratov, A. Morphology and Optical Properties of Thin Cd_3As_2 Films of a Dirac Semimetal Compound. *Metals* **2020**, *10*, 1398. [CrossRef]

Article

Effect of Substrate Temperature on the Properties of RF Magnetron Sputtered p-CuInO$_x$ Thin Films for Transparent Heterojunction Devices

Giji Skaria *, Avra Kundu and Kalpathy B. Sundaram

Department of Electrical and Computer Engineering, University of Central Florida, Orlando, FL 32816, USA; avra.kundu@ucf.edu (A.K.); kalpathy.sundaram@ucf.edu (K.B.S.)
* Correspondence: giji.skaria@ucf.edu

Abstract: Copper indium oxide (CuInO$_x$) thin films were deposited by the RF magnetron sputtering technique using a Cu$_2$O:In$_2$O$_3$ target at varying substrate temperatures up to 400 °C. Mutually exclusive requirements of having a p-type thin film along with increased conductivity and high transparency were achieved by controlling the migration of indium oxide phases during the sputtering process, as verified by the XPS studies. A morphological study performed using SEM further confirmed the crystallization and the grain growth (95–135 nm) with increasing substrate temperatures, resulting in superior conductivity and an enhanced transparency of more than 70% in the 400–700 nm range. This is due to the controlled replacement of copper sites with indium while maintaining the p-type characteristic of the thin film. Optical studies carried out on the films indicated a bandgap change in the range of 2.46–2.99 eV as a function of substrate heating. A p-CuInO$_x$/n-Si heterojunction was fabricated with a measured knee voltage of 0.85 V. The photovoltaic behavior of the device was investigated and initial solar cell parameters are reported.

Keywords: CuInO$_x$; substrate temperature; RF sputtering; XPS; transparent heterojunction; optical bandgap

1. Introduction

Transparent conducting oxides (TCOs) have unique optoelectronic properties which allow visible light to pass through while having reasonably high electrical conductivity [1–3]. TCOs have a variety of applications ranging from uses in solar cells, optical displays, reflective coatings, light emission devices, low-emissivity windows, electrochromic mirrors, UV sensors, and windows, defrosting windows, electromagnetic shielding, and transparent electronics [4,5]. The conductivity of TCOs can be tuned from insulating via semiconducting to conducting. Their transparency can also be adjusted depending on the donor/acceptor levels as well as the bandgap of the material. This enables the realization of both n-type and p-type TCOs, which make them highly attractive for transparent opto-electrical circuitries and technological applications. Most research activities have focused on the optimization of n-type TCOs [6–9], but many transparent electronic applications require the necessity of p-type TCOs as well.

Various techniques have been proposed to obtain p-TCO thin films with better electrical and optical properties. In spite of that, the conductivity of p-TCO thin films is still one or two orders of magnitude less than that in the corresponding n-TCO thin films. Increasing the conductivity of p-TCO thin films without sacrificing their visible transmittance is the most significant challenge for p-TCO technology in order to obtain high-performance active devices suitable for "Invisible Electronics" [10]. It may be noted that p-type transparent oxide semiconductors based on CuAlO$_2$, CuGaO$_2$, CuInO$_2$, SrCu$_2$O$_2$, and LaCuOCh (Ch = chalcogen) have been reported [11]. Copper indium oxide–copper aluminum oxide and copper gallium oxide-based thin-film materials have emerged as the front runner for possible p-type TCO applications [10–13]. Of the copper-based systems, the CuInO$_2$ system

is particularly interesting because it can be doped with both p-type (with Ca) and n-type (with Sn), allowing p–n homojunctions to be produced. The synthesis of $Cu_2In_2O_5$ has been reported only using either smelting [14] or chemical processes, which involve synthesizing $Cu_2In_2O_5$ from aqueous solutions of nitrates, chlorides, and sulfates of Cu, In, and Ga [15]. Therefore, the realization of p-type TCOs has been rather challenging and involves the use of chemicals that present levels of toxicity and low yield or smelting which results in a high thermal budget for the fabricated device.

However, recently, Nair et al. fabricated a transparent thin-film p–n junction consisting of Ca and tin-doped $CuInO_2$ [16] by oxygen plasma-enhanced reactive thermal evaporation. This allows for the scalable physical vapor phase deposition of the desired thin films. [10] Thus, there is a need to realize p-type TCOs which offer high yield, scalability, and low cost. RF magnetron sputtering of TCO sources, if controlled in situ, can lead to high uniformity and homogeneity, along with the ability to control the film thickness and deposition rate. It also allows for large-area deposition at a relatively low cost and optimum thermal budget [7,12,13,17–19].

In this work, we investigate the realization of p-type $CuInO_x$ thin films by RF magnetron sputtering and study the effects of substrate heating for the in situ controlling of the optoelectronic and morphological properties. It is seen that substrate heating influences the material characteristics more significantly than post-deposition annealing [13]. We can tailor the thin-film characteristics in situ to initiate crystalline growth and control the proportion of indium oxide or copper oxide phases to improve the transparency while retaining the p-type characteristics of the thin film at a reduced thermal budget as compared with post-deposition annealing [13]. Scanning electron microscopy imaging has been used to obtain the grain size of the film and UV-Vis spectral measurements have been carried out for transmittance studies. XPS analysis of the thin films has also been carried out to identify the oxidation state and bonding configurations of Cu, In, and O in copper indium oxide films. Junction studies of the p-type $CuInO_x$ have been investigated with n-Si and ITO for demonstrating heterojunction behavior, which can potentially find applications in transparent electronics, photodetectors, and solar cells.

2. Experimental Details

2.1. Deposition of Copper Indium Oxide Thin Films with Substrate Heating

Thin films of copper indium oxide ($CuInO_x$) were deposited using radio frequency (RF) magnetron sputtering. MAK 2 sputter gun (San Jose, CA, USA) was driven by a Dressler Cesar 136 FST RF Generator (Denver, CO, USA) and an Advanced Energy VarioMatch-1000 matching network (Fort Collins, CO, USA). The deposition pressure was 10 mTorr, having an argon (Ar) flow of 10 sccm. All the depositions were performed at a power of 100 W on silicon/quartz substrates. The deposition pressure, argon flow and deposition power were chosen based on the optimal deposition rate during the sputtering process. Acetone and methanol were used to clean the substrates and they were rinsed with deionized water and blow dried with a nitrogen gun prior to the deposition.

A 2" powder-pressed target of Cu_2O/In_2O_3 (1/1 mol%, 99.9% purity), ACI Alloy Inc. (San Jose, CA, USA), was used as a sputtering target. Highly dense single-phase targets for $CuInO_2$ thin-film growing are still challenging, therefore a Cu_2O–In_2O_3 composite target instead of a $CuInO_2$ single-phase target was used in this study. Furthermore, having the composite target allows for more flexibility in controlling the Cu_2O and In_2O_3 phases in the deposited thin film. The target and the substrate separation was maintained at 5 cm to achieve a uniform film thickness. Base pressure of 5×10^{-6} Torr was achieved before initiating the deposition. The deposition rate was measured for varying substrate temperatures. The thickness was measured using a Veeco Dektak-150 profilometer (Plainview, NY, USA). All films were deposited with a thickness of 2000 Å at the different substrate temperatures.

2.2. Morphological and Optical Studies of Copper Indium Oxide Thin Films

A field emission scanning electron microscope, Zeiss ULTRA-55 FEG SEM (Zeiss Microscopy, White Plains, NY, USA), was used for morphological studies. An acceleration voltage of 5 kV was used for the imaging at a working distance of 0.2 mm. A Cary 100 UV–Vis spectrophotometer (Varian Analytical Instruments, Walnut Creek, CA, USA) was used to measure the transmittance for a wavelength range of 300–800 nm. The composition of the films was determined using ESCALAB 250 Xi + X-ray photoelectron spectroscopy (XPS) by Thermo Fisher Scientific with a monochromatic Al Kα source (1486.7 eV). To remove the surface oxygen, a monatomic EX06 ion source (4 kV) was used to perform ion milling of the film surface for 30 s prior to the measurements. Thermo Fisher Scientific Avantage (software version 5.9902) was used to identify the elemental composition and fit the XPS peaks. A smart Shirley function was used to subtract the background for the XPS spectra. The peak fitting was performed by using a mixed Gaussian function after background subtraction.

2.3. Fabrication of CuInO$_x$ Heterojunctions and Electrical Studies

(a) The 1–20 Ω-cm n-type silicon was used to fabricate the Si heterojunctions. Aluminum, as the contact for the n-type silicon, was deposited on the backside of the silicon using the thermal evaporation technique. Post annealing at 400 °C for 30 min in N$_2$ ambiance was conducted before depositing the CuInO$_x$ thin films. Figure 1a shows the structure of the fabricated device.

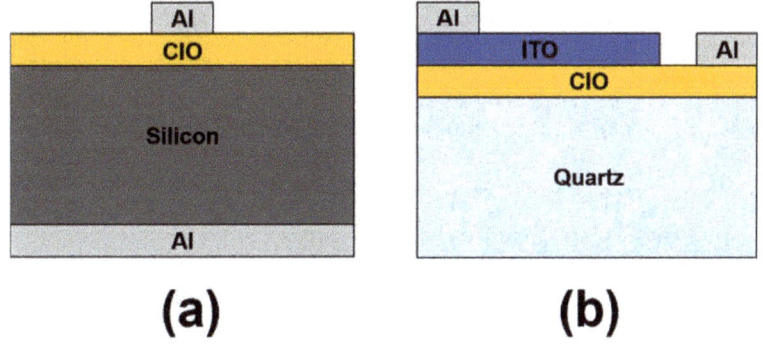

Figure 1. Device schematic for (**a**) n-Si/p-CIO heterojunction and (**b**) quartz p-CIO/n-ITO heterojunction.

(b) A quartz substrate was used to fabricate the heterojunction on a transparent platform. In total, 4000 Å of copper indium oxide films were deposited with a substrate temperature of 400 °C. A 1 cm × 1 cm area was opened and the rest of the area was masked using aluminum foil, and indium tin oxide was deposited by RF magnetron sputtering (100 W, 10 mTorr, 10 sccm Ar) at room temperature (RT) to obtain a thickness of 4000 Å. A mask with several holes (1 mm diameter) was used to deposit aluminum contacts on copper indium oxide (p-type) as well as indium tin oxide (n-type). Figure 1b shows the structure of the fabricated device.

3. Results and Discussions

3.1. Deposition and Morphological Studies

Figure 2 shows the thickness of the obtained films after 5 min of deposition at varying substrate temperatures. It is seen that the final thickness obtained at higher temperatures reduces, which may be attributed to higher thin-film densities and lower porosities [6]. It is interesting to point out here that the substrate temperature also increases due to collisions from secondary electrons [20].

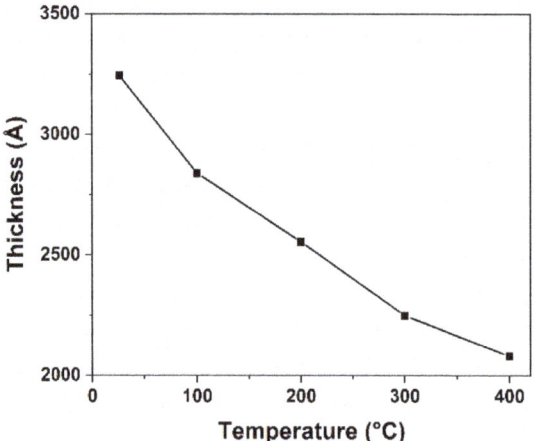

Figure 2. Thickness of CuInO$_x$ films at varying substrate temperatures after 5 min of sputtering process.

Figure 3 shows the SEM images of as-deposited CIO thin films as well as those obtained at substrate temperatures of 200, 300, and 400 °C. Changes in the morphology were identified as the substrate heating temperature was increased. Films with a substrate deposition temperature of 200 °C displayed the presence of very small grains, as shown in Figure 3b with respect to the film deposited at room temperature. Thus, a minimum of 300–400 °C is required to initiate nanocrystalline growth, as evidenced in the figure. The grain size of 10–30 nm was obtained for the films deposited at 300–400 °C. It is worthy to note that the crystallization occurs at a much lower temperature as compared with post-deposition annealing, which is typically at 500 °C and above. Therefore, a significant reduction in thermal budget is achieved by the process of substrate heating during the RF magnetron sputtering process.

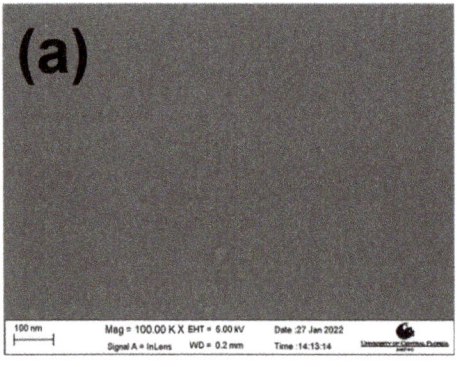
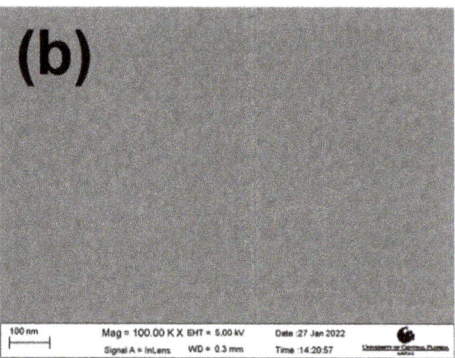

Figure 3. *Cont.*

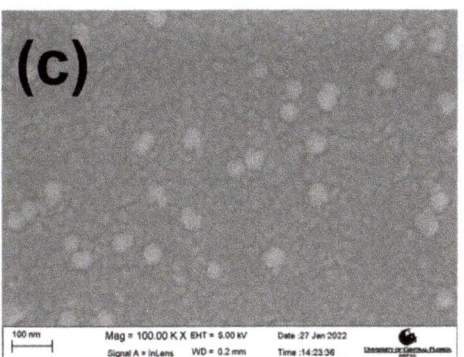
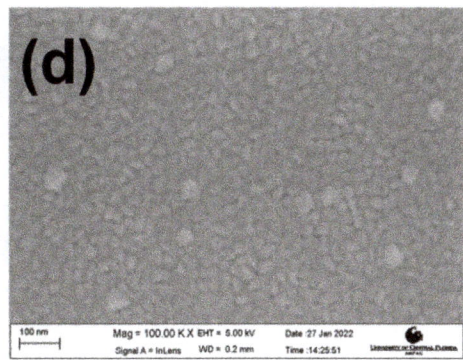

Figure 3. SEM images of CuInO$_x$ films on quartz: (**a**) room temperature, (**b**) substrate temperature of 200 °C, (**c**) substrate temperature of 300 °C, (**d**) substrate temperature of 400 °C.

3.2. XPS Analysis

Figure 4a(i–iv) shows the XPS spectra obtained from the CuInO$_x$ films sputtered at room temperature, 200, 300, and 400 °C, respectively. Peaks pertaining to Cu, In, and O were identified in the obtained XPS survey spectra shown in Figure 4a(i–iv), confirming the phase purity. From Figure 4b(i–iv), two strong peaks are obtained at 933.8 eV and 953.7 eV, belonging to Cu $2p_{3/2}$ and Cu $2p_{1/2}$, respectively [21]. The presence of Cu2+ species was confirmed by the satellite peaks between 940 and 950 eV. Figure 4c(i–iv) shows the $3d_{5/2}$ and $3d_{3/2}$ indium core levels at a binding energy of 444.1 and 451.7 eV [21]. Oxygen 1 s core level, Figure 4d(i–iv), was fitted with three Gaussian peaks located at 529.5, 531.3, and 532.5 eV (marked 1, 2, and 3), respectively. Peak 1 is attributed to oxygen attached to In (+3) while peak 2 is attributed to oxygen attached to Cu (+1). Peak 3 is attributed to strongly chemisorbed oxygen which is normally observed at 532.2 eV [21]. It is seen from Figure 4d(i–iv) that the strength of peak 2 reduces relative to that of peak 1. This indicates that there is a controlled addition of indium oxide phases to the thin film as a result of substrate heating while maintaining sufficient copper oxide phases. The addition of the indium oxide phases causes the thin film to become more transparent while the optimal presence of copper oxide phases enables the film to retain its p-type characteristics. This relative addition of indium oxide phases is due to the loss in the copper oxide phases due to the possible desorption of copper oxide at elevated temperatures. The ratio of the area of the peaks corresponding to indium oxide and copper oxide was performed for room temperature, 200, 300, and 400 °C. It is seen that the ratio of In$_2$O$_3$:Cu$_2$O increases as 0.98, 1.02, 1.06, and 1.11 with respective substrate temperatures corresponding to room temperature, 200, 300, and 400 °C. This is also validated by the transmittance measurements for transparency as well as hot probe measurements for the p-type behavior [18].

(a)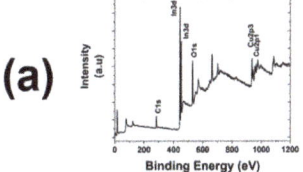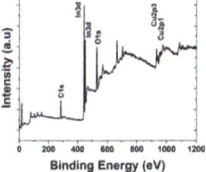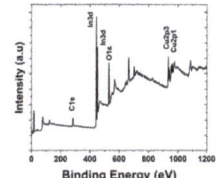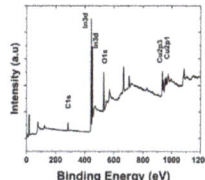

Figure 4. *Cont.*

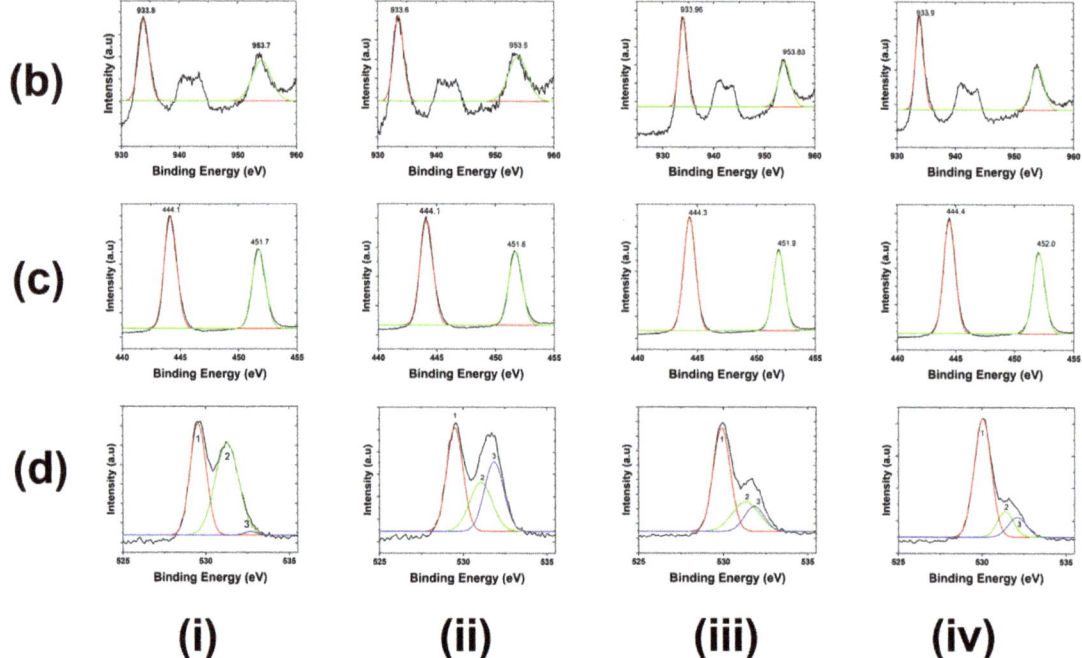

Figure 4. (a) XPS spectra of CuInO$_x$ thin films with (b) copper peaks, (c) indium peaks, and (d) oxygen peaks. (i–iv) correspond to room temperature, 200 °C, 300 °C, and 400 °C, respectively.

3.3. Optical Analysis and Bandgap

Figure 5 shows the percentage transmission for the films deposited on quartz substrates at room temperature, 200, 300, and 400 °C. Overall, the optical transmission increases for the films deposited at higher temperatures. This is because of the controlled relative increase in the indium oxide phases at the copper oxide sites as already verified by the XPS analysis. It is seen that the percentage increase in transmittance varies up to 25%–100% between wavelengths of 400 and 700 nm for a substrate temperature of 400 °C, which may be attributed to the higher transmission of indium oxide over that of copper oxide as the relative concentration of indium oxide increases with an increase in substrate temperature.

The optical bandgap of CuInO$_x$ thin films were calculated using the Tauc plot method [22,23] based on the optical transmission data. The absorption coefficient α was calculated directly from the transmission data as the reflectance was less than 5%. Absorption coefficient α was calculated using Equation (1), where d is the thickness of the film equal to 200 nm, and T is the percent of transmission. The optical bandgap (Eg) was estimated from Equation (2),

$$\alpha = \frac{1}{d} \ln\left(\frac{1}{T}\right) \quad (1)$$

$$(\alpha h\nu)^{1/n} = B\,(h\nu - E_g) \quad (2)$$

where hν is the photon energy, B is a constant, Eg is the optical bandgap, and $n = 1/2$ for the direct bandgap transition [21]. As demonstrated by Singh et al. [21], the best linear fit in the optical absorption region to the $(\alpha h\nu)^{1/n}$ versus energy curves were obtained for $n = 1/2$, implying our thin-film material has a direct band gap. Figure 6 shows the generated Tauc plot. The linear region of the curve was extrapolated to the x-axis to identify the Eg value. The extrapolated values of the bandgap are listed in Figure 6. The bandgap was in the range of 2.46–2.99 eV for temperatures ranging from room temperature to 400 °C (Figure 6). It is worth mentioning that an increase in the bandgap with temperature is also attributed to the relative increase in the indium oxide phases as it has a higher bandgap than copper oxide.

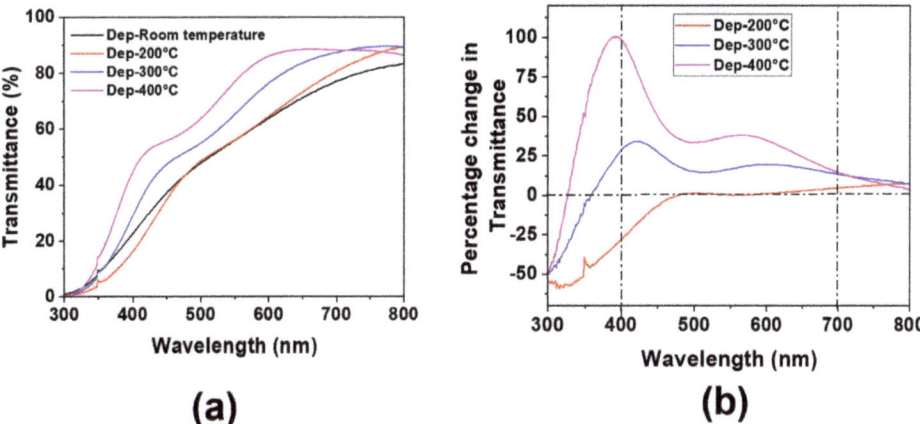

Figure 5. (**a**) Transmittance spectra of CuInO$_x$ films on quartz at different temperatures during deposition. (**b**) Percentage increase in transmittance for the deposited films due to the effect of substrate heating with respect to room temperature deposition.

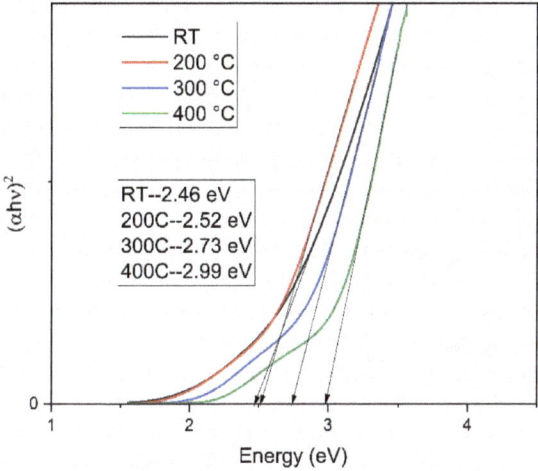

Figure 6. Tauc plots of the CuInO$_x$ thin films deposited at different substrate temperatures.

3.4. Electrical Studies

For the fabricated device structures, as outlined in Section 2.3, the I-V characteristics were performed. All the films were deposited at 400 °C as the films deposited at 400 °C had the highest transmission and also showed crystalline grains, as evidenced in the SEM images (Figure 3). For the silicon-based device, a knee voltage of 0.85 V was obtained. The value is similar to that of other such heterojunctions reported in the literature [16]. The I-V characteristics are depicted in Figure 7a,b, which shows n-Si/p-CIO heterojunction behavior. Light characteristics were also performed on the device, and an open-circuit voltage (V_{OC}) of 50 mV and short-circuit current (I_{SC}) of 5 µA were obtained. This confirms that the device is suitable for optoelectronic applications including solar cells and photodetectors. Figure 7c,d shows the I-V characteristics for the quartz-based p-CIO/n-ITO heterojunction with a knee voltage of 1.6V which is close to that of similar heterojunction devices [16,24]. The I-V characteristics again confirm a p-CIO/n-ITO heterojunction which has many potential applications in transparent electronics as both the layers have very high transmittance. Figure 8 shows the photomicrograph of the fabricated device on a quartz substrate showing a high transparency.

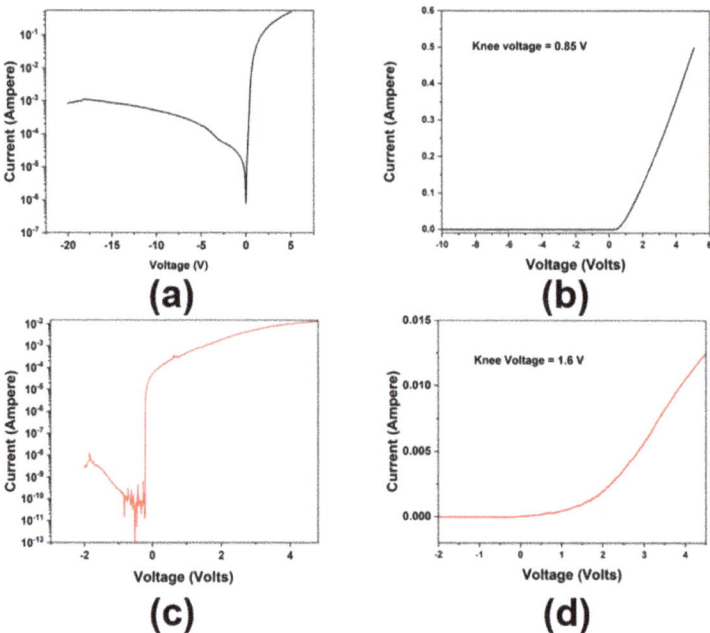

Figure 7. I-V characteristics of (**a**,**b**) n-Si/p-CIO heterojunction and (**c**,**d**) quartz p-CIO/n-ITO heterojunction with knee voltages.

Figure 8. Photomicrograph of the fabricated device on a quartz substrate showing a high transparency.

4. Conclusions

Copper indium oxide (CuInO$_x$) films, having a p-type behavior along with high transparency and superior electrical conductivity, were deposited by RF magnetron sputtering. This control over the electrical, morphological, and optical properties was obtained in situ by varying the substrate temperature during the deposition process. While the p-type nature was controlled by the presence of the copper oxide phases in the deposited thin films, the transparency was controlled by the presence of indium oxide phases. XPS results confirm that at elevated temperatures of 400 °C, the relative increase in indium oxide phases is favored, resulting in a higher optical bandgap of 2.99 eV and better transparency (>45% for 400–700 nm). It may be interesting to note here that further increasing the substrate temperature during the deposition process may increase the transparency further, as well as the crystallization, which may need to be validated by further studies. However, these advantages will be offset by the higher thermal budget with increasing temperature. Optimal optical, electrical, and morphological properties were obtained for depositions carried out at a substrate temperature of 400 °C. The heterojunction behavior of the p-type CIO was confirmed with n-Si and n-ITO, thereby laying the basis for completely transparent TCO-based optoelectronic devices.

Author Contributions: Conceptualization, K.B.S.; Data curation, G.S. and A.K.; Formal analysis, G.S. and A.K.; Investigation, G.S.; Methodology, G.S., A.K. and K.B.S.; Project administration, G.S. and K.B.S.; Software, G.S.; Supervision, K.B.S.; Validation, G.S. and A.K.; Writing—original draft, G.S.; Writing—review & editing, K.B.S. All authors have read and agreed to the published version of the manuscript.

Funding: A portion of the article processing charges were provided by the UCF College of Graduate Studies Open Access Publishing Fund.

Institutional Review Board Statement: Not applicable.

Informed Consent Statement: Not applicable.

Data Availability Statement: Not applicable.

Conflicts of Interest: The authors declare no conflict of interest.

References

1. Ginley, D.S.; Bright, C. Transparent conducting oxides. *MRS Bull.* **2000**, *25*, 15–18. [CrossRef]
2. Fortunato, E.; Ginley, D.; Hosono, H.; Paine, D.C. Transparent conducting oxides for photovoltaics. *MRS Bull.* **2007**, *32*, 242–247. [CrossRef]
3. Ye, F.; Cai, X.M.; Dai, F.P.; Zhang, D.P.; Fan, P.; Liu, L.J. Transparent Conductive Cu-In-O Thin Films Deposited by Reactive DC Magnetron Sputtering with Different Targets. In *Advanced Materials Research*; Trans Tech Publications Ltd.: Kapellweg, Switzerland, 2011.
4. Lewis, B.G.; Paine, D.C. Applications and processing of transparent conducting oxides. *MRS Bull.* **2000**, *25*, 22–27. [CrossRef]
5. Szyszka, B.; Loebmann, P.; Georg, A.; May, C.; Elsaesser, C. Development of new transparent conductors and device applications utilizing a multidisciplinary approach. *Thin Solid Film.* **2010**, *518*, 3109–3114. [CrossRef]
6. Sundaram, K.; Khan, A. Characterization and optimization of zinc oxide films by rf magnetron sputtering. *Thin Solid Film.* **1997**, *295*, 87–91. [CrossRef]
7. Shantheyanda, B.P.; Todi, V.O.; Sundaram, K.B.; Vijayakumar, A.; Oladeji, I. Compositional study of vacuum annealed Al doped ZnO thin films obtained by RF magnetron sputtering. *J. Vac. Sci. Technol. A Vac. Surf. Film.* **2011**, *29*, 051514. [CrossRef]
8. Sundaram, K.B.; Bhagavat, G.K. Chemical vapour deposition of tin oxide films and their electrical properties. *J. Phys. D Appl. Phys.* **1981**, *14*, 333. [CrossRef]
9. Sundaram, K.B.; Bhagavat, G.K. Low temperature synthesis wide optical band gap Al and (Al, Na) co-doped ZnO thin films. *Appl. Surf. Sci.* **2011**, *257*, 2341–2345.
10. Banerjee, A.; Chattopadhyay, K. Recent developments in the emerging field of crystalline p-type transparent conducting oxide thin films. *Prog. Cryst. Growth Charact. Mater.* **2005**, *50*, 52–105. [CrossRef]
11. Sheng, S.; Fang, G.; Li, C.; Xu, S.; Zhao, X. p-Type transparent conducting oxides. *Phys. Status Solidi* **2006**, *203*, 1891–1900. [CrossRef]
12. Saikumar, A.K.; Sundaresh, S.; Nehate, S.D.; Sundaram, K.B. Properties of RF Magnetron-Sputtered Copper Gallium Oxide (CuGa2O4) Thin Films. *Coatings* **2021**, *11*, 921. [CrossRef]

13. Skaria, G.; Kumar Saikumar, A.; Shivprasad, A.D.; Sundaram, K.B. Annealing Studies of Copper Indium Oxide ($Cu_2In_2O_5$) Thin Films Prepared by RF Magnetron Sputtering. *Coatings* **2021**, *11*, 1290. [CrossRef]
14. Su, C.-Y.; Mishra, D.K.; Chiu, C.-Y.; Ting, J.-M. Effects of Cu_2S sintering aid on the formation of $CuInS_2$ coatings from single crystal $Cu_2In_2O_5$ nanoparticles. *Surf. Coat. Technol.* **2013**, *231*, 517–520. [CrossRef]
15. Su, C.-Y.; Chiu, C.-Y.; Chang, C.-H.; Ting, J.-M. Synthesis of $Cu_2In_2O_5$ and $CuInGaO_4$ nanoparticles. *Thin Solid Film.* **2013**, *531*, 42–48. [CrossRef]
16. Nair, B.; Rahman, H.; Shaji, S.; Okram, G.; Sharma, V.; Philip, R.R. Calcium incorporated copper indium oxide thin films-a promising candidate for transparent electronic applications. *Thin Solid Film.* **2020**, *693*, 137673. [CrossRef]
17. Sundaresh, S.; Nehate, S.D.; Sundaram, K.B. Electrical and optical studies of reactively sputtered indium oxide thin films. *ECS J. Solid State Sci. Technol.* **2021**, *10*, 065016. [CrossRef]
18. Saikumar, A.K.; Sundaresh, S.; Nehate, S.D.; Sundaram, K.B. Preparation and Characterization of Radio Frequency Sputtered Delafossite p-type Copper Gallium Oxide (p-$CuGaO_2$) Thin Films. *ECS J. Solid State Sci. Technol.* **2022**, *11*, 023005. [CrossRef]
19. Ray, S.; Banerjee, R.; Basu, N.; Batabyal, A.K.; Barua, A.K. Properties of tin doped indium oxide thin films prepared by magnetron sputtering. *J. Appl. Phys.* **1983**, *54*, 3497–3501. [CrossRef]
20. Sundaram, K.B.; Garside, B.K. Properties of ZnO films reactively RE sputtered using a Zn target. *J. Phys. D Appl. Phys.* **1984**, *17*, 147. [CrossRef]
21. Singh, M.; Singh, V.N.; Mehta, B.R. Synthesis and properties of nanocrystalline copper indium oxide thin films deposited by RF magnetron sputtering. *J. Nanosci. Nanotechnol.* **2008**, *8*, 3889–3894. [CrossRef]
22. Ohyama, M.; Kozuka, H.; Yoko, T. Sol-gel preparation of transparent and conductive aluminum-doped zinc oxide films with highly preferential crystal orientation. *J. Am. Ceram. Soc.* **1998**, *81*, 1622–1632. [CrossRef]
23. Sundaram, K.B.; Bhagavat, G.K. Optical absorption studies on tin oxide films. *J. Phys. D Appl. Phys.* **1981**, *14*, 921. [CrossRef]
24. Yanagi, H.; Ueda, K.; Ohta, H.; Orita, M.; Hirano, M.; Hosono, H. Fabrication of all oxide transparent p–n homojunction using bipolar $CuInO_2$ semiconducting oxide with delafossite structure. *Solid State Commun.* **2001**, *121*, 15–17. [CrossRef]

Article

The Effect of RF Sputtering Temperature Conditions on the Structural and Physical Properties of Grown SbGaN Thin Film

Cao Phuong Thao [1], Dong-Hau Kuo [2,*] and Thi Tran Anh Tuan [3,*]

1 School of Engineering and Technology, Tra Vinh University, Tra Vinh 87000, Vietnam; cpthao@tvu.edu.vn
2 Department of Materials Science and Engineering, National Taiwan University of Science and Technology, Taipei 10607, Taiwan
3 School of Basic Sciences, Tra Vinh University, Tra Vinh 87000, Vietnam
* Correspondence: dhkuo@mail.ntust.edu.tw (D.-H.K.); thitrananhtuan@tvu.edu.vn (T.T.A.T.)

Abstract: By using a single ceramic SbGaN target containing a 14% Sb dopant, $Sb_{0.14}GaN$ films were successfully grown on n-Si(100), SiO_2/Si(100), and quartz substrates by an RF reactive sputtering technology at different growth temperatures, ranging from 100 to 400 °C. As a result, the structural characteristics, and optical and electrical properties of the deposited $Sb_{0.14}GaN$ films were affected by the various substrate temperature conditions. By heating the temperature deposition differently, the sputtered $Sb_{0.14}GaN$ films had a wurtzite crystal structure with a preferential ($10\bar{1}0$) plane, and these $Sb_{0.14}GaN$ films experienced a structural distortion and exhibited p-type layers. At the highest depositing temperature of 400 °C, the $Sb_{0.14}GaN$ film had the smallest bandgap energy of 2.78 eV, and the highest hole concentration of 8.97×10^{16} cm^{-3}, a conductivity of 2.1 Scm^{-1}, and a high electrical mobility of 146 cm^{2}V^{-1}s^{-1}. The p-$Sb_{0.14}$GaN/n-Si heterojunction diode was tested at different temperatures, ranging from 25 to 150 °C. The testing data showed that the change of testing temperature affected the electrical characteristics of the diode.

Keywords: acceptor Sb; temperature conditions; RF sputtering; heterojunction diode

1. Introduction

It is believed that gallium nitride and its alloys are functional semiconductor materials with a crystal wurtzite structure, wide bandgap, good thermal conductivity, high carrier mobility, and a high breakdown voltage [1,2]. In the fabrication of electronics and photoelectronic components, they have been widely applied and designated as diodes and light-emitting diodes (LED), metal-oxide semiconductor field-effect transistor (MOSFET), and hetero junction field-effect transistor (HJ-FET) [3–7]. Overall, n- and p-type Gallium nitride films can be fabricated by many processes, such as metal organic chemical vapor deposition (MOCVD) [8,9], metal organic vapor phase epitaxy (MOVPE) [10,11], low energy electron beam irradiation (LEEBI) treatment [12], or a two flow metal organic chemical vapor deposition (TF- MOCVD) [13,14]. Besides, radio frequency (RF) sputtering technology has promising properties in application, as it is more inexpensive, uncomplicated to clean, and easy to use with various sputtering conditions. Using this technique, the target can be designed according to a wide range of compositions. It is reasonable to study the doping in III–nitride compound semiconductors, and a GaN film has been easily fabricated. Recently, our group has accomplished a deposited p-GaN and n-GaN film [15–17]. In a previous experiment, we had studied the effects of an Sb dopant on the semiconductor characteristics of Sb-x-GaN films sputtered by RF reactive sputtering technology using a single ceramic Sb-x-GaN target with an Sb dopant at x = 0.0, 0.07, 0.14, and 0.2. The results determined that these deposited Sb-x-GaN films converted to a p-semiconductor at x = 0.07, 0.14, and 0.2. The p-SbGaN/n-Si(100) heterojunction diodes performed successfully and their electrical characteristics at room temperature were thoroughly investigated [18] In 2019, our previous work was performed in conditions that affected the electrical, optical, and structural properties of the GeGaN thin films grown using RF reactive sputtering [19]. At the highest temperature of 400 °C, the $Ge_{0.07}GaN$ film acted as an n-semiconductor, had a smaller bandgap energy of 3.02 eV, and achieved a higher carrier concentration of 1.30×10^{18} cm^{-3}, an electrical conductivity of 1.46 Scm^{-1}, and an electron mobility of 7 cm^{2}V^{-1}s^{-1}. However, the influences of the different heating substrate temperature conditions on the characteristics of Sb-GaN thin films have not been studied by many researchers until this work. In this work, we studied the influences of different deposition temperatures on the structural, optical, and electrical characteristics of the $Sb_{0.14}GaN$ films and the characteristics of the p-$Sb_{0.14}GaN$/n-Si heterojunction diode at different testing temperatures. Firstly, $Sb_{0.14}GaN$ thin films were deposited at various substrate heating temperatures, of 100, 200, 300, and 400 °C. Secondly, a diode was fabricated by growing the p-$Sb_{0.14}GaN$ film on an n-Si substrate with the electrodes on the top using modeling by RF sputtering, then the electrical characteristics of the diode were thoroughly investigated at various working temperatures, from 25 to 150 °C.

2. Experimental Details

To thoroughly investigate the effects of different heating substrate temperature conditions on the deposited Sb doped GaN film, the SbGaN ceramic target was prepared with Sb/(Sb+Ga) at a molar ratio of 14 at% and named $Sb_{0.14}GaN$ target. Using a single $Sb_{0.14}GaN$ target for deposition, the $Sb_{0.14}GaN$ films were successfully fabricated on n-type Si(100), SiO$_2$/Si(100), and quartz substrates by RF sputtering. The $Sb_{0.14}GaN$ films sputtered on the n-Si(100) substrates at different temperatures were used to study crystal structure properties, morphological and topographical surfaces, and analyzing the composition of the $Sb_{0.14}GaN$ films. To test the electrical properties of the films, the $Sb_{0.14}GaN$ films were deposited on an SiO$_2$/Si(100) substrate using the RF technique at various temperature ranges, 100–400 °C. Additionally, the $Sb_{0.14}GaN$ films grown on quartz substrates at different heating substrate temperatures were used to investigate the optical properties of $Sb_{0.14}GaN$ films. The n-Si(100) substrates used in this work possessed an electrical mobility of ~200 cm^{2}V^{-1}s^{-1}, an electron concentration of ~10^{15}cm^{-3}, an electrical resistivity of ~1–10 Ωcm, a diameter of 2 inches, and a thickness of ~550. In order to study the effects of deposition temperature on the SbGaN film properties, the $Sb_{0.14}GaN$ films were sputtered at substrate temperature conditions of 100, 200, 300, and 400 °C, output RF sputtering power was constant, at 120 W, and deposition time was kept at 30 min. Before depositing, the pressure of the working chamber was lowered to 1×10^{-6} Torr by the mechanical and diffusion pumps. For depositing, the reactive RF sputtering was fixed at a working pressure of 9×10^{-3} Torr and a gas mixture of Ar and N$_2$ flowed at rates of 5 sccm and 15 sccm, respectively. In this experiment, the target was located opposite the substrate at a distanced of 5 cm in the working chamber. The heterojunction diode successfully performed the reactive RF sputtering technology, and the heterojunction diode created a sputtered p-type Sb-0.14-GaN film on the n-Si substrate [18]. The heterojunction p-$Sb_{0.14}GaN$/n-Si(100) diode was made by sputtering an $Sb_{0.14}GaN$ film on an n-Si(100) substrate. The diode structure was designed in a top-top electrode model fabricated by depositing $Sb_{0.14}GaN$ film on the n-Si substrate. It is known that metals with a high work function (such as Pt, Mo, Ni, etc.) have often been applied to form Schottky contacts with an n-type semiconductor, and ohmic contacts with a p-type semiconductor in the modeling of metal-oxide-semiconductor MOS, metal-semiconductor MS, homo-, or heterojunction diodes. Besides, metals with a low work-function (such as Al, Ag, etc.) have often been employed to fabricate ohmic contacts with n-type semiconductors, and Schottky contacts with a p-type semiconductor. In this work, the modeling of Pt/p-$Sb_{0.14}GaN$/n-Si/Al was applied to Al to make an ohmic contact with n-Si and Pt to perform an ohmic contact with p-Sb-GaN. The commercially purchased Al and Pt targets (99.999%) were applied to the electrodes at a sputtering temperature of 200 °C for 30 min, on the top of the film. The physical methods for designing and operating the diodes with our sputtered III-nitride thin films are particularly described in our previous works [7,15,18,20–22]. To study the influence of the testing conditions on the diode, the p-$Sb_{0.14}GaN$/n-Si(100) diode was deployed in current–voltage tests at various testing temperatures, ranging from 25 to 150 °C.

The crystalline structure of the $Sb_{0.14}GaN$ films was studied by X–ray diffractometry (XRD, D8 Discover, Bruker, Billerica, MA, USA). Scanning electron microscopy (SEM, JSM–6500F, JEOL, Tokyo, Japan) was applied to investigate the morphological surface and cross-section images of the $Sb_{0.14}GaN$ films. The topographical surface and the film roughness formed by the root-mean-square (*rms*) value were computed by atomic force microscopy (AFM, Dimension Icon, Bruker, MA, USA). Compositional analyses of the $Sb_{0.14}GaN$ films were performed by the energy dispersive spectrometer (EDS, JSM–6500F, JEOL, Tokyo, Japan) supported on SEM. The absorption spectra of the $Sb_{0.14}GaN$ films were tested using an ultraviolet-visible (UV–Vis) spectrometer (V-670, Jasco, Tokyo, Japan), and a Hall measurement system (HMS–2000, Ecopia, Tokyo, Japan) was used to determine the electrical properties of the $Sb_{0.14}GaN$ films, consisting of hole concentration, electrical mobility, and conductivity with a maximum magnetic field of 0.51 T. A semiconductor device analyzer (Agilent, B1500A, Santa Clara, CA, USA) was used to study the electrical characteristics of the p-$Sb_{0.14}GaN$/n-Si(100) heterojunction diode at a temperature range of 25 to 150 °C.

3. Results and Discussion
3.1. Effects of Growth Temperature on the Deposited SbGaN Film Properties

The compositional films of these sputtered $Sb_{0.14}GaN$ films, grown at different temperatures in the range of 100–400 °C, were studied by the energy dispersive spectrometer supported by a SEM. Table 1 presents the EDS compositional data of the $Sb_{0.14}GaN$ films deposited at heating substrate ranges of 100–400 °C. As Ga concentration in the $Sb_{0.14}GaN$ films grown at different heating substrate temperatures, from 100 to 400 °C, was a decreasing trend, the concentration of Sb in these films increased. There was a slight reduction in Ga concentration in the $Sb_{0.14}GaN$ films, listed as 46.48, 46.31, 45.89, and 45.64 at% corresponding to substrate temperature conditions of 100, 200, 300, and 400 °C, respectively. Sb concentration in the $Sb_{0.14}GaN$ films deposited at 100–400 °C increased by 3.24, 3.91, 4.92, and 5.23 at%, respectively. Thus, the [Sb]/([Ga]+[Sb]) molar ratios were 0.065, 0.078, 0.097 and 0.103 for the $Sb_{0.14}GaN$ films at 100, 200, 300 and 400 °C, respectively. The Sb molar ratio of the sputtered $Sb_{0.14}GaN$ films increased significantly as the temperature of the heating substrate increased. As the [N]/([Ga]+[Sb]) ratio of the $Sb_{0.14}GaN$ film at 100 °C was 1.011, the [N]/([Ga]+[Sb]) ratio of the $Sb_{0.14}GaN$ films deposited at 200, 300 and 400 °C were 0.991, 0.968 and 0.966, which indicated that these $Sb_{0.14}GaN$ films had a slight nitrogen vacancy and their electrical properties could be determined by film's deficient nitrogen state. It is expected that the different temperatures of the heating substrates can explain the changes in the film properties. The morphological surfaces and cross-section patterns of $Sb_{0.14}GaN$ films grown by an RF reactive sputtering system at deposition temperatures ranging from 100 to 400 °C were figured by scanning electron microscopy. Figure 1 displays images of the SEM surface and cross sectional patterns of the $Sb_{0.14}GaN$ films deposited at a sputtering temperature in the range of 100–400 °C, with power maintained at 120 W. The surface SEM images showed that the $Sb_{0.14}GaN$ films at different heating-substrate temperatures had a continuous microstructure, film smoothness, and medium grain size in the nanometer. From the inserted cross-sectioning SEM figures in Figure 1, it can be seen that these $Sb_{0.14}GaN$ films adhered well and had good interfaces, without cracks and voids between the $Sb_{0.14}GaN$ layers and *n*-Si substrate. As shown in Table 2, these $Sb_{0.14}GaN$ films exhibited thicknesses of 1.0, 1.13, 1.17, and 1.24 µm as the heating substrate temperatures of the sputtering process were maintained at 100, 200, 300, and 400 °C, respectively. As a result, there was an incremental growth rate of 33.33, 37.67, 39.00, and 41.33 nm/min, corresponding to each sputtering temperature of the deposition system. The physical phenomenon could indicate that more Ar hit the $Sb_{0.14}GaN$ target; a higher heating substrate temperature condition corresponds to a faster sputtering rate. Additionally, Table 3 displays the sizes of wurtzite crystalline, computed by the Scherer equation from the investigated XRD data. They were found to be 24, 24, 22, and 18 nm for the Sb-014-GaN films deposited at temperatures of 100, 200, 300, and 400 °C, respectively.

The topographical surface and the root mean square (rms) roughness values of the $Sb_{0.14}GaN$ films deposited at 100, 200, 300, and 400 °C were studied by atomic force microscopy (AFM). Figure 2 presents the AFM morphologies of the $Sb_{0.14}GaN$ films after scanning dimensions of 5×5 µm^2. The roughness values of the $Sb_{0.14}GaN$ films were 1.55, 1.45, 1.26, and 1.66 nm, of the $Sb_{0.14}GaN$ films deposited at heating substrate temperatures of 100, 200, 300, and 400 °C, respectively. The GaN film sputtered using a radio frequency (RF) sputtering technique had a film roughness ranging from 0.7 to 20 nm [23], while the GaN films made with a MOCVD system had various degrees of film roughness, ranging from 0.5 to 3 nm [24]. Regarding the morphological and topographical surfaces and EDS compositional analyses of the grown $Sb_{0.14}GaN$ films, each exhibited a smooth surface and an increase in the growth rate from 33.33 to 41.33 nm/min, as the heating temperature substrate rose. It is reasonable to infer that the faster depositing of the sputtering system was formed by the strong bombardment of argon onto the $Sb_{0.14}GaN$ target at a higher RF sputtering temperature.

Figure 3 presents the XRD spectra of the $Sb_{0.14}GaN$ films deposited by RF reactive sputtering at various deposition temperatures, from 100 to 400 °C, and fixed at 120 W under an Ar/N_2 gas atmosphere. From the XRD, all the $Sb_{0.14}GaN$ films grown on the n-Si(100) substrates, whether heating substrate temperature conditions applied for the sputtering processes were 100, 200, 300, and 400 °C, had a wurtzite structure and were polycrystalline. The $(10\bar{1}0)$, $(10\bar{1}1)$, $(11\bar{2}0)$, and $(11\bar{2}2)$ peaks of diffraction were determined from those $Sb_{0.14}GaN$ films containing a growth plane with a $(10\bar{1}0)$ dominating peak; another second peak was not identified from the XRD images. At the higher sputtering temperature conditions, there was a slight shift in the position of the dominatant $(10\bar{1}0)$ peak, to a higher 2θ angle. From the tested XRD data at a 2θ angle, the positions of diffraction $(10\bar{1}0)$ peaks were discovered at 32.15°, 32.15°, 32.20°, and 32.26° on the $Sb_{0.14}GaN$ films grown at the substrate temperature conditions of 100, 200, 300, and 400 °C, respectively. Table 2 presents the statistics from the XRD analysis. The lattice constants of a, c, and the volume of the unit cell of the $Sb_{0.14}GaN$ films grown at various growth temperatures from 100 to 400 °C are shown in Table 2. The data displays that there was a slight decrease in the lattice constant of c from 5.23, 5.23, 5.22 to 5.21 Å and a from 3.212, 3.212, 3.21 to 3.20 Å for the $Sb_{0.14}GaN$ films fabricated at heating growth temperatures of 100, 200, 300 and 400 °C, respectively. Besides, the $Sb_{0.14}GaN$ films grown at 100, 200, 300, and 400 °C had cell volumes listed as 46.70, 46.69, 46.58, and 46.23 Å^3, respectively. As the $Sb_{0.14}GaN$ films were made at different RF sputtering temperatures, ranging from 100 to 400 °C, the full width at half maxima (FWHM) values of the $(10\bar{1}0)$ peaks experienced a slight increment of 0.37, 0.38, 0.42, and 0.51°, respectively. By using the Scherer equation for computing crystalline size and the XRD parameters presented in Table 2, the size of the crystallites was substantially smaller at higher heating temperatures, estimated to be 24, 24, 22, and 18 nm for the $Sb_{0.14}GaN$ films grown at 100, 200, 300, and 400 °C, respectively. It was determined that the different depositing temperatures possibly affected the crystalline structure, as the $Sb_{0.14}GaN$ films were grown at deposition temperatures of 100, 200, 300, and 400 °C. In sputtering conditions at a higher heating temperature, there was an increase in the Sb content in the as-sputtered $Sb_{0.14}GaN$ films that exhibited a smaller crystallite size and distortion in the crystal structure.

The Hall measurement system was used to measure the electrical characteristics of the RF sputtered $Sb_{0.14}GaN$ films at various temperature depositions, from 100 to 400 °C. All of these $Sb_{0.14}GaN$ films retained p-semiconductor layers, and the electrical parameters, named bulk concentration, mobility, and conductivity, of the $Sb_{0.14}GaN$ films are detailed in Figure 4. The hole concentration and electrical mobility of the $Sb_{0.14}GaN$ films grown at various heating substrate temperatures of 100, 200, 300, and 400 °C shown in Table 4 showed a significant increase, from 3.25×10^{14} cm^{-3} and 385 cm^{2}V^{-1}s^{-1}, 1.44×10^{15} cm^{-3} and 289 cm^{2}V^{-1}s^{-1}, 2.83×10^{16} cm^{-3} and 287 cm^{2}V^{-1}s^{-1}, to 8.97×10^{16} cm^{-3}

and 146 cm^{2}V^{-1}s^{-1}, respectively. This is explained by the fact that the highest Sb content of 10.3 at % appeared in the composition of the Sb$_{0.14}$GaN film grown at the highest sputtering temperature of 400 °C, correlating to the highest film hole concentration. From the experimental data in Table 4, it is supposed that the hole conductivity of a film could be calculated using hole concentration and mobility, as the Sb$_{0.14}$GaN grown at the different deposition temperatures of 100, 200, 300 and 400 °C achieved an increment of electrical conductivity, from 0.02, 0.067, and 1.3 to 2.1 Scm^{-1}, respectively. This indicates that the electrical properties of these sputtered Sb$_{0.14}$GaN films were affected by the different depositing temperature conditions.

Table 1. The Composition of EDS Analyzing Data of the Deposited Sb$_{0.14}$GaN Films at Heating Substrate Temperatures of 100, 200, 300, and 400 °C.

Temperature Deposition (°C)	Ga (at. %)	Sb (at. %)	N (at. %)	[Sb]/([Ga]+[Sb])	[N]/([Ga]+[Sb])
100	46.48	3.24	50.28	0.065	1.011
200	46.31	3.91	49.78	0.078	0.991
300	45.89	4.92	49.19	0.097	0.968
400	45.64	5.23	49.13	0.103	0.966

Table 2. The effects of Sputtering Temperature on the Structural Properties of Sb$_{0.14}$GaN Film.

Temperature Deposition (°C)	Film Thickness (μm)	Deposition Rate (nm/min)	Roughness (nm)
100	1.00	33.33	1.55
200	1.13	37.67	1.45
300	1.17	39.00	1.26
400	1.24	41.33	1.66

Table 3. Structural Properties of Sb$_{0.14}$GaN Thin Films Deposited at Temperatures Ranging from 100 to 400 °C from the X-ray Diffraction Analyses.

Temperature Deposition (°C)	2θ (1010) Peak	a (Å)	c (Å)	Volume (Å^3)	FWHM (1010) (Degree)	Crystallite Size (nm)
100	32.15	3.212	5.23	46.70	0.37	24
200	32.15	3.212	5.23	46.69	0.38	24
300	32.20	3.21	5.22	46.60	0.42	22
400	32.26	3.20	5.21	46.23	0.51	18

Table 4. The Electrical Properties of Sb$_{0.14}$GaN Films Deposited at the Different Substrate Temperatures.

Deposition Temperature (°C)	Type	Concentration N$_p$ cm^{-3}	Mobility μ cm^{2}V^{-1}s^{-1}	Conductivity σ Scm^{-1}
100	p	3.25 × 10^{14}	385	0.020
200	p	1.44 × 10^{15}	289	0.067
300	p	2.83 × 10^{16}	287	1.300
400	p	8.97 × 10^{16}	146	2.100

The optical properties of the Sb$_{0.14}$GaN films grown at different temperature depositions, from 100 to 400 °C, were measured using a UV–Vis system at room temperature. From the UV—Vis analysis data, the coefficient of optical absorption and bandgap energy,

E_g, of the $Sb_{0.14}GaN$ films could be estimated by employing the Tauc equation expressed following Equation (1):

$$(\alpha h\nu)^2 = A(h\nu - E_g) \tag{1}$$

where α is physical quantity as a coefficient of optical absorption, A is a constant, $h\nu$ is the energy value of the incident photon, and E_g is the bandgap energy of the $Sb_{0.14}GaN$ films deposited at the different growth temperatures. The plots of the $(\alpha h\nu)^2$-$h\nu$ curves and the bandgap energy of the $Sb_{0.14}GaN$ films grown at different heating substrate temperatures are shown in Figure 5. The bandgap, E_g, from the extrapolated curves were 3.01, 2.93, 2.81, and 2.78 eV for the $Sb_{0.14}GaN$ films deposited at different temperature depositions from 100 to 400 °C, respectively. In the experiment, the Sb content of the $Sb_{0.14}GaN$ films increased from 3.24 at% at a depositing temperature of 100 °C, to the highest Sb content value of 5.23% at the substrate temperature of 400 °C; thus, their bandgap E_g decreased from 3.01 to 2.78 eV. K. M. Yu et al. presented that $GaN_{1-x}Sb_x$ alloys could modify the site of absorption from 3.4 eV (GaN) to close to 1 eV for alloys composed of higher than 30 at% Sb [25]. Neugebauer et al. illustrated that the smallest formation of energy remained in p-GaN in the state of vacancy-nitrogen (a donor), and in the n-GaN of an vacancy-gallium state (an acceptor) [26]. Mattila clearly confirmed a sophisticated formation between donors of positively charged and negatively charged cation vacancies [27]. As the $Sb_{0.14}GaN$ thin films were sputtered at different temperature depositions, from 100 to 400 °C, there was smaller absorption energy and lower defect levels caused by presenting the solid Sb solution into the GaN, and Sb formed as an acceptor in the GaN.

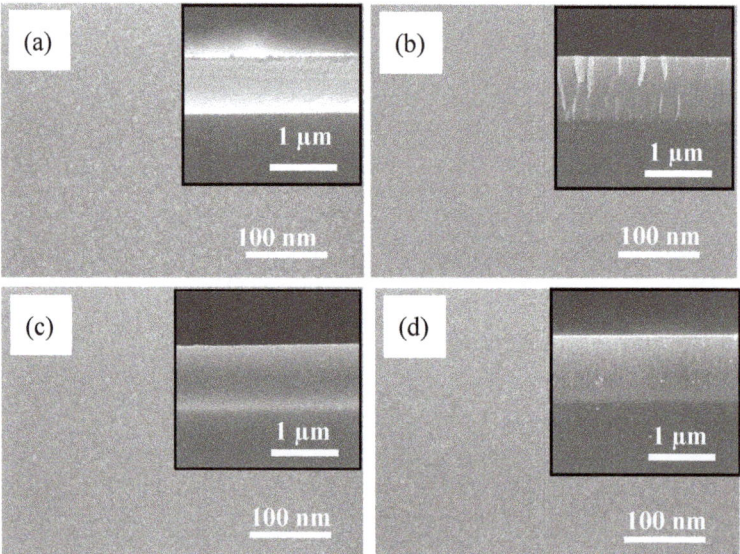

Figure 1. SEM images of Sb-0.14-GaN films sputtered at different heating substrate temperatures: (**a**) 100 °C, (**b**) 200 °C, (**c**) 300 °C and (**d**) 400 °C.

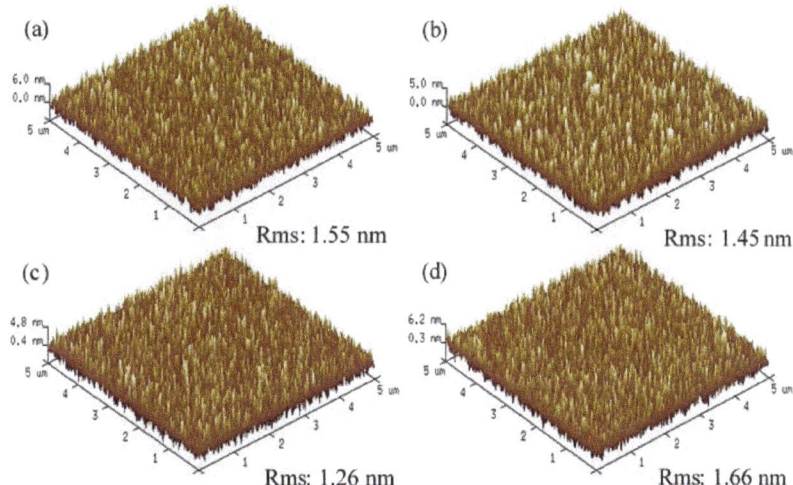

Figure 2. 3D AFM of SbGaN films at different sputtering temperatures of (**a**) 100 °C, (**b**) 200 °C, (**c**) 300 °C, and (**d**) 400 °C.

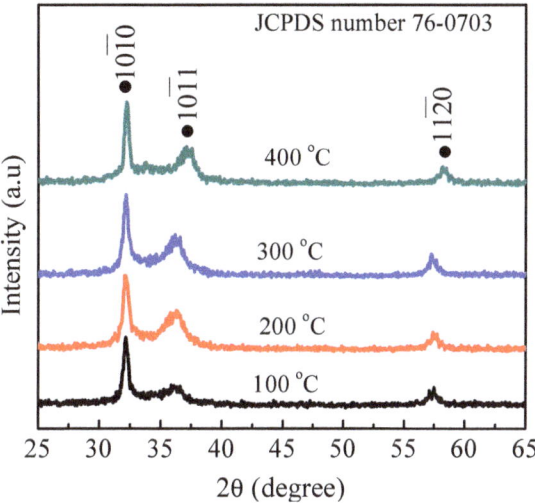

Figure 3. XRD patterns of SbGaN films deposited at heating substrate temperatures ranging from 100 to 400 °C.

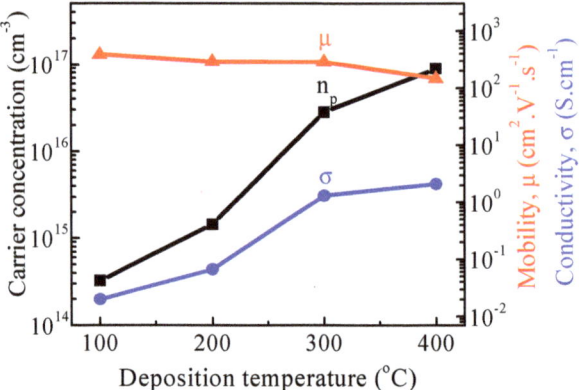

Figure 4. Electrical properties of $Sb_{0.14}GaN$ films deposited at different substrate temperature ranges of 100–400 °C.

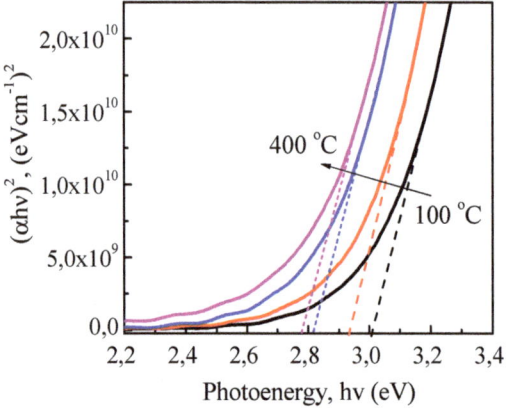

Figure 5. Plots of $(\alpha h\nu)^2$ vs. photon energy (hν) for the optical band gap determination of the $Sb_{0.14}GaN$ films sputtered at different substrate temperatures.

3.2. The p-$Sb_{0.14}$GaN/n-Si heterojunction Diode

In our previous work, we have successfully tested the I–V characteristics of the p-$Sb_{0.14}$GaN/n-Si(100) heterojunction diode at room temperature by using the RF sputtering technique [18]. In this experiment, the p-$Sb_{0.14}$GaN/n-Si(100) diode was investigated at various testing temperatures, ranging from 25 to 150 °C. Figure 6 presents the current density and voltage characteristic of the p-$Sb_{0.14}$GaN/n-Si diode tested at the different temperature ranges of 25–150 °C. The p-$Sb_{0.14}$GaN/n-Si(100) diode was tested at a range of voltages from −20 to +20 V, and applied at the different working temperatures of 25, 75, 100, 125, and 150 °C. As listed in Table 5, the leakage currents of this diode were 5.13×10^{-5}, 1.26×10^{-4}, 5.98×10^{-4}, 1.53×10^{-2}, and 2.96×10^{-2} Acm^{-2} at the reverse bias of −5 V, respectively. The J–V curves in Figure 6 present that the p-$Sb_{0.14}$GaN/n-Si diode at a working temperature range 25–150 °C had the same turnon voltages at 1.25 V, while the diode experienced breakdown voltages at 20 V. There was a relationship between the current density of the diode and testing temperature. As the testing temperature rose from 25, 75, 100, 125 to 150 °C at the forward bias of 20 V, the current density of the p-$Sb_{0.14}$GaN/n-Si diode increased from 0.139, 0.461, 0.663, 0.886 to 1.34 A/cm^2, respectively. It is supposed that the electrical properties of the sputtered p-type $Sb_{0.14}$GaN at different heating substrate temperatures presented in Figure 4 applied to the changes in leakage current density and forward current density values of the diode in Figures 6 and 7.

Table 5. The Parameters and the Electrical Characteristics of the p-Sb$_{0.14}$GaN/n-Si Diode Tested at Different Temperatures.

Samples	Leakage Current Density (A/cm^2) at −5 V	Barrier Height (eV)	I-V		Cheungs' dV/dlnI Versus I	
			n	R$_S$ (kΩ)		n
25 °C	5.13 × 10^{-5}	0.45	5.60	7.51		5.59
75 °C	1.26 × 10^{-4}	0.58	4.00	6.72		3.93
100 °C	5.98 × 10^{-4}	0.56	4.76	1.37		4.78
125 °C	1.53 × 10^{-2}	0.54	5.31	0.53		5.36
150 °C	2.96 × 10^{-2}	0.52	5.71	0.38		5.74

By applying a standard thermionic-emission (TE) model displayed in the below Equation (2), the electrical properties of the diode can be calculated as qV > 3 kT [6,11,28]:

$$I = I_0 \exp\left[\frac{q}{nKT}(V - IR_s)\right] \quad (2)$$

where q is the carrier charge (1.60 × 10^{-19} C), V is the applied voltage, I$_0$ is the value of the saturation current, n is the ideality factor, I$_0$ the saturation current, the series resistance is named R$_s$, K displays the Boltzmann constant (1.38 × 10^{-23} JK^{-1}), and T is the investigating temperature in Kelvin. With the loop of *ln*I versus V figured by using Equation (2), the intersection of the interpolated straight line from the linear region of the semilog chart determined that the saturation current (I$_0$) of the diode can be given by the intersection of the interpolated straight line from the linear region of the semilog chart. The I$_0$ values were 7.60 × 10^{-5}, 1.7 × 10^{-4}, 3.3 × 10^{-4}, 5.6 × 10^{-4}, and 1.3 × 10^{-3} A for the *p*-Sb$_{0.14}$GaN/*n*-Si diode at a testing temperature ranging from 25 to 150 °C. Additionally, φ$_B$, the barrier height, and n, the ideality factor, can be calculated from the retrieved I$_0$ data by applying Equations (3) and (4) [15,17,28].

Figure 6 also determined the electrical characteristics of the p-Sb$_{0.14}$GaN/n-Si diode tested at different working temperatures. From the data for various investigating temperatures of 25, 75, 100, 125, and 150 °C in Table 5, the ideality factor *n* can be figured to be 5.6, 4.0, 4.76, 5.31, and 5.71. In addition, by testing within the temperature range of 25–150 °C, φ$_B$ the barrier height can be computed to be greater than the values estimated in Figure 7, e.g., 0.45, 0.58, 0.56, 0.54, and 0.52 eV, respectively.

$$\phi_B = \frac{KT}{q}\ln\left(\frac{AA^*T^2}{I_0}\right) \quad (3)$$

$$n = \left(\frac{q}{KT}\right) \times \left(\frac{dV}{d(\ln I)}\right) \quad (4)$$

By using Cheungs' method presented in Equation (5), the R$_s$ series resistance and n, ideality factor, were calculated for the *p*-Sb$_{0.14}$GaN/*n*-Si diode worked at the different temperatures and shown in Figure 8 [17,28–30]. The n ideality factors were found to be 5.59, 3.93, 4.78, 5.36, and 5.74. The R$_S$ series resistances were 7.51, 6.72, 1.37, 0.53, and 0.38 kΩ for the testing temperatures of 25, 75, 100, 125, and 150 °C, respectively (Table 5).

$$\left(\frac{dV}{d(\ln I)}\right) = \frac{nKT}{q} + IR_s \quad (5)$$

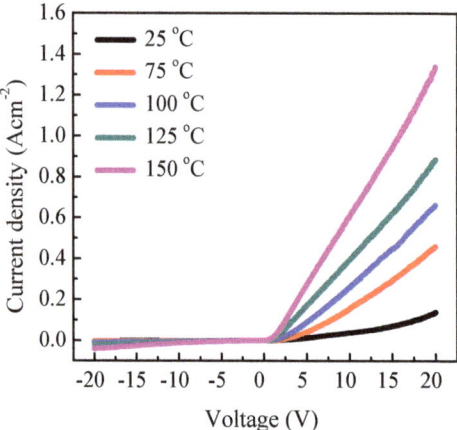

Figure 6. The current density–voltage characteristic of the p-Sb$_{0.14}$GaN/n-Si heterojunction diode investigated (a) at room temperature and (b) at the different temperature ranges from 25 °C to 150 °C.

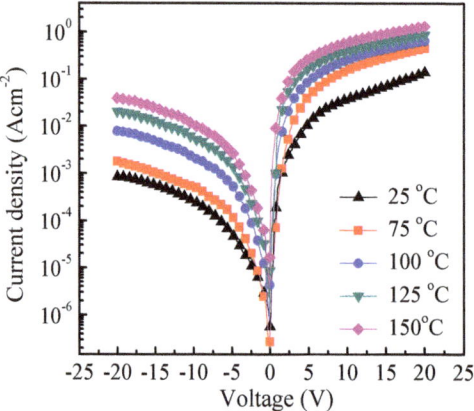

Figure 7. The reverse and forward current density–voltage (I–V) plots of the p-Sb$_{0.14}$GaN/n-Si heterojunction diodes measured at the different temperatures, ranging from 25 °C to 150 °C.

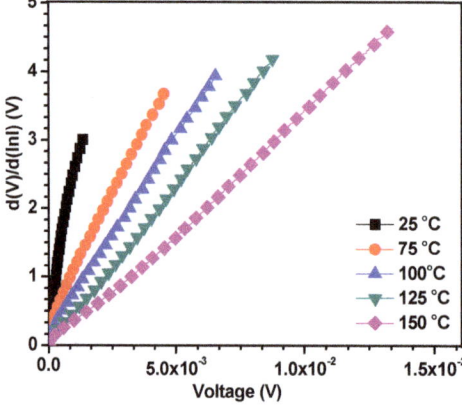

Figure 8. Plots of dV/dln(I) versus current density for p-Sb$_{0.14}$GaN/n-Si diodes measured at the different temperature.

From the surveyed data, it can be seen that as the heating substrate temperature of the sputtering process increases, the $Sb_{0.14}GaN$ film possessed more electrical conductivity (σ) and carrier concentration (n_e), as shown in Table 4. This means that the $Sb_{0.14}GaN$ device had a higher leakage current density and forward current density at higher testing temperatures.

At this time, there is no report on the SbGaN/Si heterojunction device. Our diode showed an excellent breakdown voltage of 20 V compared with some devices based on GaN and its alloys, which often display breakdown voltages of ~ 13V [30–34]. Similar results were also determined in the investigation of heterojunction diodes with different approaches. Mohd Yusoff et al. studied the p–n junction diode based on GaN grown on Si(111) substrate at room temperature (RT). Their ideality factors (n) decreased from 15.14 to 19.68, with an increase in the testing temperature within the range of 30–104 °C [35]. Chirakkara et al. fabricated a n–ZnO/p–Si(100) hetero diode by pulsed laser deposition. Their barrier heights increased from 0.6 eV (at 300 K) to 0.76 eV (at 390 K) when testing at various temperatures, ranging from 300 to 390 K [36]. Lin et al. reported leakage current densities of 6.65×10^{-7} A/cm^2 at a reverse voltage of 5 V and a large turn-on voltage of 9.2 V, which was measured for I–V curves in the sputtered-n-$Al_{0.05}In_{0.075}GaN$/p-Si heterodiode. In another work, they also identified a turnon voltage of 2.70 V in the sputtered n-$Al_{0.075}In_{0.25}GaN$/p-Si diode when testing at RT [37,38].

In addition, although the reported III–V diodes made by MOCVD displayed the epitaxial growth of their semiconductors, their deposition was conducted at high temperatures of 900 °C for GaN [9,10,28]. The p-$Sb_{0.14}GaN$/n-Si diodes had shown the special properties shown in the $I-V$ measurements, which could be attributed to the low temperature process only at 400 °C. Our diode with stable electrical properties up to test temperatures of 150 °C has great potential for application in power diodes and electronic devices.

4. Conclusions

$Sb_{0.14}GaN$ films were successfully fabricated on n-type Si(100) substrates by applying radio frequency (RF) reactive sputtering technology at various heating substrate temperatures, from 100 to 400 °C. The investigated data indicated that all the $Sb_{0.14}GaN$ films retained the structure of the polycrystalline and electrical conductivity at the different depositing temperature conditions. At the various sputtering temperatures of the deposition process, the $Sb_{0.14}GaN$ films were affected by the film properties and experienced a significant structural distortion. At the highest heating substrate temperature of 400 °C, the $Sb_{0.14}GaN$ film behaved as a p-semiconductor layer and had the highest bulk hole concentration and electrical conductivity, of 8.97×10^{16} cm^{-3}, and 2.1 Scm^{-1}, respectively. The electrical characteristics of the p-$Sb_{0.14}GaN$ layer/n-Si substrate diode were measured and computed at the different testing temperature ranges of 25–150 °C using semiconductor device analyzer. These results proved that the testing temperature conditions affected the electrical characteristics of the p-$Sb_{0.14}GaN$/n-Si device. Forward bias analysis at 20 V revealed the highest current density of 1.34 A/cm^2 to be for the diode operating at 150 °C and it is believed that there is a relationship between a working temperature decrement of 125, 100, 75, and 25 °C and a significant current density decrease from 0.886, 0.663, 0.461 to 0.139 A/cm^2, respectively.

Author Contributions: Data curation: C.P.T. and T.T.A.T.; methodology: C.P.T. and T.T.A.T.; writing—original draft and investigation: C.P.T. and T.T.A.T.; formal analysis, funding acquisition and writing—review & editing: C.P.T., T.T.A.T.; supervision: D.-H.K. All authors have read and agreed to the published version of the manuscript.

Funding: This research received no external funding.

Institutional Review Board Statement: Not Applicable.

Informed Consent Statement: Not Applicable.

Data Availability Statement: Data is contained within the article or supplementary material.

Acknowledgments: This work was supported by the Ministry of Science and Technology of the Republic of China under grant numbers 109-2222-E-011-006.

Conflicts of Interest: The authors declare no conflict of interest.

References

1. Akasaki, I.; Amano, H. Crystal Growth and Conductivity Control of Group III Nitride Semiconductors and Their Application to Short Wavelength Light Emitters. *Jpn. J. Appl. Phys.* **1997**, *36*, 5393–5408. [CrossRef]
2. Jain, S.C.; Willander, M.; Narayan, J.; Overstraeten, R.V. III-nitrides-Growth, characterization, and properties. *J. Appl. Phys.* **2000**, *87*, 965–1006. [CrossRef]
3. Fujii, T.; Gao, Y.; Sharma, R.; Hu, E.L.; DenBaars, S.P.; Nakamura, S. Increase in the extraction efficiency of GaN-based light-emitting diodes via surface roughening. *Appl. Phys. Lett.* **2004**, *84*, 855–857. [CrossRef]
4. Pearton, S.J.; Ren, F.; Zhang, A.P.; Lee, K.P. Fabrication and performance of GaN electronic devices. *Mater. Sci. Eng. R Rep.* **2000**, *30*, 205–212. [CrossRef]
5. Rajan, S.; Chini, A.; Wong, M.H.; Speck, J.S.; Mishra, U.K. N-polar GaN/AlGaN/GaN high electron mobility transistors. *J.Appl. Phys.* **2007**, *102*, 044501. [CrossRef]
6. Tuan, T.T.A.; Kuo, D.-H.; Li, C.C.; Yen, W.-C. Schottky barrier characteristics of Pt contacts to all sputtering-made n-type GaN and MOS diodes. *J. Mater. Sci. Mater. Electron.* **2014**, *25*, 3264–3270. [CrossRef]
7. Tuan, T.T.A.; Kuo, D.-H.; Saragih, A.D.; Li, G.-Z. Electrical properties of RF-sputtered Zn-doped GaN films and p-Zn-GaN/n-Si hetero junction diode with low leakage current of 10^{-9} A and a high rectification ratio above 10^5. *Mater. Sci. Eng. B* **2017**, *222*, 18–25. [CrossRef]
8. Shuji, N. GaN Growth Using GaN Buffer Layer. *Jpn. J. Appl. Phys.* **1991**, *30*, L1705.
9. Guarneros, C.; Sánchez, V. Magnesium doped GaN grown by MOCVD. *Mater. Sci. Eng. B* **2010**, *174*, 263–265. [CrossRef]
10. Akasaki, I.; Amano, H.; Koide, Y.; Hiramatsu, K.; Sawaki, N. Effects of ain buffer layer on crystallographic structure and on electrical and optical properties of GaN and $Ga_{1-x}Al_xN$ (0 < x < 0.4) films grown on sapphire substrate by MOVPE. *J. Crystal. Growth.* **1989**, *98*, 209–219.
11. Shota, K.; Kazumasa, H.; Nobuhiko, S. Fabrication of GaN Hexagonal Pyramids on Dot-Patterned GaN/Sapphire Substrates via Selective Metalorganic Vapor Phase Epitaxy. *Jpn. J. Appl. Phys.* **1995**, *34*, L1184.
12. Hiroshi, A.; Masahiro, K.; Kazumasa, H.; Isamu, A. P-type Conduction in Mg-Doped GaN Treated with Low-Energy Electron Beam Irradiation (LEEBI). *Japan. J. Appl. Phys.* **1989**, *28*, L2112.
13. Nakamura, S.; Harada, Y.; Seno, M. Novel metalorganic chemical vapor deposition system for GaN growth. *Appl. Phys. Lett.* **1991**, *58*, 2021–2023. [CrossRef]
14. Shuji, N.; Takashi, M.; Masayuki, S.; Naruhito, I. Thermal Annealing Effects on P-Type Mg-Doped GaN Films. *Jpn. J. Appl. Phys.* **1992**, *31*, L139.
15. Li, C.-C.; Kuo, D.-H. Material and technology developments of the totally sputtering-made p/n GaN diodes for cost-effective power electronics. *J. Mater. Sci. Mater. Electron.* **2014**, *25*, 1942–1948. [CrossRef]
16. Thao, C.P.; Kuo, D.H. Electrical and structural characteristics of Ge-doped GaN thin films and its hetero-junction diode made all by RF reactive sputtering. *Mater. Sci. Semicond. Process* **2018**, *74*, 336–341. [CrossRef]
17. Kuo, D.H.; Tuan, T.T.A.; Li, C.C.; Yen, W.C. Electrical and structural properties of Mg-doped $In_xGa_{1-x}N$ (x ≤ 0.1) and p-InGaN/n-GaN junction diode made all by RF reactive sputtering. *Mater. Sci. Eng. B* **2015**, *193*, 18–25. [CrossRef]
18. Thao, C.P.; Tuan, T.T.A.; Kuo, D.H.; Ke, W.C. Reactively Sputtered Sb-GaN Films and its Hetero-Junction Diode: The Exploration of the n-to-p Transition. *Coatings* **2020**, *10*, 210. [CrossRef]
19. Thao, C.P.; Kuo, D.H.; Tuan, T.T.A.; Tuan, K.A.; Vu, N.H.; Via Sa Na, T.T.; Nhut, K.V.; Sau, N.V. The Effect of RF Sputtering Conditions on the Physical Characteristics of Deposited GeGaN Thin Film. *Coatings* **2019**, *9*, 645. [CrossRef]
20. Li, C.C.; Kuo, D.H. Effects of growth temperature on electrical and structural properties of sputtered GaN films with a cermet target. *J. Matter. Sci Mater. Electron.* **2014**, *25*, 1404–1409. [CrossRef]
21. Tuan, T.T.A.; Kuo, D.H. Characteristics of RF reactive sputter-deposited Pt/SiO_2/n-InGaN MOS Schottky diodes. *Mater. Sci. Semicond Process* **2015**, *30*, 314–320. [CrossRef]
22. Li, C.C.; Kuo, D.H.; Hsieh, P.W.; Huang, Y.S. Thick In_xGa_{1-x} N Films Prepared by Reactive Sputtering with Single Cermet Targets. *J. Electron. Mater.* **2013**, *42*, 2445–2449. [CrossRef]
23. Kim, H.W.; Kim, N.H. Preparation of GaN films on ZnO buffer layers by RF magnetron sputtering. *Appl. Sur. Sci.* **2004**, *236*, 192–197. [CrossRef]
24. Chyr, I.; Lee, B.; Chao, L.C.; Steckl, A.J. Damage generation and removal in the Ga^+ focused ion beam micromachining of GaN for photonic applications. *J. Vacuum. Sci. Tech. B* **1999**, *17*, 3063–3067. [CrossRef]
25. Yu, K.M.; Sarney, W.L.; Novikov, S.V.; Segercrantz, N.; Ting, M.; Shaw, M.; Svensson, S.P.; Martin, R.W.; Walukiewicz, W.; Foxon, C.T. Highly mismatched $GaN_{1-x}Sb_x$ alloys: Synthesis, structure and electronic properties. *Semicond. Sci. Tech.* **2016**, *31*, 083001. [CrossRef]
26. Neugebauer, J.; Van de Walle, C.G. Atomic geometry and electronic structure of native defects in GaN. *Phys. Rev. B* **1994**, *50*, 8067–8070. [CrossRef] [PubMed]

27. Mattila, T.; Nieminen, R.M. Point-defect complexes and broadband luminescence in GaN and AlN. *Phys. Rev. B* **1997**, *55*, 9571–9576. [CrossRef]
28. Reddy, V.R.; Prasanna, B.P.; Padma, R. Electrical Properties of Rapidly Annealed Ir and Ir/Au Schottky Contacts on n-Type InGaN. *J. Metall.* **2012**, *1*, 1–9. [CrossRef]
29. Cheung, S.K.; Cheung, N.W. Extraction of Schottky diode parameters from forward current-voltage characteristics. *Appl. Phys. Lett.* **1986**, *49*, 85–87. [CrossRef]
30. Padma, R.; Reddy, V.R. Electrical Properties of Ir/n-InGaN/Ti/Al Schottky Barrier Diode in a Wide Temperature Range. *Adv. Mater. Lett.* **2014**, *5*, 31–38. [CrossRef]
31. Lee, S.Y.; Kim, T.H.; Suh, D.I.; Park, J.E. An electrical characterization of a hetero-junction nanowire (NW) PN diode (n-GaN NW/pSi) formed by dielectrophoresis alignment. *Phys. E* **2007**, *36*, 194–198. [CrossRef]
32. Cao, X.A.; Lachode, J.R.; Ren, F. Implanted p–n junction in GaN. *Solid State Electron.* **1999**, *43*, 1235–1238. [CrossRef]
33. Hickman, R.; Vanhove, J.M.; Chow, P.P.; Klaassen, J.J. GaN PN junction issues and developments. *Solid. State Electron.* **2000**, *44*, 377–381. [CrossRef]
34. Baik, K.H.; Irokawa, Y.; Ren, F.; Pearton, S.J.; Park, S.S.; Park, S.J. Temperature dependence of forward current characteristics of GaN junction and Schottky rectifiers. *Solid State Electron.* **2003**, *47*, 1533–1538. [CrossRef]
35. Mohd, M.Z.; Baharin, A.; Hassan, Z.; Abu, H.; Abdullah, M.J. MBE growth of GaN pn-junction photodetector on AlN/Si substrate with Ni/Ag as Ohmic contact. *Superlattices Microstruct.* **2013**, *56*, 35–44. [CrossRef]
36. Chirakkara, S.; Krupanidhi, S.B. Study of n-ZnO/p-Si(100) thin film heterojunctions by pulsed laser deposition without buffer layer. *Thin Solid Films* **2012**, *520*, 5894–5899. [CrossRef]
37. Lin, K.; Kuo, D.H. Characteristics and electrical properties of reactively sputtered AlInGaN films from three different $Al_{0.05}In_xGa_{0.95-x}N$ targets with $x = 0.075, 0.15,$ and 0.25. *Mater. Sci. Semicond. Process.* **2017**, *57*, 63–69. [CrossRef]
38. Lin, K.; Kuo, D.H. Characterization of quaternary AlInGaN films obtained by incorporating Al into InGaN film with the RF reactive magnetron sputtering technology. *J. Matter. Sci. Mater. Electron.* **2017**, *28*, 43–51. [CrossRef]

Article

Improved Mechanical Properties and Corrosion Resistance of Mg-Based Bulk Metallic Glass Composite by Coating with Zr-Based Metallic Glass Thin Film

Pei-Hua Tsai [1], Chung-I Lee [1], Sin-Mao Song [1], Yu-Chin Liao [2], Tsung-Hsiung Li [1], Jason Shian-Ching Jang [1,2,*] and Jinn P. Chu [3]

1. Institute of Materials Science and Engineering, National Central University, Taoyuan 32001, Taiwan; peggyphtsai@gmail.com (P.-H.T.); worm30221@gmail.com (C.-I.L.); bear82112760103@gmail.com (S.-M.S.); pshunterbabu@gmail.com (T.-H.L.)
2. Department of Mechanical Engineering, National Central University, Taoyuan 32001, Taiwan; llllurker@gmail.com
3. Department of Materials Science and Engineering, National Taiwan University of Science and Technology, Taipei 10607, Taiwan; jpchu@mail.ntust.edu.tw
* Correspondence: jscjang@ncu.edu.tw

Received: 12 November 2020; Accepted: 10 December 2020; Published: 12 December 2020

Abstract: Mg-based bulk metallic glass (BMG) and its composite (BMGC) can be excellent candidates as lightweight structure materials, but lack of anti-corrosion ability may restrict their application. In order to enhance the natural weak point of Mg-based BMGC, a 200-nm thick Zr-based metallic glass thin film (MGTF) (($Zr_{53}Cu_{30}Ni_9Al_8)_{99.5}Si_{0.5}$) was applied and its mechanical properties as well as its corrosion resistance were appraised. The results of a 3-point bending test revealed that the flexural strength of the Mg-based BMGC with 200-nm thick Zr-based MGTF coating can be greatly enhanced from 180 to 254 MPa. We propose that the Zr-based MGTF coating can help to cover any small defects of a substrate surface, provide a protecting layer to prevent stress concentration, and cease crack initiation from the specimen surface during bending tests. Moreover, the results of anti-corrosion behavior analysis revealed a similar trend between the Mg-based BMG, Mg-based BMGC, and Mg-based BMGC with Zr-based MGTF coating in 0.9 wt.% sodium chloride solution. The readings show a positive effect with the Zr-based MGTF coating. Therefore, the 200-nm thick Zr-based MGTF coating is a promising solution to provide protection for both mechanical and anti-corrosion behaviors of Mg-based BMGC and reinforce its capability as structure material in island environments.

Keywords: Mg alloy; bulk metallic glass; composites; thin film coating; mechanical properties

1. Introduction

Zr-, Ti-, Ni-, and Fe-based alloys bulk metallic glasses (BMGs) have been well studied in the past few decades [1–3]. Ti-based BMGs are an excellent candidate for bio-application due to their relatively low density and much more compatible Young's modulus in comparison with stainless steel or Co-Cr-Mo alloy. Several toxic-element-free Ti-based BMG alloy systems have been developed with good glass-forming ability [4,5]. Ni- and Fe-based BMGs generally possesses excellent mechanical properties and can be promising structure materials [6–8]. In addition, Fe-based BMG exhibits extremely high hardness around 1200 Hv and excellent anti-wear resistance ability, meaning it can be used in medical tool parts and for surgical blades with better durability [9]. Mg-based BMG possesses the low-density advantage and can be designed as a lightweight component for the automotive, aerospace, and 3C industries [10–13]. Yet, monolithic Mg-based BMGs show a very brittle behavior and will break into pieces before yielding [10,11]. To conquer the problem of brittleness, extensive efforts have been devoted to develop Mg-based metallic glass composites (BMGCs) with homogeneous micro- or nano-scaled second-phase dispersion in a BMG matrix in the past decade. These include the incorporation of in-situ precipitation of micro- or nano-crystalline phases and ex-situ added micro-sized refractory ceramics or ductile metal particles in the Mg-based BMGCs [14–19]. Mg-based metallic glass composite reinforced with Nb particles can reach the high strength of 900 MPa and large plasticity of 12.1% ± 2% [14]. In situ addition of Mg flakes into Mg–Cu–Y–Zn BMG alloy can significantly improve mechanical properties such as compressive plastic strain up to 18% and ultimate strength up to 1.2 GPa [15]. Moreover, ex-situ addition of 40 vol.% Ti spherical powder improved the ductility from 0% (monolithic glass) to 41% plastic deformation for the composite [16,17]. Thereafter, many Mg-based BMGCs have been developed and all exhibit significant improvements in plasticity as well as toughness. Among these developed Mg-based BMGCs, one special Mg-based BMGC with porous Mo [18,19] performs the optimum combination of yield strength (1100 MPa) and plasticity (>25%), which was chosen as the substrate for further study. Nevertheless, inheriting the reactive characteristics of Mg-based alloys, this Mg-based BMGC is still concerning as it cannot sustain corrosion attacks due to the salty atmosphere of island environments. Therefore, surface treatment for protecting the Mg-based BMGC from the corrosion of a salty atmosphere is essential in the island environment. For conventional Mg alloys, anodic surface treatment [20–22] and micro-arc surface treatment [23,24] are commonly applied to form a protective oxide layer on the Mg alloy surface to prevent the attack of a salty atmosphere in the island environment. However, these two treatments belong to wet processes with alkaline electrolyte solution and have the pollution concern of wastewater. Therefore, a dry process, i.e., sputtering coating treatment [25], is believed to be a better green process for using on the surface treatment of Mg-based alloys. In parallel, Zr-based metallic glasses are reported not only to possess good mechanical properties but also to have excellent corrosion resistance in salty aqueous solutions [26–28]. Hence, coating a thin layer of Zr-based metallic glass thin film (MGTF) on the surface of Mg-based BMGC by sputtering is suggested as an effective and green approach to improve its corrosion resistance in the salty atmosphere.

Accordingly, to further investigate the effect of Zr-based MGTF coating on the corrosion resistance as well mechanical properties of Mg-based BMGC [16], the Zr-based BMG (($Zr_{48}Cu_{36}Al_8Ag_8)_{99.5}Si_{0.5}$) was selected as the target material for the sputtering materials due to its high glass-forming ability (GFA), high corrosion resistance, and good mechanical properties. The as-polished specimens of $Mg_{58}Cu_{28.5}Gd_{11}Ag_{2.5}$ BMGC with 25 vol.% Mo particle additions were chosen as the substrate to coat with a 200-nm thick Zr-based MGTF coupled with different thin film buffer layers by the sputtering method. Then, the microstructures, mechanical properties, and corrosion resistances of these MGTF-coated samples were systematically investigated.

2. Materials and Methods

2.1. Preparation of Sputtering Target

First, we carefully measured pure elements of Zr, Cu, Ni, Al, and Si based on the composition of $(Zr_{53}Cu_{30}Ni_9Al_8)_{99.5}Si_{0.5}$. Then, we prepared the ingot via arc-melting 4 times in a Ti-gettered argon atmosphere furnace to assure homogeneity. The final product, a plate with dimensions of 30 mm in *width*, 80 mm in *length* and 8 mm in *thickness*, was produced via casting the ingot into a water-cooling copper mold by suction. The plates were cut with wire-cut Electrical Discharge Machining (EDM) and then assembled into the target with 8 mm in thickness and 3 inches in diameter, as shown in Figure 1. Table 1 shows the chemical composition of the Zr-based target which was firstly examined by EDS before the sputtering process.

Figure 1. The appearance of the assembled Zr-based target with dimensions of 76 mm in *diameter* and 8 mm in *thickness* for the sputtering process.

Table 1. Chemical composition of Mg-based bulk metallic glass (BMG) and Mg-based bulk metallic glass composite (BMGC) analyzed by EDS.

Material	(at.%)	Mg	Cu	Gd
Mg-based BMG	Design composition	58	31	11
	Average	51.94	34.78	13.28
	Deviation	0.3	0.38	0.02
Mg-based BMGC	Design composition	58	31	11
	Average	53.7	34.09	12.22
	Deviation	0.82	0.58	0.02

2.2. Sample Preparation of Mg-Based BMGC

The composition of $Mg_{58}Cu_{28.5}Gd_{11}Ag_{2.5}$ was selected as the base alloy for preparing the BMGC with the addition of 25% porous Mo particles (with average particle size 25 ± 4 μm). The composite master alloy ingots were prepared by following the process procedure from our previous report [16]. Then, these composite alloy ingots were further re-melted by induction melting in a quartz tube and injected into a water-cooled Cu mold by argon pressure to obtain the BMGC plates with dimensions of 50 mm L × 15 mm W × 3 mm T. The temperature of the Cu mold was kept at 8 °C to reach a cooling rate of 63 K/s for 2-mm thick plates and to obtain a BMGC plate with residual porosity less than 0.15 vol.% [17]. Samples for the three-point bending tests were taken from the as-cast Mg-based BMG and BMGC plates with sample dimensions of 4 mm W × 3 mm T × 35 mm L (of B (thickness) = 2.5 mm, W (width) = 7.5 mm, and S (span) = 36 mm). The as-machined and fine polished BMGC samples were then deposited with two different combinations of thin film coating; Film A: 50-nm thick Cu buffer layer plus 200-nm thick Zr-based MGTF, and Film B: 25-nm thick Al/25-nm thick Ti buffer layer plus 200-nm thick Zr-based MGTF with DC sputtering system (MDX1000, Advanced Energy Industries, Denver, CO, USA). The operating parameters of the DC sputtering procedure were set as follows: the distance between the specimen and target was 10 cm with a base pressure of 10^{-5} Pa and working pressure at 0.5 Pa. In parallel, the Ar flow was set at 5.4 sccm, with a sputtering time period of 30 min and 20 W of sputtering power. In addition, an attached test piece for coating thickness examination was coated at the same sputtering conditions as the specimen of the bending test.

2.3. Characterization of Microstructure and Properties

The amorphous states of the as-cast Mg-based BMG and BMGC were examined by X-ray diffraction (XRD, Bruker D8A, Cu-Kα radiation, Billerica, MA, USA) and the amorphous state of Zr-based MGTF coating was examined by grazing incident X-ray diffraction analysis (GIXRD, Philips Xpert-Pro PW-3040, Amsterdam, The Netherlands, operated at 40 kV and a 0.5-degree incident angle) with mono-chromatic Cu-Kα radiation. The thickness and composition of the MGTF coating were examined by field emission scanning electron microscopy (FESEM, FEI INSPECT F50, Waltham, MA, USA, operated at 20 kV) with energy dispersion spectroscopy (EDS) at the cross-section of a coupled test specimen. The adhesion capability between the thin film coating and substrate was evaluated by tape testing, which follows the standard of ASTM D3359-09 Test Method B [29]—Cross cut. The hardness of the Mg-based BMG and BMGC was tested by Vickers' hardness tester and that of the Zr-based MGTF coating was checked by a nano-indenter (Hystron, TI 950 Tribo-Indenter, St, Eden Prairie, MN, USA). Following the standard of ASTM E855-08 [30], three-point bending tests were conducted by a universal testing system (MTS Criterion Modle42, Eden Prairie, MN, USA) equipped with a bending gauge, as shown schematically in Figure 2. Before the bending test, the average values of surface roughness of all specimens were confirmed to be less than 10 nm by examination with an atomic force microscope (AFM, Bruker Dimension edge, Billericacity, MA, USA). The morphologies of fractured surfaces after the bending test were examined by FESEM. To further investigate the electrochemical behavior and corrosion properties, 0.9 wt.% NaCl solution was chosen to be the corrosion environment. The Mg-based BMG, Mg-based BMGC, and Mg-based BMGC with Film B were studied by potential dynamic polarization measurements which were conducted by the Autolab PGStat 302 potentiostat (Utrecht, The Netherlands) in a three-electrode cell. The counter and reference electrodes (Saturated Calomel Electrode, SCE) were platinum wire and Ag/AgCl, and specimens with an immersion area of about 25 mm^2 were used as a working electrode. The polarization scan was started from −1.5 to 1.5 V with a scan rate 20 mV/s. Corrosion behavior indicators such as corrosion current density (I_{corr}), corrosion potential (E_{corr}), and corrosion rate can be obtained by the Tafel extrapolation method from anodic polarization curves.

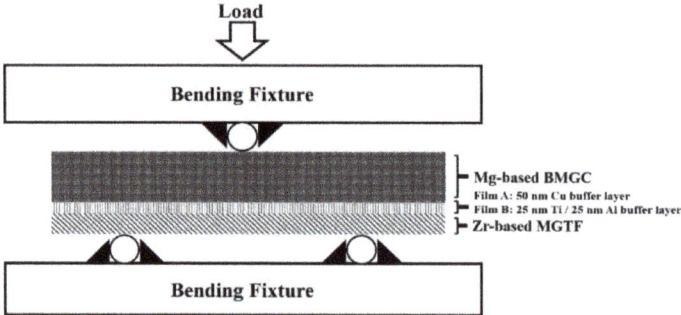

Figure 2. Schematic illustration of three-point bending and sample dimension.

3. Results

The Zr-based MGTF coatings were firstly examined using EDS to confirm their composition in comparison with the Zr-based MG target. The results of EDS confirmed that the composition of the thin film was close to its pre-set composition, as shown in Table 2. The coating thickness of attached test pieces for different combinations of buffer layer and Zr-based MGTF was found very close to the preset thickness (they are around 50 and 200 nm, respectively), as shown in Figure 3. In addition, the XRD patterns reveal that the Zr-based MGTF coating, the Mg-based BMG, and the Mg-based BMGC all present the amorphous state (typical broadened and diffused humps around 30–50 degrees of 2θ), except the high-intensity crystalline peaks resulting from the Mo particles embedded in the

Mg-based BMGC samples, as shown in Figure 4. In parallel, the average surface roughness can be decreased from 10 (bare Mg-based BMGC substrate) to 4 nm by coating with Zr-based MGTF. This is similar to the results of a published report [31].

Table 2. Chemical composition of Zr-based MGTF coating analyzed by EDS.

Material	(at.%)	Zr	Cu	Ni	Al	Si
	Design composition	52.73	29.85	8.96	7.96	0.5
Zr-based MGTF	Average	52.44	31.69	12.05	3.69	0.13
	Deviation	0.59	0.19	0.02	0.25	0.01

 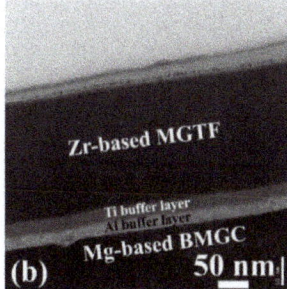

Figure 3. Cross-sectional TEM images of the Mg-based BMGC coated with a buffer layer and Zr-based metallic glass thin film (MGTF); (**a**) with Film A coating; (**b**) with Film B coating.

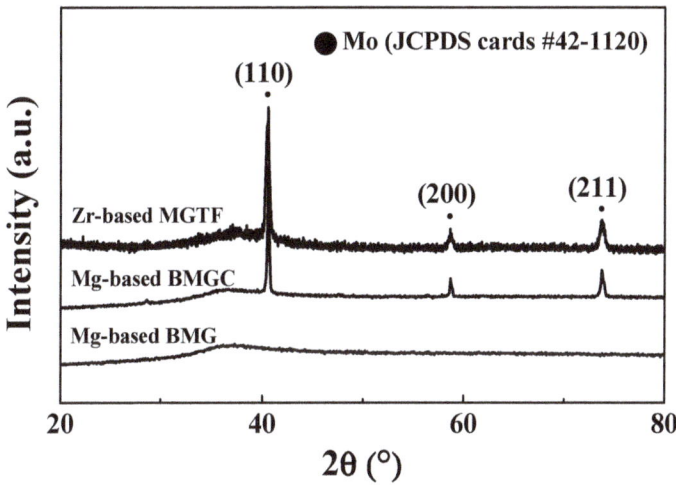

Figure 4. X-ray diffraction patterns of Mg-based BMG, Mg-based BMGC, and Mg-based BMGC coated with Zr-based MGTF.

The results of adhesion testing show that the buffer layer coating does affect the adhesion capability of the Zr-based MGTF coatings on the Mg-based BMGC substrate, as shown in Figure 5. The buffer layer of 25-nm thick Al/25-nm thick Ti (Film B) possesses much better adhesion capability than the buffer layer of 50-nm thick Cu (Film A), and only Film B can reach a 4B grade and meet the industrial requirement. This is presumed to be attributed to the large atomic size misfit at the interface of Mg (rMg = 1.60 nm)/Cu (rCu = 1.28 nm) and results in a weak adhesion. On the contrary, the atomic size misfits at the interfaces of Mg (rMg = 1.60 nm)/Al (rAl = 1.43 nm) and Al (rAl = 1.43 nm)/Ti (rAl = 1.47 nm) are much smaller than the Mg/Cu one.

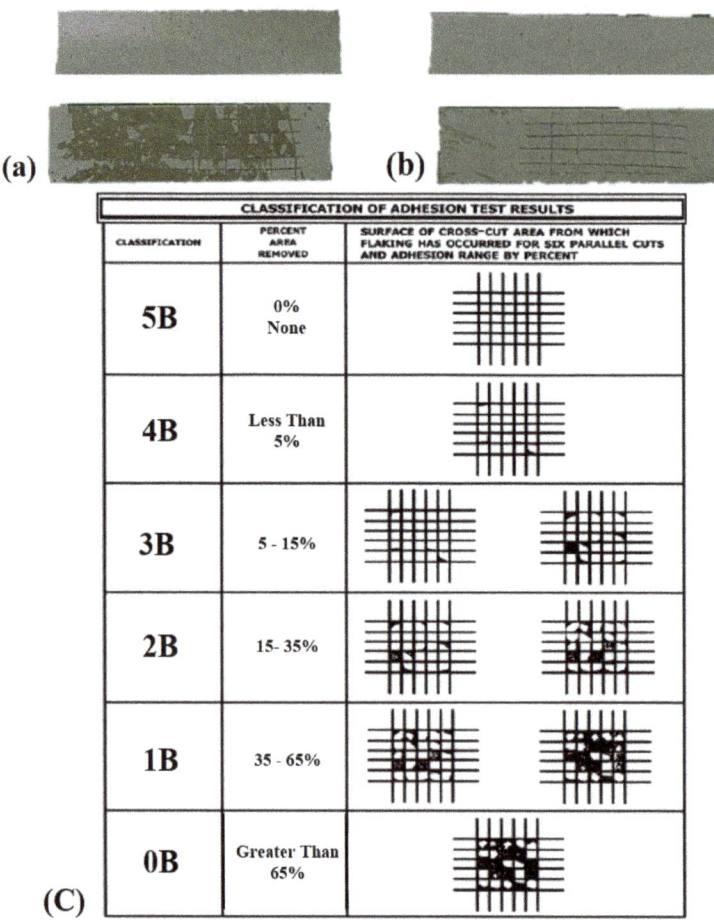

Figure 5. Results of adhesion test of the Zr-based MGTF coating on the Mg-based BMGC with different buffer layers; (**a**) with Film A coating; (**b**) with Film B coating. (**c**) Classification of adhesion test results (ASTM D3359-09) [29].

In addition, the results of the bending tests also reveal the significant improvement in the bending fracture strength after coating with Zr-based MGTF, as shown in Figure 6. The bending fracture strength can be improved from 180 (the bare Mg-based BMGC) to 254 MPa (Mg-based BMGC coated with Film B). However, the Mg-based BMGC coated with Film A only presents a slight increase in bending fracture strength (189 MPa) compared with the bare Mg-based BMGC (180 MPa) due to the weak adhesion of the buffer layer between the Zr-based MGTF and Mg-based BMGC.

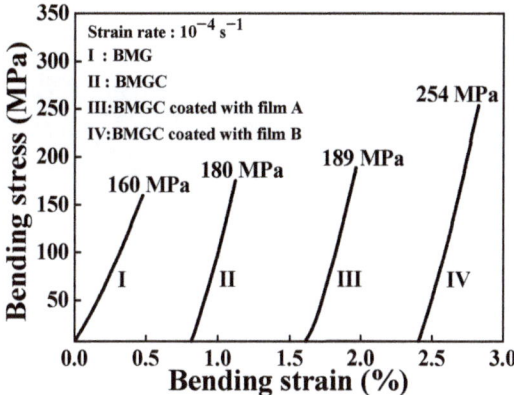

Figure 6. Stress-strain curves of bending tests for the samples of Mg-based BMG, Mg-based BMGC, and Mg-based BMGC with Zr-based MGTF coating.

According to the SEM observations, it is revealed that the fracture edge and fracture surface for each sample after bending test exhibit distinct different morphologies, as illustrated in Figure 7. The Mg-based BMG sample only presents very brittle fracture behavior, a sharp fracture edge, and a flat fracture surface. However, the Mg-based BMGC sample shows some cracking traces near the fracture edge with little rough fracture surface, indicating that it has better toughness than the Mg-based BMG, which is in agreement with the previous report [18]. Moreover, the sample of Mg-based BMGC with Film B coating shows a more different morphology near the fracture edge, with many cracks accompanied by several remelting-like traces on the surface of the Zr-based MGTF and relatively rough fracture surface.

In the literature, the improvement in the bending strength and ductility of the MGTF-coated BMG sample has been proposed to be attributed to several major factors [32]: (1) mechanical properties of the thin film coating; (2) surface roughness of the coating; (3) adhesion capability of the coating on the substrate; (4) the flexibility of thin film coating. Accordingly, the significant improvement in the bending strength of the Mg-based BMGC with a Zr-based MGTF coating in this study is suggested to be due to the high strength and good flexibility of the Zr-based MGTF coating [33–35] and the good adhesion of MGTF/Ti/Al buffer-layer coating to substrate [25].

Figure 8 shows the results of the potentiodynamic polarization test in the 0.9 wt.% NaCl solutions of the Mg-based BMG, Mg-based BMGC, and Mg-based BMGC with Zr-based MGTF coating. The free corrosion potential (E_{corr}) and the corrosion current density (I_{corr}) can be measured from the polarization curves and are listed in Table 3. Values of the corrosion voltage readings for the Mg-based BMG, Mg-based BMGC, and Mg-based BMGC with Zr-based MGTF coating are about −1.04, −1.06, and 0.8 V, respectively. This indicates that the Mg-based BMGC with Zr-based MGTF coating needs to be polarized further before it starts to corrode. In addition, the corrosion current densities for the Mg-based BMG, Mg-based BMGC, and Mg-based BMGC with Zr-based MGTF coating estimated by the Tafel slope method are about 9.07×10^{-5}, 5.38×10^{-4}, and 9.06×10^{-6} A/cm^2, respectively, in 0.9 wt.% NaCl solution, as listed in Table 3. In a comparison of the polarization curves in a 0.9 wt.% NaCl solution amount of the three samples, the Zr-based MGTF coating provided much better corrosion resistance than the bare substrate of Mg-based BMG and BMGC due to a relatively small corrosion current density (I_{corr}).

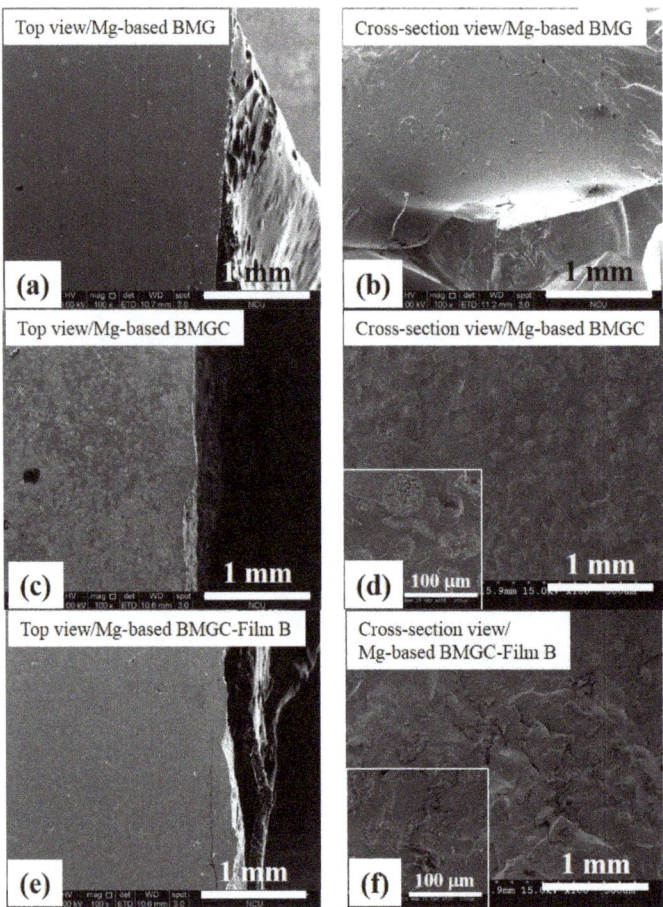

Figure 7. SEM images of the fracture edges and fracture surfaces for the samples after bending tests. (**a**) and (**b**) are the Mg-based BMG; (**c**) and (**d**) are the Mg-based BMGC; (**e**) and (**f**) are the Mg-based BMGC with Zr-based MGTF coating. Each insert in (**d**,**f**) is the enlarged image from the circle area of each figure.

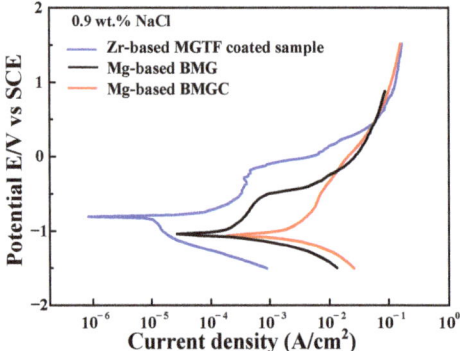

Figure 8. Potential dynamic polarization curves of Mg-based BMG, Mg-based BMGC, and Mg-based BMGC with Zr-based MGTF coating in 0.9 wt.% NaCl solution.

Table 3. Electrochemical parameters of the polarization test for Mg-based BMG, Mg-based BMGC, and Mg-based BMGC with Zr-based MGTF coating in 0.9 wt.% NaCl solution.

Material	E_{corr} (V)	I_{corr} (A/cm^2)
Mg-based BMG	−1.04	9.07×10^{-5}
Mg-based BMGC	−1.06	5.38×10^{-4}
Zr-based MGTF	−0.8	9.06×10^{-6}

The potential is related to SCE (Ag/AgCl).

4. Conclusions

This study revealed that the smooth surface, the excellent adhesion, and the high strength of the Zr-based MGTF have a significant effect on improving the bending strength of Mg-based BMGC. By means of a thin-layer Zr-based MGTF coating accompanied with the Ti/Al buffer-layered coating, the bending strength of Mg-based BMGC could be increased from 180 (bare substrate) to 254 MPa (MGTF-coated sample), which is a 41% improvement. The superior mechanical properties of the Zr-based MGTF such as the high strength and great flexibility, accompanied with the good adhesion to the substrate by Ti/Al buffer-layered coating are the major factors to improve the bending strength of Mg-based BMGC. In addition, the Zr-based MGTF exhibits much better corrosion resistance in 0.9 wt.% sodium chloride solution than the Mg-based BMGC. Therefore, adding a 200-nm thick Zr-based MGTF coating on the Mg-based BMGC by sputtering is believed to be a promising method to protect the Mg-based BMGC from the island environment for many industrial applications.

Author Contributions: Conceptualization, J.S.-C.J.; methodology, J.S.-C.J. and P.-H.T.; formal analysis, C.-I.L. and P.-H.T.; investigation, C.-I.L., Y.-C.L., S.-M.S., and T.-H.L.; resources, J.P.C.; data curation, P.-H.T. and T.-H.L.; writing—original draft preparation, P.-H.T.; writing—review and editing, J.S.-C.J.; supervision, J.S.-C.J. All authors have read and agreed to the published version of the manuscript.

Funding: This research was funded by the Ministry of Science and Technology of ROC, under the Project MOST 103-2221-E-008 -028 -MY3 and MOST 105-2918-I-008-001.

Acknowledgments: The authors gratefully acknowledge the support from the Ministry of Science and Technology of Taiwan, ROC, under the projects MOST 103-2221-E-008-028-MY3 and MOST 105-2918-I-008-001, and the support for analysis from the Precision Instrument Center of National Central University. We also acknowledge the support for analysis from J.P.Chu (National Taiwan University of Science and Technology) and J.W.Lee (Ming-Chi University of Technology).

Conflicts of Interest: The authors declare no conflict of interest.

References

1. Johnson, W.L. Bulk Glass-forming metallic alloys: Science and technology. *MRS Bull.* **1999**, *24*, 42–56. [CrossRef]
2. Inoue, A. Stabilization of metallic supercooled liquid and bulk amorphous alloys. *Acta Mater.* **2000**, *48*, 279–306. [CrossRef]
3. Oak, J.J.; Louzguine, D.V.; Inoue, A. Investigation of glass-forming ability, deformation and corrosion behavior of Ni-free Ti-based BMG alloys designed for application as dental implants. *Mater. Sci. Eng. C* **2009**, *29*, 322–327. [CrossRef]
4. Zhang, T.; Inoue, A. New bulk glassy Ni-based alloys with high strength of 3000 MPa. *Mater. Trans.* **2002**, *43*, 708–711. [CrossRef]
5. Suryanarayana, C.; Inoue, A. *Bulk Metallic Glasses*; CRC Press/Taylor & Francis Group: Boca Raton, FL, USA, 2011.
6. Zhu, S.L.; Wang, X.M.; Qin, F.X.; Inoue, A. A new Ti-based bulk glassy alloy with potential for biomedical application. *Mater. Sci. Eng. A* **2007**, *459*, 233–237. [CrossRef]

7. Inoue, A.; Shen, B.L.; Chang, C.T. Super-high strength of over 4000 MPa for Fe-based bulk glassy alloys in $((Fe_{1-x}Co_x)_{0.75}B_{0.2}Si_{0.05})_{96}Nb_4$ system. *Acta Mater.* **2004**, *52*, 4093–4099. [CrossRef]
8. Lu, Z.P.; Liu, C.T.; Thompson, J.R.; Porter, W.D. Structural amorphous steels. *Phys. Rev. Lett.* **2004**, *92*, 245503. [CrossRef]
9. Tsai, P.H.; Xiao, A.C.; Li, J.B.; Jang, J.S.C.; Chu, J.P.; Huang, J.C. Prominent Fe-based bulk amorphous steel alloy with large supercooled liquid region and superior corrosion resistance. *J. Alloys Compd.* **2014**, *586*, 94–98. [CrossRef]
10. Amiya, K.; Inoue, A. Thermal stability and mechanical properties of Mg–Y–Cu–M (M = Ag, Pd) bulk amorphous Alloys. *Mater. Trans. JIM* **2000**, *41*, 1460–1462. [CrossRef]
11. Xu, Y.K.; Ma, H.; Xu, J.; Ma, E. Mg-based bulk metallic glass composites with plasticity and gigapascal strength. *Acta Mater.* **2005**, *53*, 1857–1866. [CrossRef]
12. Ma, H.; Shi, L.L.; Xu, J.; Li, Y.; Ma, E. Discovering inch-diameter metallic glasses in three-dimensional composition space. *Appl. Phys. Lett.* **2005**, *87*, 181915. [CrossRef]
13. Park, E.S.; Kyeong, J.S.; Kim, D.H. Enhanced glass forming ability and plasticity in Mg-based bulk metallic glasses. *Mater. Sci. Eng. A* **2007**, *449*, 225–229. [CrossRef]
14. Pan, D.G.; Zhang, H.F.; Wang, A.M.; Hu, Z.Q. Enhanced plasticity in Mg-based bulk metallic glass composite reinforced with ductile Nb particles. *Appl. Phys. Lett.* **2006**, *89*, 261904. [CrossRef]
15. Hui, X.; Dong, W.; Chen, G.L.; Yao, K.F. Formation, microstructure and properties of long-period order structure reinforced Mg-based bulk metallic glass composites. *Acta Mater.* **2007**, *55*, 907–920. [CrossRef]
16. Jang, J.S.C.; Ciou, J.Y.; Hung, T.H.; Huang, J.C.; Du, X.H. Enhanced mechanical performance of Mg metallic glass with porous Mo particles. *Appl. Phys. Lett.* **2008**, *92*, 011930. [CrossRef]
17. Jang, J.S.C.; Li, T.H.; Jian, S.R.; Huang, J.C.; Nieh, T.G. Effects of characteristics of Mo dispersions on the plasticity of Mg-based bulk metallic glass composites. *Intermetallics* **2011**, *19*, 738–743. [CrossRef]
18. Jang, J.S.C.; Li, J.B.; Lee, S.L.; Chang, Y.S.; Jian, S.R.; Huang, J.C.; Nieh, T.G. Prominent plasticity of Mg-based bulk metallic glass composites by ex-situ spherical Ti particles. *Intermetallics* **2012**, *30*, 25–29. [CrossRef]
19. Kinaka, M.; Kato, H.; Hasegawa, M.; Inoue, A. High specific strength Mg-based bulk metallic glass matrix composite highly ductilized by Ti dispersoid. *Mater. Sci. Eng. A* **2008**, *494*, 299–303. [CrossRef]
20. Mizutania, Y.; Kim, S.J.; Ichino, R.; Okido, M. Anodizing of Mg alloys in alkaline solutions. *Surf. Coat. Tech.* **2003**, *169–170*, 143–146. [CrossRef]
21. Abulsain, M.; Berkani, A.; Bonilla, F.A.; Liu, Y.; Arenas, M.A.; Skeldon, P.; Thompson, G.E.; Bailey, P.; Noakes, T.C.Q.; Shimizu, K.; et al. Anodic oxidation of Mg–Cu and Mg–Zn alloys. *Electrochim. Acta* **2004**, *49*, 899–904. [CrossRef]
22. Zhang, Y.; Yan, C. Development of anodic film on Mg alloy AZ91D. *Surf. Coat. Tech.* **2006**, *201*, 2381–2386. [CrossRef]
23. Guo, H.F.; An, M.Z.; Huo, H.B.; Xu, S.; Wu, L.J. Microstructure characteristic of ceramic coatings fabricated on magnesium alloys by micro-arc oxidation in alkaline silicate solutions. *Appl. Surf. Sci.* **2006**, *252*, 7911–7916. [CrossRef]
24. Chen, F.; Zhou, H.; Yao, B.; Qin, Z.; Zhang, Q. Corrosion resistance property of the ceramic coating obtained through microarc oxidation on the AZ31 magnesium alloy surfaces. *Surf. Coat. Tech.* **2007**, *201*, 4905–4908. [CrossRef]
25. Chu, J.P.; Jang, J.S.C.; Huang, J.C.; Chou, H.S.; Yang, Y.; Ye, J.C.; Wang, Y.C.; Lee, J.W.; Liu, F.X.; Liaw, P.K.; et al. Thin film metallic glasses: Unique properties and potential applications. *Thin Solid Film.* **2012**, *520*, 5097–5122. [CrossRef]
26. Chieh, T.C.; Chu, J.; Liu, C.T.; Wu, J.K. Corrosion of $Zr_{52.5}Cu_{17.9}Ni_{14.6}Al_{10}Ti_5$ bulk metallic glasses in aqueous solutions. *Mater. Lett.* **2003**, *57*, 3022–3025. [CrossRef]
27. Lin, C.H.; Huang, C.H.; Chuang, J.F.; Lee, H.C.; Liu, M.C.; Du, X.H.; Huang, J.C.; Jang, J.S.C.; Chen, C.H. Simulated body-fluid tests and electrochemical investigations on biocompatibility of metallic glasses. *Mater. Sci. Eng. C* **2012**, *32*, 2578–2582. [CrossRef]
28. Deng, Y.L.; Lee, J.W.; Lou, B.S.; Duh, J.G.; Chu, J.P.; Jang, J.S.C. The fabrication and property evaluation of Zr–Ti–B–Si thin film metallic glass materials. *Surf. Coat. Tech.* **2014**, *259*, 115–122. [CrossRef]
29. ASTM D3359-09 Standard Test Methods for Measuring Adhesion by Tape Test; ASTM: PA, USA. Available online: https://www.astm.org/DATABASE.CART/HISTORICAL/D3359-09.htm (accessed on 12 November 2020).

30. ASTM E855-08 Standard Test Methods For Bend Testing Of Metallic Flat Materials For Spring Applications Involving Static Loading; ASTM: PA, USA. Available online: https://webstore.ansi.org/standards/astm/astme85508 (accessed on 12 November 2020).
31. Chang, Y.Z.; Tsai, P.H.; Li, J.B.; Lin, H.C.; Jang, J.S.C.; Li, C.; Chen, G.J.; Chen, Y.C.; Chu, J.P.; Liaw, P.K. Zr-based metallic glass thin film coating for fatigue-properties improvement of 7075-T6 aluminum alloy. *Thin Solid Films* **2013**, *544*, 331–334. [CrossRef]
32. Chu, J.P.; Greene, J.E.; Jang, J.S.C.; Huang, J.C.; Shen, Y.L.; Liaw, P.K.; Yokoyama, Y.; Inoue, A.; Nieh, T.G. Bendable bulk metallic glass: Effects of a thin, adhesive, strong, and ductile coating. *Acta Mater.* **2012**, *60*, 3226–3238. [CrossRef]
33. Chu, J.P.; Huang, J.C.; Jang, J.S.C.; Wang, Y.C.; Liaw, P.K. Thin film metallic glasses: Preparations, properties, and applications. *JOM* **2010**, *62*, 19–24. [CrossRef]
34. Schroers, J.; Johnson, W.L. Ductile bulk metallic glass. *Phys. Rev. Lett.* **2004**, *93*, 255506. [CrossRef] [PubMed]
35. Conner, R.D.; Li, Y.; Nix, W.D.; Johnson, W.L. Shear band spacing under bending of Zr-based metallic glass plates. *Acta Mater.* **2004**, *52*, 2429–2434. [CrossRef]

Publisher's Note: MDPI stays neutral with regard to jurisdictional claims in published maps and institutional affiliations.

© 2020 by the authors. Licensee MDPI, Basel, Switzerland. This article is an open access article distributed under the terms and conditions of the Creative Commons Attribution (CC BY) license (http://creativecommons.org/licenses/by/4.0/).

Article

Synthesis and Electron-Beam Evaporation of Gadolinium-Doped Ceria Thin Films

Fariza Kalyk [1,2,*], Artūras Žalga [3], Andrius Vasiliauskas [2], Tomas Tamulevičius [1,2], Sigitas Tamulevičius [1,2] and Brigita Abakevičienė [1,2]

[1] Department of Physics, Kaunas University of Technology, Studentų St. 50, 51368 Kaunas, Lithuania; tomas.tamulevicius@ktu.lt (T.T.); sigitas.tamulevicius@ktu.lt (S.T.); brigita.abakeviciene@ktu.lt (B.A.)
[2] Institute of Materials Science, Kaunas University of Technology, K. Baršausko St. 59, 51423 Kaunas, Lithuania; andrius.vasiliauskas@ktu.lt
[3] Department of Applied Chemistry, Faculty of Chemistry and Geosciences, Vilnius University, Naugarduko St. 24, 03225 Vilnius, Lithuania; arturas.zalga@chf.vu.lt
* Correspondence: fariza.kalyk@ktu.lt; Tel.: +37-062903862

Abstract: Gadolinium-doped ceria (GDC) nanopowders, prepared using the co-precipitation synthesis method, were applied as a starting material to form ceria-based thin films using the electron-beam technique. The scanning electron microscopy (SEM) analysis of the pressed ceramic pellets' cross-sectional views showed a dense structure with no visible defects, pores, or cracks. The AC impedance spectroscopy showed an increase in the total ionic conductivity of the ceramic pellets with an increase in the concentration of Gd_2O_3 in GDC. The highest total ionic conductivity was obtained for $Gd_{0.1}Ce_{0.9}O_{2-\delta}$ (σ_{total} is 11×10^{-3} S·cm^{-1} at 600 °C), with activation energies of 0.85 and 0.67 eV in both the low- and high-temperature ranges, respectively. The results of the X-ray photoelectron spectroscopy (XPS) and inductively coupled plasma optical emission spectrometer (ICP-OES) measurements revealed that the stoichiometry for the evaporated thin films differs, on average, by ~28% compared to the target material. The heat-treatment of the GDC thin films at 600 °C, 700 °C, 800 °C, and 900 °C for 1 h in the air had a minor effect on the surface roughness and the morphology. The results of Raman spectroscopy confirmed the improvement of the crystallinity for the corresponding thin films. The optimum heat-treating temperature for thin films does not exceed 800 °C.

Keywords: gadolinium-doped ceria; GDC; co-precipitation synthesis; electron-beam evaporation; thin films; SOFC; impedance spectroscopy

1. Introduction

Miniaturized solid-oxide fuel cells (μ-SOFCs), constructed using thin-film technologies, can achieve high specific energy and energy density and may, one day, partially replace Li batteries in portable devices [1–5]. However, the initial materials used in the fabrication of the μ-SOFC process should fully satisfy their requirements. Recently, the thickness of the μ-SOFC three-layered structure (anode-electrolyte-cathode) has been reduced to a one-micron size. Thus, the thickness of the electrolyte thin film in μ-SOFC becomes thinner, e.g., ~600 nm, compared to conventional SOFC (~1 μm) [6–8]. This reduced thickness can minimize the ionic transport path and significantly reduce the ohmic resistance [9]. The development of thin-film ceramic electrolytes over the past several decades has led to reduced operating temperatures for SOFCs [10]. Conventional materials, such as ceria or zirconia-based ceramics, are still widely used as electrolytes [2,11]. Due to their superior properties, such as high ionic conductivity and low activation energy [12], gadolinium-doped ceria (GDC) ceramics are widely applied in the production of μ-SOFC as an electrolyte [4], interlayer [9], or in the composition of an anode [13]. GDC is one of the most promising electrolytes for μ-SOFC, with only one condition: that the operating temperature should be below 650 °C [12], due to the reduction of Ce^{4+} to a Ce^{3+}, resulting in the failure of the electrolyte material at temperatures higher than 750 °C [14].

The properties of μ-SOFC electrolyte thin films primarily depend on the initial materials and their characteristics; therefore, the choice of synthesis method, processing stages, and conditions are particularly important. The synthesis of ceria-based electrolytes with the desired properties can be carried out using wet chemical synthesis routes, such as the sol-gel process [15], combustion synthesis [16], hydrothermal synthesis [17], polyol [12,18], the acetic acrylic method [19], and the co-precipitation method using nitrates [20–23], oxalates [24,25], and acetates [26,27]. Compared to other techniques, the co-precipitation (CP) method has the advantages of good control of the starting material and the processing parameters, the low temperature of the process, and the high purity and homogeneity of the product due to the possibility of controlling the starting solution [28]. The control of the initial material synthesis process is essential for the preparation of electrolyte materials with the required properties: crystalline structure, crystallite size, and even distribution of grains. This can be achieved by changing the synthesis parameters, such as the deposition rate and duration, precipitation, material, concentration, stirring speed, residue solubility, ambient pH, temperature, etc. [28].

Many studies have been carried out employing vacuum deposition techniques for the preparation of the electrolyte thin films used in μ-SOFC [29–35]. Although physical vapor deposition (PVD) techniques present some challenges, such as complexity and relatively high cost, they enable the production of very thin and dense films on either porous or dense substrates [34]. PVD processes can be used to deposit films of elements and alloys, as well as compounds employing reactive deposition processes. Moreover, films can be formed at temperatures that are much lower than those required in traditional ceramic processing [31]. Unfortunately, the film's stoichiometry is found to be difficult to control, due to the deposition behavior of the composite material [4].

The most common techniques for the formation of μ-SOFC electrolytes are ion beam sputtering and pulsed laser deposition (PLD) [4]. However, compared to the other PVD methods, the e-beam evaporation technique has the advantages of a high deposition rate and large deposition area [34,36].

The electrolyte of μ-SOFC should have a high density and demonstrate good ionic conductivity [1,4,29,37,38]. These requirements can be achieved by controlling the concentration of impurities in the sample, e.g., mol% of gadolinia in GDC thin films [14], and selecting the appropriate method for the deposition of the thin film—in all cases the optimal chemical composition of elements in the films has to be ensured. Saporiti F. et al. [39] showed that the pulsed laser deposition technique is well suited for the formation of thin films that have adhered well to the electrolyte substrate, enabling the production of thin films with the same stoichiometry as the target. However, in this method, it is difficult to control the surface morphology as well as the porosity of the film. Uhlenbruck S. et al. [40] employed magnetron sputtering and the e-beam evaporation technique (EB-PVD) for the fabrication of GDC electrolyte thin films. The authors summarized that irrespective of deposition temperatures, the measured ratio of Ce and Gd of the GDC thin film corresponds almost exactly to the theoretical value of the target composition. While Sanghoon Ji et al. [38] showed that the chemical composition of the deposited thin film depends on the target material and substrate temperature during the deposition process. In addition, Wibowo R.A. et al. [41] explained that the deviation of the chemical composition is due to the different sputtering yield or evaporation process during the sputtering and EB-PVD deposition processes, respectively. However, the stoichiometry of electrolyte thin films produced by the EB-PVD technique has not been sufficiently studied.

This research aimed to synthesize and characterize the initial/target material, form dense ceria-based thin films using the EB-PVD technique, determine the stoichiometric deviation in the evaporated thin films compared to the target/initial material, and investigate the influence of additional heat treatment on the formed thin films. Thus, a co-precipitation synthesis route was employed for the preparation of GDC ceramic powders with different concentrations of Gd (10, 15, and 20 mol%), which were further used as target materials in the EB-PVD process. Since the target may influence the chemical composition of the film, the chemical composition of the evaporated thin films was estimated using XPS measurement. GDC thin films were annealed at various temperatures to study the effect on the structural properties of the films and to find the optimal annealing temperature. The obtained experimental results will help to select the optimal conditions for the formation of thin films with the desired properties, using the electron-beam evaporation technique, which can be used as an electrolyte for SOFC.

2. Materials and Methods

2.1. Synthesis of the Target Material and the Formation of Thin Films

Gadolinium (III) ($Gd(NO_3)_3 \cdot 6H_2O$, 99.9%, Sigma Aldrich, St. Louis, MO, USA) and cerium (III) nitrate hexahydrates ($Ce(NO_3)_3 \cdot 6H_2O$, 99.0%, Fluka, Charlotte, NC, USA) were used as metal precursors for the preparation of gadolinium-doped ceria (GDC) $Ce_{1-x}Gd_xO_{2-\delta}$ (where x = 0.1, 0.15, and 0.2 mol%) ceramic powders using the co-precipitation (CP) synthesis method. According to the concentration of Gd, the ceramic powders and pellets were denoted as 10-GDC, 15-GDC, and 20-GDC, respectively. A stoichiometric amount of gadolinium and cerium nitrate hexahydrates were dissolved in distilled water. The obtained solution of Gd and Ce salts was poured dropwise into an aqueous solution of oxalic acid under active stirring at 50 °C for 30 min, resulting in the formation of a white opaque colloidal solution. Ammonium hydroxide (NH_4OH, 25%, Sigma Aldrich, St. Louis, MO, USA) was used to adjust the pH ratio to ~8–9 and to promote sedimentation. The precipitate was filtered by vacuum filtration using a Büchner funnel, washed, and dried for 24 h at room temperature in air. Finally, the synthesized powders were ground, milled in an agate mortar, and calcined at different temperatures (200, 400, 600, 800, 900, 1000, 1100, and 1200 °C) for 5 h (5 °C/min) in air. These calcination temperatures were chosen to study the crystalline phases and the changes in crystallinity, and to verify at which temperature the oxides are completely formed.

The GDC powders, synthesized and calcined at 900 °C for 5 h, were pressed into pellets with a diameter of 10 mm and a thickness of ~1.5 mm (for impedance measurements) and ~3.7 mm (for electron-beam evaporation) using uniaxial compression at 200 MPa. Subsequently, the pellets were annealed at 1200 °C for 5 h in the air (5 °C/min). The density of the pellets was measured by a weight-volume method, using the theoretical density of 7.235 g/cm^3 [41]. For impedance spectroscopy measurements, platinum paste (conductive paste Lot No. 13032810, Mateck, Jülich, Germany) was applied on both parallel sides of the polished GDC pellets and then dried at 300 °C for 2 h.

GDC ceramics with a thickness of ~3.7 mm were used as the target material for evaporation on Si (thickness of the films: ~800 nm) using a UVN-71P3 electron-beam evaporation system. The evaporation process was carried out at a pressure of 0.7 Pa, with an evaporation rate of ~2 nm/s; the distance between the electron gun (power: 10 kW) and the substrate was 250 mm. The temperature of the substrate was kept at 200 °C during the evaporation process. The evaporated thin films are denoted as 10-GDC, 15-GDC, and 20-GDC, respectively.

2.2. Characterization Techniques

The thermal decomposition of the synthesized powders was analyzed using thermogravimetric (TGA) and differential thermal (DTA) analyses (PerkinElmer STA 6000, Shelton, CT, USA). Dried but not calcined synthesized GDC powders (5–10 mg) were heated from 25 to 950 °C (heating rate 10 °C/min) in dry flowing air (20 mL/min). To define the elemental compositions, 100 mg of powders were dissolved in concentrated sulfuric acid, and the diluted solutions were analyzed with an inductively coupled plasma optical emission spectrometer (ICP-OES, Vista-Pro, Varian, Mulgrave, Victoria, Australia). The Brunauer–Emmett–Teller (BET) surface area analyzer (Sorptometer KELVIN 1042, Ithaca, NY, USA) was used to determine the bulk surface area of the powders calcined at 900 °C.

The BET surface area and the equivalent particle size (D_{BET}) were calculated using the following equation [42]:

$$D_{BET} = \frac{6 \times 10^3}{d_{th} S_{BET}} \quad (1)$$

where S_{BET} is the specific surface area (m^2/g) and d_{th} is the theoretical density of the solid solution oxide (g/cm^3), calculated according to the following equation:

$$d_{th} = \frac{4[(1-x)M_{Ce} + xM_{Gd} + (2-x/2)M_O]}{a^3 N_A} \quad (2)$$

where x is the dopant concentration, N_A is the Avogadro constant, M is the atomic weight, and a represents the lattice parameters of the solid solution.

The crystal structure of the synthesized GDC powders was determined using a D8 Discover X-ray diffractometer (Bruker AXS GmbH, Karlsruhe, Germany) with a $Cu\ K_\alpha$ (λ = 1.5418 Å) radiation source and parallel beam geometry with a 60 mm Göbel mirror. A Soller slit with an axial divergence of 2.5° was utilized on the primary side. The diffraction patterns were recorded using a fast-counting LynxEye (0D mode) silicon strip detector with a 2.475° opening angle and a 6 mm slit opening. The peak intensities were scanned over the range of 20–90° (coupled 2θ-θ scans), with a 0.02° step size and time per step of 0.2 s.

The microstructure and elemental composition of the samples were estimated using scanning electron microscopy (SEM) (FEI Quanta 200 FEG, Hillsboro, Oregon, USA) in a low-vacuum mode, equipped with an energy-dispersive X-ray spectrometer (EDS). To obtain the EDS spectra, the accelerating voltage was 5 kV.

The electrical properties of GDC pellets were investigated via impedance measurements, using an impedance analyzer Alpha-AK (Novocontrol Technologies, Montabaur, Germany) in the temperature range of 200–800 °C, and from 1 Hz to 1 MHz of the frequency range. The obtained impedance spectra were fitted using the equivalent circuits by the Zview2 software. The plots of σ vs. $1000/T$ were analyzed and the activation energy was obtained using the Arrhenius plot, according to the following equation:

$$\sigma_{b,\ gb} = \sigma_0 exp \frac{-\Delta E_{b,\ gb}}{kT} \quad (3)$$

where σ_0 is the pre-exponential factor, k is the Boltzmann constant (0.86 × 10^{-4} eV K^{-1}), T is the temperature, and $\Delta E_{b,gb}$ represents the activation energies of bulk and grain boundary conductivity.

X-ray photoelectron spectroscopy (XPS) was used to study the atomic composition of the as-deposited GDC electrolyte thin films. A Thermo Scientific ESCALAB 250Xi spectrometer (Thermo Fisher, 2013, Waltham, MA, USA). with monochromatized AlK_α radiation ($h\nu$ = 1486.6 eV) was used for the surface analysis. The base pressure in the analytical chamber was 2 × 10^{-7} Pa, the x-ray spot size was 0.3 mm, and 40 eV pass energy was used during the spectra acquisition. The energy scale of the system was calibrated according to the peak positions of Au 4f7/2, Ag 3d5/2, and Cu 2p3/2. The GDC thin films were analyzed without a surface-cleaning procedure, and the calculations of atomic concentration were performed using the original ESCALAB 250Xi Avantage software.

For the characterization of the bonding structure of GDC thin films on a Si (100) substrate, a Renishaw inVia Raman spectrometer (Renishaw, Wotton-under-Edge, UK)) equipped with a wavelength of 532 nm, a 45 mW excitation laser power, a 50× objective (NA = 0.75, Leica Microsystems, Wetzlar, Germany), and an integration time of 10 s was used. The measurements were performed in the 100–800 cm^{-1} spectral range, with an exposure time of 10 s and 1% of the laser power. The background was subtracted from the obtained Raman spectra and was fitted by Lorentzian-shaped lines in the spectral range of 440–490 cm^{-1}.

The morphology, topography, and surface roughness parameters (R_q, R_{sk}, Z_{mean}, and R_{ku}) of GDC and SDC thin films on different substrates were analyzed using an NT-206

(Microtestmachines Co., Gomel, Belarus) atomic force microscope and the SPM-data processing software, SurfaceXplorer (Ultrafast Systems, Sarasota, FL, USA). The measurements were performed at room temperature in the air. A silicon cantilever with a tip curvature radius of 10 nm, spring constant of 3 N/m, and cone angle of 20° was operating in a contact scanning mode, with a 12 µm × 12 µm field of view.

3. Results

3.1. Characterization of the Powders

In this work, thermal analysis was applied to show the most important differentiating features of the thermal decomposition of intermediate products. The thermal decomposition process revealed and characterized the individual peculiarities of each sample, which was prepared according to the co-precipitation synthesis method. The analysis was carried out for all concentrations of GDC nanopowders. However, the obtained results showed the same trend regardless of the concentration; thus, only the 10-GDC results are presented (Figure 1a).

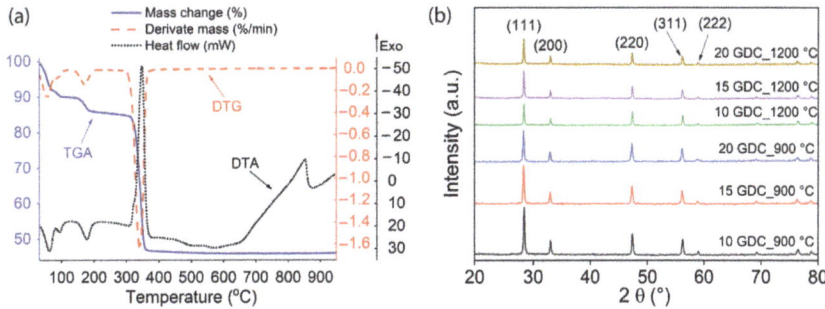

Figure 1. Thermal analysis curves (**a**) for 10-GDC ceramic powders and (**b**) the X-ray diffraction patterns of GDC nanopowders, calcined at 900 °C and 1200 °C.

The decomposition of CP precursor is closely related to the degradation behavior of hydrated oxalic acid. In the first temperature range from 30 °C to 100 °C, the evaporation of water molecules and a corresponding mass change of about 12–13% were identified. This effect was confirmed by a strong endothermic peak in the DTA curve. During the further increase in temperature to 190 °C, a bright endothermic effect was observed on the DTA curve that corresponds to the melting and partial decomposition of oxalic acid, and a mass change in the sample of about 4–5%. The last mass change in the range from 310 °C to 370 °C was attributed to the final decomposition of the initial metal oxalate precursor. The mass change of about 38% and the strong exothermic peak on the DTA curve suggest the release of carbon dioxide, the formation of which was promoted by the redox properties of ceria. There was also a slight increase in mass (0.16%) above 800 °C. This effect reflects an endothermic peak on the DTA curve at temperatures from 850 °C to 950 °C. In conclusion, the final tendency of the crystallization of double oxide at elevated temperatures depended only on the initial sizes of the crystallites, which were formed at a temperature of about 400 °C.

The X-ray diffraction patterns of 10-, 15-, and 20-GDC powders, calcined at 900 and 1200 °C for 5 h, are presented in Figure 1b. The obtained results show that GDC powders have a cubic fluorite crystal structure with an $Fm\bar{3}m$ space group and with the dominating (111) crystallographic plane; all positions of the diffraction peaks match the standard XRD data (PDF- 4 database: 011-7336, 006-3415, 013-6571). The crystallite size, D, was calculated from the X-ray broadening, using Scherrer's equation [12]. With increasing calcining temperature, the crystallite size increased and was in the range of from 5 nm to 42–48 nm for the different concentrations of GDC. According to the obtained results, the tendency of the growth of crystallite sizes for different concentrations of GDC powders, annealed at the same temperatures, was the same. The equivalent crystallite size, D_{BET}, was calculated from the BET analysis, according to Equation (1). Comparing the results of the crystallite size, D, and the equivalent size of the GDC ceramic powders calcined at 900 °C, the D_{BET} was higher within the range of 70–108 nm (Table 1). The difference between D and D_{BET} could be caused by the occurrence of crystalline nanodomains within the individual nanocrystals in the calcined powders. The XRD technique registered these nanodomains as individual crystals, while the BET technique measured only the surface area of the parent nanocrystal. Moreover, the nanocrystals could be clustered together to form agglomerates of the crystalline nanoparticles. To probe the agglomeration extent of the particles, the factor ϕ was defined using:

$$\phi = D_{BET}/D \quad (4)$$

where D_{BET} is the specific surface area determined by BET analysis, and D is calculated according to Scherrer's equation, using the X-ray diffraction peak broadening data [42]. This ratio is well known as a factor that reflects the agglomeration extent of the primary crystallites and is an indicator of their porous agglomerate nature; a value of 1.0 specifies their complete dispersion [42]. The related results are presented in Table 1, where the ϕ factor increased with the increase in the molar concentration of Gd_2O_3 in GDC. The obtained results are in good agreement with the SEM results; those synthesized using co-precipitation synthesis and calcined at 900 °C have agglomerated features (Figure 2).

Table 1. The summary of the elemental analysis results and the physical properties of the GDC ceramic powders, calcined at 900 °C.

Sample	Expected Molar Ratio of Gd to Ce in GDC	Gd Content in GDC from ICP-OES (r.u.)	Gd Content in GDC from EDS (r.u.)	S_{BET} (m^2/g)	D (nm)	d_{th} (g/cm^3)	D_{BET} (nm)	ϕ (r.u.)
10-GDC	0.18: 0.82	0.18	0.16	11.7	32.6	7.235	70.3	2.156
15-GDC	0.26: 0.74	0.26	0.25	10.19	32.7	7.244	81.4	2.489
20-GDC	0.33: 0.67	0.33	0.32	7.7	31.1	7.251	108.0	3.472

Figure 2. SEM images of 10-GDC ceramic powders calcined at (**a**) 800 °C, (**b**) 900 °C, and (**c**) 1200 °C for 5 h.

The lattice parameter (*a*) was calculated according to Bragg's Law:

$$a = \frac{\lambda}{2sin\theta_{hkl}}\sqrt{h^2 + k^2 + l^2} \quad (5)$$

where h, k, l are the Miller indices of the crystallographic plane (in the calculations, the (111) plane was used), θ_{hkl} is the Bragg's angle, and λ is the wavelength of X-ray radiation.

The changes in GDC lattice parameters depend on both the concentration of Gd_2O_3 and the calcination temperature. The lattice parameter increases with increases in the molar concentration of Gd_2O_3 since the ionic radius of the gadolinium cation, Gd^{3+}, is larger than the ionic radius of the cerium cation, Ce^{4+} (r_{Gd}^{3+} = 0.1053 nm and r_{Ce}^{4+} = 0.097 nm) [42]. Moreover, the growth of the lattice parameter with the addition of Gd confirms the incorporation of gadolinium ions into the lattice.

The morphology of GDC ceramic powders calcined at 800 °C, 900 °C, and 1200 °C for 5 h was investigated using scanning electron microscopy (Figure 2). It can be seen that the structure of nanopowders changes with the increase in the calcination temperature. For example, as the temperature rose from 800 °C to 900 °C, the formation of a structure with fragments and protrusions (~400 nm in width) was observed, which is typical for the co-precipitation synthesis method (Figure 2b). Therefore, further powder treatment was necessary, as pellets pressed from such uncrushed powders have a relatively low density. At a maximum calcination temperature of 1200 °C (Figure 2c), the formation of grains was observed; that is, the structure had acquired its final form. The same tendency was observed for other concentrations of GDC. Similar results were obtained by Zha S. et al. [43], who determined the coarse structure of the GDC powders, synthesized using oxalic acid co-precipitation synthesis and sintering at 750 °C for 1 h.

When summarizing the TG-DTA, SEM, and XRD results, the calcination of GDC ceramic powders at 900 °C ensures the full formation of a cubic fluorite crystal structure with an $Fm\bar{3}m$ space group and helps to achieve thermal stability in the material.

The elemental analysis of the synthesized GDC powders (concentration of gadolinium) was measured using two different methods: ICP-OES and SEM/EDS analyses (Table 1). The obtained results indicate that the chemical composition of the material is controlled by the composition of the synthesis solution. Furthermore, the properties of the ceramics can be influenced by the compaction of the powders during the calcination process.

3.2. Characterization of the Pellets

To characterize the ionic conductivity of $Ce_{1-x}Gd_xO_{2-\delta}$ ceramics, the distribution of grains and the density of the pellets were calculated. The sintering quality of the ceramics and their microstructure are important factors for the analysis of ionic conductivity using impedance spectroscopy. Burcu et al. [44] found that the grain boundary resistance increased due to low sinterability and the increment of porosity in samarium-doped ceria electrolytes synthesized by the electrospinning method. The sinterability depends on the sintering temperature and time period, the diffusion coefficient of the atoms, and the dispersivity of the particles. Calcination of the fine-dispersive powders, with individual grains growing together, helps to reduce the porosity of the pellets from 30–50% to a few percent. Figure 3 presents SEM images of the cross-section of GDC pellets annealed at 1200 °C for 5 h. To perform SEM measurement, the pellets were broken in half. Due to the roughness of the surface after breaking the pellets, the sides were polished, cleaned with ethanol, and thermally etched at 1100 °C for 1 h.

Figure 3. SEM images of the cross-section view of (**a**) 10-GDC, (**b**) 15-GDC, and (**c**) 20-GDC ceramic pellets at 100 k magnification.

The SEM results of the GDC pellets showed that they consisted of grains of different sizes, without visible defects or cracks. The ceramics were dense and with no porosity; the average grain sizes of the pellets were about 320 nm (10-GDC), 310 nm (15-GDC), and

302 nm (20-GDC). All synthesized GDC ceramics were investigated using AC impedance spectroscopy; the obtained complex impedance plots at different temperature and frequency ranges are presented in Figure 4. After measuring the complex resistance over a wide frequency range of electric field, it was possible to separate the different conductivity processes, such as the bulk, grain boundary, and total ionic conductivity.

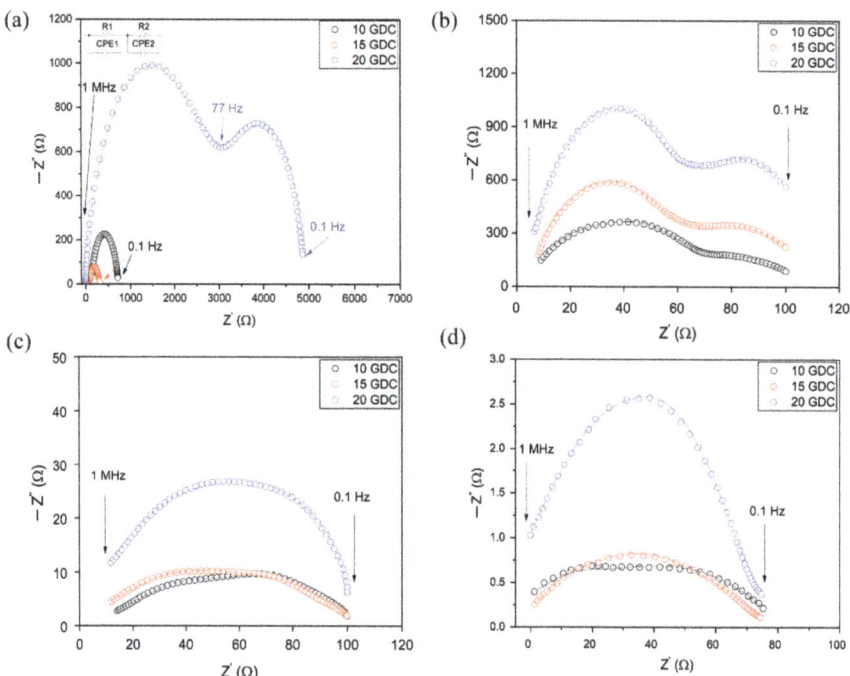

Figure 4. Complex impedance plots at: (**a**) 250 °C, (**b**) 400 °C, (**c**) 600 °C, and (**d**) 800 °C of the different compositions of GDC pellets annealed at 1200 °C. An example of an equivalent circuit, where CPE is the constant phase element, and R is the resistance for bulk, grain boundary, and electrode has been inserted in the top left-hand corner of (**a**).

The semicircle in the high-frequency range is related to the oxygen ion relaxation in the grain interior, while the semicircle in the medium-frequency range can be attributed to the ionic migration in the grain boundaries, and the semicircle in the low-frequency range corresponds to the electrode polarization (the frequency increase from right to left in the impedance spectra) [45–47]. From Figure 4, it can be seen that with the increase in the concentration of gadolinium, the resistance of the GDC pellets increased. The 20-GDC pellets showed the highest resistance at all temperature ranges, resulting in the lowest ionic conductivity, while the 10-GDC pellets had the lowest resistance and the highest conductivity. With the increasing temperature, we observed a decrease in the resistance, thus increasing the conductivity. Such results allow us to conclude that the conductivity of GDC ceramics depends on the concentration of Gd_2O_3. Koettgen et al. [48] showed the bulk and grain boundary conductivity values of Sm- and Gd-doped ceria, synthesized by the sol-gel method as a function of the dopant fraction. Comparing the different dopant concentrations of Gd-doped ceria with Sm-doped ceria, the largest bulk conductivity was reported for $Ce_{0.93}Sm_{0.07}O_{1.965}$, leading to maximum total conductivity. The authors of an earlier study stated that the conductivity gained in value with increasing dopant and, subsequently, decreased with the further increment of dopant fraction [48]. In our own previous work, we compared the different concentrations (x = 0.1, 0.2, and

0.3) of SDC (samarium-doped ceria) synthesized using different synthesis methods and obtained the highest conductivity values for SDC synthesized by the combustion method, with x = 0.2 [49]. However, Wattanathana et al. [50] synthesized SDC (x = 0.1, 0.15, and 0.2) via the thermal decomposition of metal-organic complexes and reported the highest conductivity for 15 SDC, determining that the Sm^{3+} ions were replaced at the Ce^{4+} sites within the ceria structure. The authors stated that due to the formation of the amorphous Sm_2O_3 phase, increasing the doping concentration led to a lower conductivity value.

The temperature dependencies of the total ionic conductivity obey Arrhenius' law [51] and its plot is presented in Figure 5.

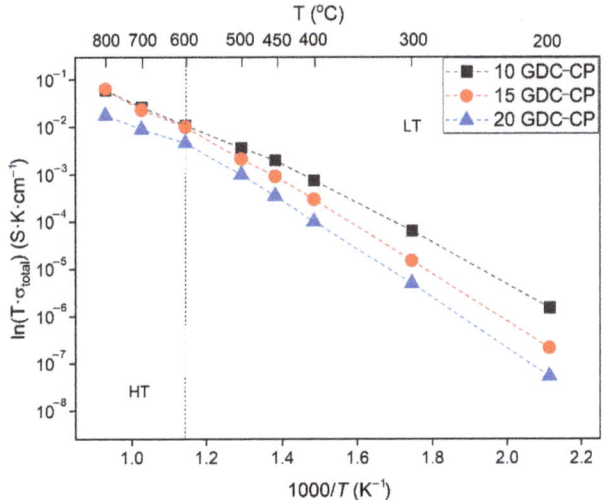

Figure 5. Temperature dependence of total conductivity of GDC ceramic pellets. The dashed line shows the separation of high- (HT) and low-temperature (LT) ranges.

Different temperature ranges (low- (LT) and high-temperature (HT) ranges), separated by the dashed line in Figure 5, occur due to the thermodynamics between defect species and their interactions, as well as the redox reactions, followed by the formation of polarons in the ceria lattice [51]. From the obtained results (Figure 5 and Table 2), we can observe the decrease in the activation energy in the high-temperature range for all concentrations of GDC. In the high-temperature range, the charge-carrying defects are dictated by intrinsic defects and no association enthalpy is present; thus, the activation energy decreases [52]. The activation energies determined from the Arrhenius plots, as a function of the molar concentration in GDC ceramic pellets, and the values of bulk, grain boundary, and total conductivity at 600 °C are presented in Table 2.

Table 2. Values of the bulk (ΔE_b), grain boundary (ΔE_{gb}), and total (ΔE_{total}) conductivity and activation energy of GDC ceramic pellets at low- (LT) and high-temperature (HT) ranges.

Sample	ΔE_a (eV)		Total Conductivity (S·cm^{-1})		
	LT	HT	400 °C	600 °C	800 °C
10-GDC	0.85	0.67	0.7×10^{-3}	11×10^{-3}	5.9×10^{-2}
15-GDC	0.95	0.80	0.3×10^{-3}	10×10^{-3}	6.3×10^{-2}
20-GDC	0.99	0.85	0.1×10^{-3}	4.7×10^{-3}	1.8×10^{-2}

With an increase in the molar concentration, the total conductivity decreases, and the activation energy increases (Table 2). A similar tendency was sustained for the bulk and grain boundary conductivity. Thus, the highest total ionic conductivity of 11×10^{-3} S·cm^{-1} at 600 °C, with the lowest activation energies at both LT and HT (0.85 and 0.67 eV, respectively), was found for 10-GDC. Similar results were obtained by Fuentes and Baker [53], who synthesized $Gd_{0.1}Ce_{0.9}O_{1.95}$ by the sol-gel technique and reported a conductivity value of 11×10^{-3} S·cm^{-1} (at 600 °C) for the samples sintered at 1300 °C for 8 h. Öksüzömer M.A.F. et al. [12] synthesized $Gd_{0.1}Ce_{0.9}O_{1.95}$ and $Gd_{0.2}Ce_{0.8}O_{1.9}$ powders through the polyol process and obtained higher conductivity values at 800 °C, with low activation energies for 10-GDC compared to 20-GDC (2.11×10^{-2} and 2.01×10^{-2} S·cm^{-1}, respectively). Jaiswal N. et.al. [54] reported values of ionic conductivity of 0.01 S cm^{-1} for $Ce_{0.9}Gd_{0.1}O_{1.95}$ and 3.02×10^{-3} S cm^{-1} for $Ce_{0.85}Gd_{0.15}O_{1.925}$ at 500 °C. Murutoglu M. et al. [55] used the cold sintering-assisted densification of GDC and reported activation energy of 0.69 eV at a high temperature range. Zhang J. et. al. [56] compared the electrolytes for SOFC and found that, compared to ScSZ and YSZ at 500 °C, GDC had the highest ionic conductivity at 5.8×10^{-3} S cm^{-1}.

3.3. Characterization of Thin Films

The synthesized $Ce_{1-x}Gd_xO_{2-\delta}$ pellets were evaporated on Si substrates using the electron beam evaporation (EB-PVD) technique. To study the effect of additional heat treatment on the properties of the formed GDC thin films, the samples were annealed at 600, 700, 800, and 900 °C for 1 h. The results determining the concentration of Gd_2O_3 in GDC thin films are summarized in Table 3.

Table 3. The chemical composition of evaporated 10-, 15-, and 20-GDC thin films, obtained from XPS and ICP-OES measurements.

Notation	Gd Content in GDC from ICP-OES (r.u.)	Gd Content in Thin Film from XPS (r.u.)	Molar Content of Gd_2O_3 in Thin Film (mol%)	Decrease in the Molar Content of Gd_2O_3 in Thin Film (%)
10-GDC	0.13	0.13	6.90	31.0
15-GDC	0.21	0.20	11.3	24.7
20-GDC	0.27	0.26	14.4	28.0

From the chemical composition results, we can see that the molar content of Gd_2O_3 in all GDC thin films is lower by an average of 28% (31% for 10-GDC, 24.7% for 15-GDC, and 28% for 20-GDC) than the target material used in the EB-PVD process (Table 3). As we can see, thin films prepared using the 15-GDC ceramic pellets as a target material ensured the formation of 10-GDC thin films, the target of which (10-GDC ceramic pellet) had the highest ionic conductivity [14]. It should be noted that the evaporation process of the solid solution is complex, and many factors may influence the change in the chemical composition of thin films compared to the target material [57]. The changes occur due to the different evaporation temperatures and evaporation rates of the individual components, which experience the same process temperature. Metals evaporate as a function of temperature and vacuum level; high vacuum conditions lead to a greater evaporation rate. As a result, different elements evaporate at different evaporation rates, and the resultant chemical composition of the condensed material may vary, compared to the target material. Therefore, the composition of the deposited thin films using the evaporation technique may vary, compared to the target material.

The surface roughness and morphology of GDC thin films were studied by SEM and AFM and are presented in Figure 6. Annealing the films had an insignificant effect on the morphology and surface roughness of the films. GDC thin films at all temperatures were dense and quite rough, containing nanoscale grains. The thickness values fluctuated slightly (~800 nm), depending on the annealing temperature.

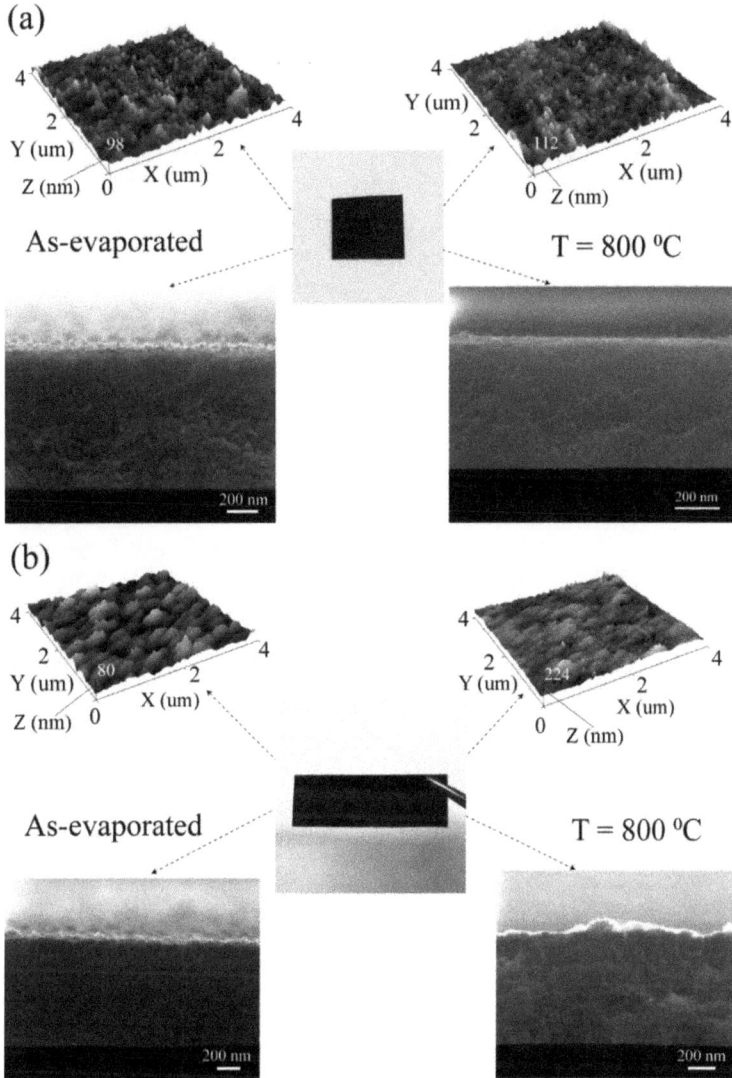

Figure 6. Photo, SEM cross-section view, and 3D AFM images of (**a**) 10-GDC and (**b**) 15-GDC thin films: as-evaporated (left) and annealed at 800 °C (right).

The roughness parameters were calculated from AFM images and the results are shown in Figure 7a. The roughness kurtosis (R_{ku}) value showed that all films had a spiky surface (above 3 nm) and the highest values of 6.23 nm (10-GDC) and 7.48 nm (15-GDC) were in the films annealed at 800 °C, while the smallest values were obtained for the as-evaporated samples and the films annealed at 900 °C of all concentrations.

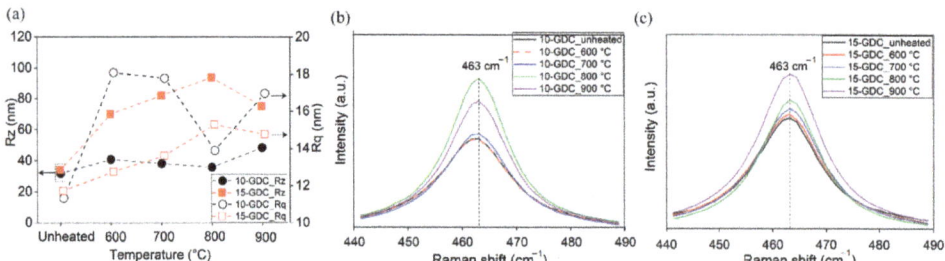

Figure 7. (**a**) Surface roughness parameters calculated from the AFM images of 10-GDC and 15-GDC thin films: as-evaporated, and after additional thermal treatment, where R_q is the root mean square and R_z is the average height. Raman spectra of (**b**) 10-GDC and (**c**) 15-GDC thin films: unheated, and after additional thermal treatment at 600, 700, 800, and 900 °C for 1 h.

The root mean square values of the surface roughness (R_q) of 10-GDC fluctuated with the annealing temperature and had a maximum value of 18.08 nm at 600 °C, with the lowest value of 11.32 nm for as-evaporated films. The R_q values of 15-GDC showed a different tendency and increased with the increase in annealing temperature, reaching the maximum value of 15.29 nm at 800 °C. The calculated average height (R_z) gradually increased for 10-GDC and had a maximum of 48.44 nm at 900 °C, while for 15-GDC, the rapid increase in R_z had a maximum of 112.61 nm at 800 °C (Figure 7a). All recorded roughness parameters increased with the annealing temperature, since the mobility of atoms increased, which led to the agglomeration of particles and an increase in their size [58,59].

The results from the Raman spectroscopy measurements showed that all GDC thin films had a major band at 463 cm^{-1} (Figure 7b,c). Considering the fact that CeO$_2$, annealed at 600 °C, shows a band at 475 cm^{-1} due to the F$_{2g}$ symmetric vibration of the cubic phase, while Gd$_2$O$_3$ shows a peak at ~360–370 cm^{-1}, the absence of these features confirms the formation of a single phase [60–63]. According to Prasad D.H. et al. [22] the formation of a single cubic phase of GDC happens when Gd^{+3} ions partially occupy the interstitial spaces of the ceria lattice, which leads to a shift in the F$_{2g}$ symmetry. The obtained GDC peak at 463 cm^{-1} (Figure 7b,c) can be related to the symmetric vibration of Ce-O, with a shift of ~12 cm^{-1} toward a lower wavenumber. As can be seen from the obtained results, the intensity of the peaks increased with the increase in the annealing temperature of the films. This can be attributed to the improvement in the crystallinity of GDC thin films [64–66]. Moreover, with an increase in the doping concentration, the peaks became wider (from 17.8 cm^{-1} to 28.4 cm^{-1}), which might be associated with the crystal size [64]. Regardless of the concentration of GDC films, the Raman peak shifted to lower frequencies with the increase in the annealing temperature. This behavior is a size-induced phenomenon observed in nanoscale systems, explained by the combined effects of lattice strain and associated with defect species and phonon confinement [64–67]. Kosacki et al. [66] and Weber et al. [67] stated that the width of the Raman peak has a linear dependence on the reciprocal of the crystal size. Similar behavior was reported by other authors [64–67]. El-Habib A. et. al. [68] presented Gd-doped CeO$_2$ nanocrystalline thin films using spray pyrolysis and reported that the peak asymmetry and broadening of Raman spectra could be attributed to the existence of an oxygen vacancy and Ce^{3+}, which changed with the addition of Gd^{3+}. With a gradual increase in crystal size, there may be a simultaneous enhancement of defects in the thin film and the growth of other crystalline phases [67]. For the 10-GDC film, the peak intensities increased and showed their maximum position when annealed at 800 °C. However, the intensity dropped in the samples annealed at 900 °C. Different results were obtained for 15-GDC thin films, where the intensity had its maximum at 900 °C. This difference in results was associated with different growth rates depending on the concentration, and, at higher temperatures, these growths were more intense. Moreover, the changes in the Raman spectrum, depending on the concentration of gadolinium happen due to a reduction in the number of Ce–O$_8$ vibrational units. The symmetrical stretching mode of the Ce–O$_8$ is sensitive to the substitution of Ce^{4+} by Gd^{3+}, and the changes in Raman spectrum shape and position are related to the lattice expansion induced by the substitution of Ce^{4+} (97 pm) by Gd^{3+} (105.3 pm) and the presence of oxygen vacancies [64,67]. Thus, the optimal annealing temperature for GDC thin films is up to 800 °C since, at this temperature, the overall crystallinity and roughness of the films are improved.

4. Conclusions

Gadolinium-doped ceria ceramics, synthesized using a co-precipitation synthesis method at different Gd$_2$O$_3$ concentrations (10, 15, and 20 mol%), were used as a target material for the formation of ceria-based thin films by the EB-PVD technique. From the impedance spectroscopy measurements, we observed an increase in activation energy (E_a) and a decrease in total ionic conductivity (σ_{total}) with the increasing molar concentration of the material. Of the pellets, the 10 mol% Gd$_2$O$_3$-containing GDC ceramic pellet had the highest conductivity (where σ_{total} was 11×10^{-3} S·cm^{-1} at 600 °C) and the lowest activation energy (where E_a was 0.85 eV (low-temperature region) and 0.67 eV (high-temperature region)). During the evaporation, a deviation in the individual component in the stoichiometry of the evaporated thin films occurred, while the decrease in the molar content of Gd$_2$O$_3$ in ~28% was lower than in the target material used for the process. Thus, to produce thin films by e-beam evaporation, one should have a target material with a higher concentration of gadolinium than is required for the desired concentration of the final material. This additional heat treatment resulted only in insignificant changes in the morphology of the films; the thickness and roughness values fluctuated, depending on the temperature. The Raman spectroscopy confirmed the improvement in the crystallinity of GDC thin films and the decrease in the grain boundary phase volume during the thermal annealing. The optimal annealing temperature for GDC thin films was found to be up to 800 °C as, at this temperature, the overall crystallinity and roughness of the films were improved.

Author Contributions: F.K.: Conceptualization, investigation, formal analysis, data curation, writing—original draft. A.Ž.: Visualization, investigation. A.V.: investigation. T.T.: Review, investigation. S.T.: Review and editing. B.A.: Writing—original draft, investigation, methodology, supervision. All authors have read and agreed to the published version of the manuscript.

Funding: The research was funded by project No P-LL-21-124 of a bilateral research funding program by the Research Council of Lithuania (LMT) and Narodowe Centrum Nauki (NCN) DAINA-2.

Institutional Review Board Statement: Not applicable.

Informed Consent Statement: Not applicable.

Data Availability Statement: Not applicable.

Conflicts of Interest: The authors declare no conflict of interest.

References

1. Evans, A.; Bieberle-Hütter, A.; Rupp, J.L.M.; Gauckler, L.J. Review on microfabricated micro-solid oxide fuel cell membranes. *J. Power Sources* **2009**, *194*, 119–129. [CrossRef]
2. Kim, K.J.; Park, B.H.; Kim, S.J.; Lee, Y.; Bae, H.; Choi, G.M. Micro solid oxide fuel cell fabricated on porous stainless steel: A new strategy for enhanced thermal cycling ability. *Sci. Rep.* **2016**, *6*, 22443. [CrossRef] [PubMed]
3. Rupp, J.L.M.; Gauckler, L.J. Microstructures and electrical conductivity of nanocrystalline ceria-based thin films. *Solid State Ion.* **2006**, *177*, 2513–2518. [CrossRef]
4. Beckel, D.; Bieberle-Hütter, A.; Harvey, A.; Infortuna, A.; Muecke, U.P.; Prestat, M.; Rupp, J.L.M.; Gauckler, L.J. Thin films for micro solid oxide fuel cells. *J. Power Sources* **2007**, *173*, 325–345. [CrossRef]
5. Bieberle-Hütter, A.; Beckel, D.; Infortuna, A.; Muecke, U.P.; Rupp, J.L.M.; Gauckler, L.J.; Rey-Mermet, S.; Muralt, P.; Bieri, N.R.; Hotz, N.; et al. A micro-solid oxide fuel cell system as battery replacement. *J. Power Sources* **2008**, *177*, 123–130. [CrossRef]
6. Yamamoto, O. Solid oxide fuel cells: Fundamental aspects and prospects. *Electrochim. Acta* **2000**, *45*, 2423–2435. [CrossRef]

7. Wachsman, E.D.; Singhal, S.C. Solid oxide fuel cell commercialization, research, and challenges. *Electrochem. Soc. Interface* **2009**, *18*, 38–43. [CrossRef]
8. van Gestel, T.; Sebold, D.; Buchkremer, P.H. Processing of 8YSZ and CGO thin film electrolyte layers for intermediate- and low-temperature SOFCs. *J. Eur. Ceram. Soc.* **2015**, *35*, 1505–1515. [CrossRef]
9. Bae, J.; Lee, D.; Hong, S.; Yang, H.; Kim, Y.B. Three-dimensional hexagonal GDC interlayer for area enhancement of low-temperature solid oxide fuel cells. *Surf. Coat. Technol.* **2015**, *279*, 54–59. [CrossRef]
10. Ji, S.; An, J.; Jang, D.Y.; Jee, Y.; Shim, J.H.; Cha, S.W. On the reduced electrical conductivity of radio-frequency sputtered doped ceria thin film by elevating the substrate temperature. *Curr. Appl. Phys.* **2016**, *16*, 324–328. [CrossRef]
11. Jiang, S.P. Thermally and Electrochemically Induced Electrode/Electrolyte Interfaces in Solid Oxide Fuel Cells: An AFM and EIS Study. *J. Electrochem. Soc.* **2015**, *162*, F1119–F1128. [CrossRef]
12. Öksüzömer, M.A.F.; Dönmez, G.; Sariboğa, V.; Altinçekiç, T.G. Microstructure and ionic conductivity properties of gadolinia doped ceria (GdxCe1-xO2-x/2) electrolytes for intermediate temperature SOFCs prepared by the polyol method. *Ceram. Int.* **2013**, *39*, 7305–7315. [CrossRef]
13. Wang, Y.; Mori, T.; Li, J.G.; Yajima, Y. Low-temperature fabrication and electrical property of 10 mol% Sm 2O3-doped CeO2 ceramics. *Sci. Technol. Adv. Mater.* **2003**, *4*, 229–238. [CrossRef]
14. Chourashiya, M.G.; Jadhav, L.D. Synthesis and characterization of 10%Gd doped ceria (GDC) deposited on NiO-GDC anode-grade-ceramic substrate as half cell for IT-SOFC. *Int. J. Hydrogen Energy* **2011**, *36*, 14984–14995. [CrossRef]
15. Fuentes, R.O.; Baker, R.T. Synthesis and properties of Gadolinium-doped ceria solid solutions for IT-SOFC electrolytes. *Int. J. Hydrog. Energy* **2008**, *33*, 3480–3484. [CrossRef]
16. Zarkov, A.; Stanulis, A.; Salkus, T.; Kezionis, A.; Jasulaitiene, V.; Ramanauskas, R.; Tautkus, S.; Kareiva, A. Synthesis of nanocrystalline gadolinium doped ceria via sol-gel combustion and sol-gel synthesis routes. *Ceram. Int.* **2016**, *42*, 3972–3988. [CrossRef]
17. Dell'Agli, G.; Spiridigliozzi, L.; Marocco, A.; Accardo, G.; Ferone, C.; Cioffi, R. Effect of the mineralizer solution in the hydrothermal synthesis of gadolinium-doped (10% mol Gd) ceria nanopowders. *J. Appl. Biomater. Funct. Mater.* **2016**, *14*, e189–e196. [CrossRef]
18. Dönmez, G.; Sariboıa, V.; Altinçekiç, T.G.; Öksüzömer, M.A.F. Polyol synthesis and investigation of Ce1-535 xRExO2-x/2 (RE = Sm, Gd, Nd, La, $0 \leq x \leq 0.25$) electrolytes for IT-SOFCs. *J. Am. Ceram. Soc.* **2014**, *98*, 501–509. [CrossRef]
19. Liu, A.Z.; Wang, J.X.; He, C.R.; Miao, H.; Zhang, Y.; Wang, G.W. Synthesis and characterization of Gd0.1Ce0.9O 1.95 nanopowder via an acetic-acrylicmethod. *Ceram. Int.* **2013**, *39*, 6229–6235. [CrossRef]
20. Biesuz, M.; Dell'Agli, G.; Spiridigliozzi, L.; Ferone, C.; Sglavo, V.M. Conventional and field-assisted sintering of nanosized Gd-doped ceria synthesized by co-precipitation. *Ceram. Int.* **2016**, *42*, 11766–11771. [CrossRef]
21. Hsieh, T.H.; Ray, D.T.; Fu, Y.P. Co-precipitation synthesis and AC conductivity behavior of gadolinium-doped ceria. *Ceram. Int.* **2013**, *39*, 7967–7973. [CrossRef]
22. Prasad, D.H.; Kim, H.R.; Park, J.S.; Son, J.W.; Kim, B.K.; Lee, H.W.; Lee, J.H. Superior sinterability of nano-crystalline gadolinium doped ceria powders synthesized by co-precipitation method. *J. Alloys Compd.* **2010**, *495*, 238–241. [CrossRef]
23. Ikuma, Y.; Takao, K.; Kamiya, M.; Shimada, E. X-ray study of cerium oxide doped with gadolinium oxide fired at low temperatures. *Mater. Sci. Eng. B Solid-State Mater. Adv. Technol.* **2003**, *99*, 48–51. [CrossRef]
24. Zha, S.; Xia, C.; Meng, G. Effect of Gd (Sm) doping on properties of ceria electrolyte for solid oxide fuel cells. *J. Power Sources* **2003**, *115*, 44–48. [CrossRef]
25. Tianshu, Z.; Hing, P.; Huang, H.; Kilner, J. Ionic conductivity in the CeO2-Gd2O3 system ($0.05 \leq Gd/Ce \leq 0.4$) prepared by oxlate coprecipitation. *Solid State Ion.* **2002**, *148*, 567–573. [CrossRef]
26. Arabaci, A.; Solak, N. High Temperature—FTIR Characterization of Gadolinia Doped Ceria. *Adv. Sci. Technol.* **2010**, *72*, 249–254. [CrossRef]
27. Arabac, A.; Öksüzömer, M.F. Preparation and characterization of 10 mol% Gd doped CeO 2 (GDC) electrolyte for SOFC applications. *Ceram. Int.* **2012**, *38*, 6509–6515. [CrossRef]
28. Shao, Z.; Zhou, W.; Zhu, Z. Advanced synthesis of materials for intermediate-temperature solid oxide fuel cells. *Prog. Mater. Sci.* **2012**, *57*, 804–874. [CrossRef]
29. Paek, J.Y.; Chang, I.; Park, J.H.; Ji, S.; Cha, S.W. A study on properties of yttrium-stabilized zirconia thin films fabricated by different deposition techniques. *Renew. Energy* **2014**, *65*, 202–206. [CrossRef]
30. Huang, H.H.; Diao, C.C.; Yang, C.F.; Huang, C.J. Effects of substrate temperatures on the crystallizations and microstructures of electron beam evaporation YSZ thin films. *J. Alloys Compd.* **2010**, *500*, 82–86. [CrossRef]
31. Laukaitis, G.; Dudonis, J. Microstructure of gadolinium doped ceria oxide thin films formed by electron beam deposition. *J. Alloys Compd.* **2008**, *459*, 320–327. [CrossRef]
32. Galdikas, A.; Čerapaite-Trušinskiene, R.; Laukaitis, G.; Dudonis, J. Real-time kinetic modeling of YSZ thin film roughness deposited by e-beam evaporation technique. *Appl. Surf. Sci.* **2008**, *62*, 941–946. [CrossRef]
33. Sakaliuniene, J.; Čyviene, J.; Abakevičiene, B.; Dudonis, J. Investigation of structural and optical properties of GDC thin films deposited by reactive magnetron sputtering. *Acta Phys. Pol. A* **2011**, *120*, 63–65. [CrossRef]
34. Hong, Y.S.; Kim, S.H.; Kim, W.J.; Yoon, H.H. Fabrication and characterization GDC electrolyte thin films by e-beam technique for IT-SOFC. *Curr. Appl. Phys.* **2011**, *11*, S163–S168. [CrossRef]

35. Hartmanová, M.; Jergel, M.; Thurzo, I.; Kundracik, F.; Gmucová, K.; Chromik, S.; Ortega, L. Thin Film Electrolytes: Yttria Stabilized Zirconia and Ceria. *Russ. J. Electrochem.* **2003**, *39*, 478–486. [CrossRef]
36. Laukaitis, G.; Virbukas, D. The structural and electrical properties of GDC10 thin films formed by e-beam technique. *Solid State Ion.* **2013**, *247–248*, 41–47. [CrossRef]
37. Chandran, P.R.; Arjunan, T.V. A review of materials used for solid oxide fuel cell. *Int. J. ChemTech Res.* **2015**, *7*, 488.
38. Ji, S.; Chang, I.; Lee, Y.H.; Park, J.; Paek, J.Y.; Lee, M.H.; Cha, S.W. Fabrication of low-temperature solid oxide fuel cells with a nanothin protective layer by atomic layer deposition. *Nanoscale Res. Lett.* **2013**, *8*, 48. [CrossRef]
39. Saporiti, F.; Juarez, R.E.; Audebert, F.; Boudard, M. Yttria and ceria doped zirconia thin films grown by pulsed laser deposition. *Mater. Res.* **2013**, *16*, 655–660. [CrossRef]
40. Uhlenbruck, S.; Jordan, N.; Sebold, D.; Buchkremer, H.P.; Haanappel, V.A.C.; Stöver, D. Thin film coating technologies of (Ce,Gd)O2-δ interlayers for application in ceramic high-temperature fuel cells. *Thin Solid Film.* **2007**, *515*, 4053–4060. [CrossRef]
41. Wibowo, R.A.; Kim, W.S.; Lee, E.S.; Munir, B.; Kim, K.H. Single step preparation of quaternary Cu2 ZnSnSe4 thin films by RF magnetron sputtering from binary chalcogenide targets. *J. Phys. Chem. Solids* **2007**, *68*, 1908–1913. [CrossRef]
42. Li, J.G.; Wang, Y.; Ikegami, T.; Mori, T.; Ishigaki, T. Reactive 10 mol% RE2O3 (RE = Gd and Sm) doped CeO2 nanopowders: Synthesis, characterization, and low-temperature sintering into dense ceramics. *Mater. Sci. Eng. B Solid-State Mater. Adv. Technol.* **2005**, *121*, 54–59. [CrossRef]
43. Zha, S.; Moore, A.; Abernathy, H.; Liu, M. GDC-Based Low-Temperature SOFCs Powered by Hydrocarbon Fuels. *J. Electrochem. Soc.* **2004**, *151*, A1128–A1133. [CrossRef]
44. Aygün, B.; Özdemir, H.; Öksüzömer, M.A.F. Structural, morphological and conductivity properties of samaria doped ceria (SmxCe1-xO2-x/2) electrolytes synthesized by electrospinning method. *Mater. Chem. Phys.* **2019**, *232*, 82–87. [CrossRef]
45. Zhan, Z.; Wen, T.-L.; Tu, H.; Lu, Z.-Y. AC Impedance Investigation of Samarium-Doped Ceria. *J. Electrochem. Soc.* **2001**, *148*, A427–A432. [CrossRef]
46. Toor, S.Y.; Croiset, E. Reducing sintering temperature while maintaining high conductivity for SOFC electrolyte: Copper as sintering aid for Samarium Doped Ceria. *Ceram. Int.* **2020**, *46*, 1148–1157. [CrossRef]
47. Huang, Q.A.; Liu, M.; Liu, M. Impedance Spectroscopy Study of an SDC-based SOFC with High Open Circuit Voltage. *Electrochim. Acta* **2015**, *177*, 227–236. [CrossRef]
48. Koettgen, J.; Martin, M. The ionic conductivity of Sm-doped ceria. *J. Am. Ceram. Soc.* **2020**, *103*, 3776–3787. [CrossRef]
49. Kalyk, F.; Stankevičiūtė, A.; Budrytė, G.; Gaidamavičienė, G.; Žalga, A.; Kriūkienė, R.; Kavaliauskas, Ž.; Leszczyńska, M.; Abakevičienė, B. Comparative study of samarium-doped ceria nanopowders synthesized by various chemical synthesis routes. *Ceram. Int.* **2020**, *46*, 24385–24394. [CrossRef]
50. Wattanathana, W.; Veranitisagul, C.; Wannapaiboon, S.; Klysubun, W.; Koonsaeng, N.; Laobuthee, A. Samarium doped ceria (SDC) synthesized by a metal triethanolamine complex decomposition method: Characterization and an ionic conductivity study. *Ceram. Int.* **2017**, *43*, 9823–9830. [CrossRef]
51. Anantharaman, S.B.; Bauri, R. Effect of sintering atmosphere on densification, redox chemistry and conduction behavior of nanocrystalline Gd-doped CeO2 electrolytes. *Ceram. Int.* **2013**, *39*, 9421–9428. [CrossRef]
52. Mokkelbost, T.; Kaus, I.; Grande, T.; Einarsrud, M.A. Combustion synthesis and characterization of nanocrystalline CeO2-based powders. *Chem. Mater.* **2004**, *16*, 5489–5494. [CrossRef]
53. Fuentes, R.O.; Baker, R.T. Structural, morphological and electrical properties of Gd0.1Ce0.9O1.95 prepared by a citrate complexation method. *J. Power Sources* **2009**, *186*, 268–277. [CrossRef]
54. Murutoglu, M.; Ucun, T.; Ulasan, O.; Buyukaksoy, A.; Tur, Y.K.; Yilmaz, H. Cold sintering-assisted densification of GDC electrolytes for SOFC applications. *Int. J. Hydrog. Energy* **2022**, *47*, 19772–19779. [CrossRef]
55. Zhang, J.; Lenser, C.; Menzler, N.H.; Guillon, O. Comparison of solid oxide fuel cell (SOFC) electrolyte materials for operation at 500 °C. *Solid State Ion.* **2020**, *344*, 115138. [CrossRef]
56. Jaiswal, N.; Tanwar, K.; Suman, R.; Kumar, D.; Upadhyay, S.; Parkash, O. A brief review on ceria based solid electrolytes for solid oxide fuel cells. *J. Alloys Compd.* **2019**, *781*, 984–1005. [CrossRef]
57. Mackay, D.; van Wesenbeeck, I. Correlation of chemical evaporation rate with vapor pressure. *Environ. Sci. Technol.* **2014**, *48*, 10259–10263. [CrossRef]
58. Hajakbari, F.; Ensandoust, M. Study of thermal annealing effect on the properties of silver thin films prepared by DC magnetron sputtering. *Acta Phys. Pol. A* **2016**, *129*, 680–682. [CrossRef]
59. Otieno, F.; Airo, M.; Erasmus, R.M.; Quandt, A.; Billing, D.G.; Wamwangi, D. Annealing effect on the structural and optical behavior of ZnO:Eu3+ thin film grown using RF magnetron sputtering technique and application to dye sensitized solar cells. *Sci. Rep.* **2020**, *10*, 8557. [CrossRef]
60. Medisetti, S.; Ahn, J.; Patil, S.; Goel, A.; Bangaru, Y.; Sabhahit, G.V.; Babu, G.U.B.; Lee, J.H.; Dasari, H.P. Synthesis of GDC electrolyte material for IT-SOFCs using glucose & fructose and its characterization. *Nano-Struct. Nano-Objects* **2017**, *11*, 7–12. [CrossRef]
61. Khalipova, O.S.; Lair, V.; Ringuedé, A. Electrochemical synthesis and characterization of Gadolinia-Doped Ceria thin films. *Electrochim. Acta* **2014**, *116*, 183–187. [CrossRef]

62. Matta, J.; Courcot, D.; Abi-Aad, E.; Aboukaïs, A. Identification of vanadium oxide species and trapped single electrons in interaction with the CeVO4 phase in vanadium-cerium oxide systems. 51V MAS NMR, EPR, Raman, and thermal analysis studies. *Chem. Mater.* **2002**, *14*, 4118–4125. [CrossRef]
63. Jadhav, L.D.; Patil, S.P.; Jamale, A.P.; Chavan, A.U. Solution combustion synthesis: Role of oxidant to fuel ratio on powder properties. *Mater. Sci. Forum* **2013**, *757*, 85–98. [CrossRef]
64. Zarkov, A.; Stanulis, A.; Mikoliunaite, L.; Katelnikovas, A.; Jasulaitiene, V.; Ramanauskas, R.; Tautkus, S.; Kareiva, A. Chemical solution deposition of pure and Gd-doped ceria thin films: Structural, morphological and optical properties. *Ceram. Int.* **2017**, *43*, 4280–4287. [CrossRef]
65. Saitzek, S.; Blach, J.F.; Villain, S.; Gavarri, J.R. Nanostructured ceria: A comparative study from X-ray diffraction, Raman spectroscopy and BET specific surface measurements. *Phys. Status Solidi Appl. Mater. Sci.* **2008**, *205*, 1534–1539. [CrossRef]
66. Taniguchi, T.; Watanabe, T.; Sugiyama, N.; Subramani, A.K.; Wagata, H.; Matsushita, N.; Yoshimura, M. Identifying defects in ceria-based nanocrystals by UV resonance Raman spectroscopy. *J. Phys. Chem. C* **2009**, *113*, 19789–19793. [CrossRef]
67. Chin, H.S.; Chao, L.S. The effect of thermal annealing processes on structural and photoluminescence of zinc oxide thin film. *J. Nanomater.* **2013**, *2013*, 424953. [CrossRef]
68. El-Habib, A.; Addou, M.; Aouni, A.; Diani, M.; Zimou, J.; Bouachri, M.; Brioual, B.; Allah, R.F.; Rossi, Z.; Jbilou, M. Oxygen vacancies and defects tailored microstructural, optical and electrochemical properties of Gd doped CeO2 nanocrystalline thin films. *Mater. Sci. Semicond. Process.* **2022**, *145*, 106631. [CrossRef]

Article

Improving Transport Properties of GaN-Based HEMT on Si (111) by Controlling SiH₄ Flow Rate of the SiN$_x$ Nano-Mask

Jin-Ji Dai [1,2], Cheng-Wei Liu [1], Ssu-Kuan Wu [1], Sa-Hoang Huynh [1], Jhen-Gang Jiang [1], Sui-An Yen [1], Thi Thu Mai [1], Hua-Chiang Wen [1], Wu-Ching Chou [1,*], Chih-Wei Hu [2] and Rong Xuan [2]

[1] Department of Electrophysics, National Chiao Tung University, Hsinchu 30010, Taiwan; jinjidai@gmail.com (J.-J.D.); william798424@gmail.com (C.-W.L.); wusykuann@gmail.com (S.-K.W.); hoangsa1429@gmail.com (S.-H.H.); agon810025@gmail.com (J.-G.J.); blue361993@gmail.com (S.-A.Y.); maithucs@gmail.com (T.T.M.); a091316104@gmail.com (H.-C.W.)

[2] Technology Development Division, Episil-Precision Inc., Hsinchu 30010, Taiwan; lf8825@gmail.com (C.-W.H.); protonesprotones@gmail.com (R.X.)

* Correspondence: wuchingchou@mail.nctu.edu.tw; Tel.: +886-3-5712121 (ext. 56129)

Citation: Dai, J.-J.; Liu, C.-W.; Wu, S.-K.; Huynh, S.-H.; Jiang, J.-G.; Yen, S.-A.; Mai, T.T.; Wen, H.-C.; Chou, W.-C.; Hu, C.-W.; et al. Improving Transport Properties of GaN-Based HEMT on Si (111) by Controlling SiH₄ Flow Rate of the SiN$_x$ Nano-Mask. *Coatings* 2021, *11*, 16. https://dx.doi.org/10.3390/coatings11010016

Received: 30 November 2020
Accepted: 22 December 2020
Published: 25 December 2020

Publisher's Note: MDPI stays neutral with regard to jurisdictional claims in published maps and institutional affiliations.

Copyright: © 2020 by the authors. Licensee MDPI, Basel, Switzerland. This article is an open access article distributed under the terms and conditions of the Creative Commons Attribution (CC BY) license (https://creativecommons.org/licenses/by/4.0/).

Abstract: The AlGaN/AlN/GaN high electron mobility transistor structures were grown on a Si (111) substrate by metalorganic chemical vapor deposition in combination with the insertion of a SiN$_x$ nano-mask into the low-temperature GaN buffer layer. Herein, the impact of SiH₄ flow rate on two-dimensional electron gas (2DEG) properties was comprehensively investigated, where an increase in SiH₄ flow rate resulted in a decrease in edge-type threading dislocation density during coalescence process and an improvement of 2DEG electronic properties. The study also reveals that controlling the SiH₄ flow rate of the SiN$_x$ nano-mask grown at low temperatures in a short time is an effective strategy to overcome the surface desorption issue that causes surface roughness degradation. The highest electron mobility of 1970 cm^2/V·s and sheet carrier concentration of 6.42×10^{12} cm^{-2} can be achieved via an optimized SiH₄ flow rate of 50 sccm.

Keywords: GaN HEMT; SiN$_x$ nano-mask; edge threading dislocation; V-defects; 2DEG

1. Introduction

GaN-based high electron mobility transistors (HEMTs) have attracted much intense research because of its high-voltage operation, high-frequency switching, and high thermal conductivity for high-power and high-frequency device applications [1–3]. However, the growth of GaN-based HEMTs faces the problem of non-commercial nitride substrates, where sapphire (Al$_2$O$_3$) and silicon carbide (SiC) are two commonly used substrates. In recent years, to realize a compromise between high performance and wafer-size scalability, low cost, and integrated compatibility with the conventional Si-based technology, many research groups have grown GaN-based HEMT on a Si (111) substrate. Even so, the large lattice mismatch of 17% along with a thermal expansion coefficient mismatch of 54% between the Si (111) substrate and GaN layer is the most challenging of any epitaxial growth methods to achieve high-quality GaN layers. High-density dislocation (~10^{10} cm^{-2}) and surface cracks due to thermal expansion during the cooling process are generally observed in the GaN/Si (111) layers, contributing to the degradation of the GaN-based HEMT performance. In general, the mainstream of research has focused on reducing the current collapse, which is attributed to both the interface trapping of AlGaN barrier/passivation layers and the bulk trapping of defects/dislocations [4–10]. To mitigate the bulk crystalline defects/dislocations, several buffer structural designs have been employed such as the graded/stepped AlGaN layer configurations [11–13], AlN/GaN or AlGaN/GaN superlattices (SL) [14–16], epitaxial lateral overgrowth (ELOG) technique [17], and three-dimensional (3D)-to-two-dimensional (2D) growth mode [18,19]. Especially, inserting a SiN$_x$ nano-mask into the matrix of GaN layer to promote the growth mode transition in

GaN from 3D island-like overgrowth to 2D coalescence has been becoming a widely used technique [19–33], in which the threading dislocations (TDs) are efficiently bent or even annihilate each other. There are several approaches to investigate the effect of SiN_x nano-masks on the TD reduction such as varying either the SiN_x deposition temperature or time to modulate the SiN_x nano-mask configuration [19–22], using different growth conditions of the (Al)GaN layer overgrown on the SiN_x nano-mask [23–25], or inserting multi-SiN_x nano-mask into the (Al)GaN buffer layers [26–28]. Among these approaches, configuring a single SiN_x nano-mask is considered the simplest strategy. By adjusting the growth time and temperature, the grain size and distribution of the SiN_x nano-mask are modulated in such a way that the 3D-to-2D growth mode transformation is promoted [20,32]. However, any excess of the growth temperature and growth time will enhance the surface desorption of the GaN underlayer, leading to the aggregation of Ga atoms [33]. As a result, the threading dislocation density (TDD), etch pit density (EPD), and surface roughness of the GaN layers would increase, causing detrimental effects on the HEMT device performance. On the other hand, if the SiN_x nano-mask is grown at too low temperature or short duration, its grain size would be tiny and ineffective to generate the transition of 3D-to-2D growth mode as well as to suppress the TDD. It is therefore essential to study an alternative method to solve the surface desorption issue, while retaining the efficiency of the inserted-SiN_x nano-mask, in terms of suppressing TDs and improving two-dimensional electron gas (2DEG) properties of the GaN HEMT structure.

For this conversation, we present inserting a SiN_x nano-mask into the low-temperature GaN (LT-GaN) layer grown by metalorganic chemical vapor deposition (MOCVD), where the nano-mask formation was controlled via the SiH_4 flow rate variation instead of either the growth temperature or the growth time. Thus, this novel method allows easily controlling the SiN_x configuration, enhancing the annihilation of TDs. Importantly, its low temperature and short-time growth were expected to highly limit the surface desorption of GaN underlayer. The results in this study show that the crystalline quality and the transport properties of GaN-based HEMT structure were strongly governed by varying the SiH_4 flow rate.

2. Experimental

The full structures of GaN-based HEMT were grown on 6" Si (111) substrates by the G4 MOCVD system (Aixtron, Herzogenrath, Germany). The conventional precursors including TMGa, TMAl, NH_3, and silane (1% diluted-SiH_4) were used for the Ga, Al, N, and Si source, respectively. In order to grow the lattice-matched GaN layer on the Si (111) substrate, an AlN nucleation layer of about 300 nm was initially grown to prevent Ga–Si melt-back etching, followed by a transition layer of ~1.8 μm. This is to effectively modulate stress and prevent misalignment. A semi-insulating LT-GaN buffer layer with a thickness of 1.8 μm was then deposited at 990 °C, followed by a 2DEG heterostructure. This 2DEG structure was composed of a 300 nm HT-GaN (1020 °C) channel layer, 0.5 nm AlN spacer layer, and 10 nm $Al_{0.25}Ga_{0.75}N$ barrier layer as shown in Figure 1. Finally, the structure was capped by a 4 nm thin GaN layer to protect the wafer surface from oxidation. To study the effect of the SiN_x nano-mask on reducing the stress and TDD, the SiN_x nano-mask was inserted at 300 nm above the LT-GaN/transition layer interface as can be seen in Figure 1. The growth time of the nano-mask was 1 min, while the growth temperature and the ammonia gas flow were kept at the same with that of the LT-GaN layer. Herein, a highly concentrated 1%-SiH_4 source was employed to conduct a high deposition rate and low surface desorption. Five samples with different SiH_4 source flow rates of 0, 25, 50, 75, and 100 sccm were used and denoted as samples A, B, C, D, and E, respectively. Sample A is a control sample, where no SiN_x nano-mask was inserted. The in situ reflectivity and curvature measurement were monitored simultaneously by the LayTec EpiCurve-TT (LayTec, Berlin, Germany) unit integrated inside the MOCVD chamber. The 633 nm light source and 650 nm laser beam were used for reflectivity and curvature characterization, respectively. The experimental etching pit density (EPD) was performed in molten-KOH

(85%-KOH:H$_2$O = 1:5) in 6 min at 220 °C. The pit density was then calculated from atomic force microscopes (AFM) (NT-MDT Spectrum Instruments, Moscow, Russia) imaging on scan areas of (5 × 5) μm^2. The surface morphology of the samples before and after the etching process was carefully observed by JSM7001F (JEOL, Tokyo, Japan) scanning electron microscope (SEM) and AFM. The TDD was evaluated from the full width at half maximum (FWHM) XRD-PANalytical X'Pert PRO MRD (Malvern Panalytical, Almelo, The Netherlands), scanned on (002) and (102) planes, the etch pit density (EPD) technique, and G2 F-20 STEM (FEI Tecnai^{TM}, Hillsboro, OR, USA). The wafer bowing measurement was also carried out after the growth by ADE 9700 equipment. To investigate the 2DEG properties, the standard Hall effect measurement (BioRad HL5500, Hercules, CA, USA) was conducted at room temperature (RT) under a magnetic field of 0.57 T.

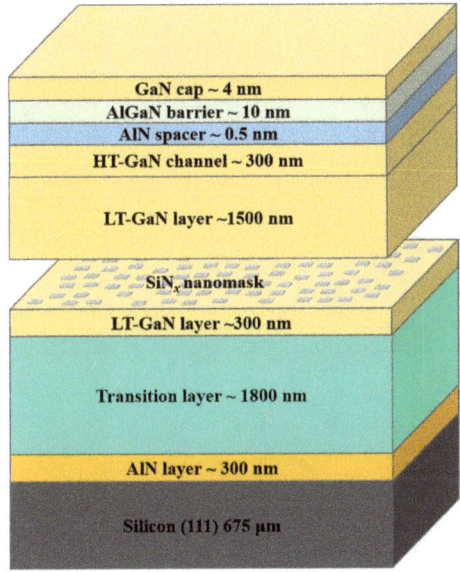

Figure 1. Schematic structure of GaN-based HEMT with inserted SiN$_x$ nano-masks grown at different SiH$_4$ flow rates.

3. Results and Discussion

The in situ monitoring of the reflectivity and surface curvature of all five samples is shown in Figure 2. The growth rate and surface morphology of the samples during the epitaxial growth could be evaluated from the reflectivity measurement (Figure 2a). Before the SiN$_x$ nano-mask growth, the growth rate of ~4.2 μm/h of the 300 nm thick LT-GaN initial layer of all samples was indicated from the dependence of reflectance oscillation on the growth time. During the SiN$_x$ nano-mask insertion, the reflectance intensity of the samples was extinguished and flattened because of depositing 3D SiN$_x$ nanocrystals. Then, the maximum intensity of the oscillation peaks was gradually restored during the 1.5 μm LT-GaN overlayer growth, where the required time to recover the reflectance peak intensity to maximum, as before the SiN$_x$ deposition, was different between the inserted SiN$_x$ samples. Obviously, the recovering time, marked as the length of colored arrows in Figure 2a, increased with increasing the SiH$_4$ flow rate. In other words, the time required for the growth mode transition from 3D-island to 2D-coalescence of the LT-GaN overlayer strongly depends on the SiH$_4$ flux, in which transition time increases with the amount of SiH$_4$ flux used. It is also emphasized that this transition could not complete if the SiH$_4$ flow rate exceeds a threshold of 75 sccm, as in the case of sample E. Indeed, the reflectivity at the end of the LT-GaN overlayer of this sample only equals ~40% as compared to that at end of the LT-GaN underlayer. However, the following high-temperature GaN (HT-GaN) channel layer growth of this sample exhibited a fast restoration in the reflectivity only after three oscillation cycles (see blue-dash arrow in Figure 2a, sample E). This result suggests that the SiH_4 flux modulation during inserting SiN_x interlayer is a considerable approach in terms of recovering the 2D growth mode of the HT-GaN layer even if a SiH_4 flux as high as 100 sccm is used.

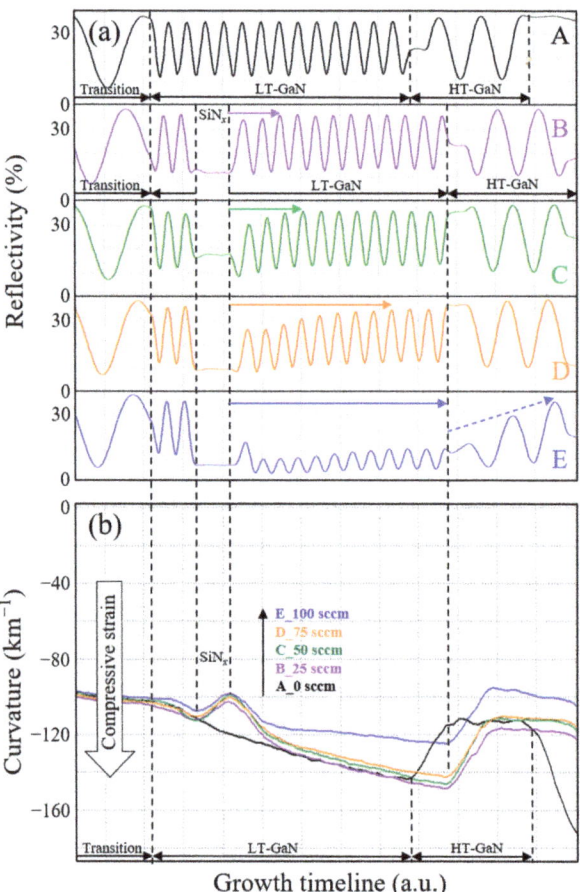

Figure 2. In situ (**a**) reflectivity and (**b**) curvature monitoring of sample surfaces.

To study the surface morphology with varying SiH_4 flow rate, scanning electron microscopy (SEM) and atomic force microscopy (AFM) were employed as shown in Figure 3a–j. No V-defects were observed on the surfaces of samples B and C, while a few defects were visible on both sample D and sample E. The SEM image (Figure 3d) of sample D exposes several small pits. These pits are well known as stacking mismatch boundary (SMB) defects, attributed to the atom alignment faults during the coalescence process. Meanwhile, the surface of sample E displays the additional appearance of hexagonal micro-pit (HMP) defects, which are caused by the incomplete coalescence process as mentioned in the reflectivity discussion (Figure 2a) [34]. The presence of SMB and HMP defects on these samples is expected to decline the transport properties of 2DEG, which is discussed later. On the other hand, the surface morphologies of the inserted SiN_x samples were preserved as the control sample with a root mean square (RMS) roughness of around 0.22 nm as shown in

Figure 3f–j. This result indicates clearly that the surface desorption of LT-GaN underlayer in our study was restrained efficiently during SiN$_x$ interlayer growth.

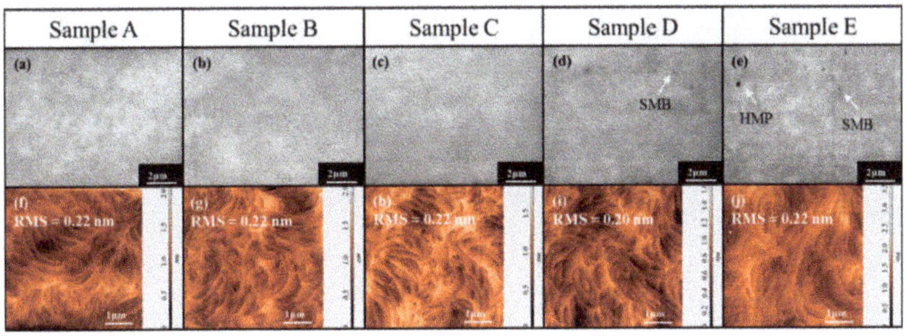

Figure 3. (**a–e**) SEM and (**f–j**) AFM images of the inserted SiN$_x$ nano-mask samples with various SiH$_4$ flow rates.

Figure 2b shows that all five samples were suffering compressive strains, where the strain in the transition layer and 300 nm LT-GaN underlayer increased slowly with the growth time. After inserting the SiN$_x$ nano-mask, the compressive strain in the LT-GaN overlayer of the SiN$_x$ samples (samples B–D) increased in the same fashion, quickly approaching the strain magnitude of sample A. This is in agreement with the reflectivity results, where a transition from 3D island-like to 2D growth mode in the following LT-GaN layers was taking place. Furthermore, note that the strain increasing rate of sample E was much lower than that of other samples as can be seen in Figure 2b. Thus, the compressive strain of sample E (highest SiH$_4$ flow rate of 100 sccm) was relieved more effectively than in other samples. That is because the transformation from 3D to 2D growth mode of the LT-GaN overlayer is not completed in this sample, as the reflectivity results indicated (Figure 2a). Indeed, it is directly related to the morphology as revealed by SEM (Figure 3e), where the long 3D growth mode duration of the LT-GaN overlayer and the incomplete coalescence of the HT-GaN channel layer resulted in the formation of HMP defects. Moreover, measurements of curvature showed that strain of layers is strongly affected by SiH$_4$ flow rate that is also in agreement with the ex situ wafer bow measurements, in which the positive warp of sample A, B, C, D, and E were 22.0, 15.7, 15.7, 14.8, and 9.6 μm, respectively. Obviously, the warp of 9.6 μm presented in sample E is much smaller than that of other samples.

In order to investigate the relationship between the TDD and the SiH$_4$ flow rate, the EPD and X-ray diffraction (XRD) techniques were employed, and the results are summarized in Table 1. The omega XRD scan of the sample near (002) and (102) plane could be found in Figure S1. The EPD slightly decreased as the SiH$_4$ flow rate was increased from 25 to 75 sccm. However, it rapidly reduced to 3.24×10^8 cm^{-2} when the SiH$_4$ flow rate increased further to 100 sccm (sample E). Of course, the EPD directly relates to the TDD in the epilayers. Herein, the TD in GaN growth could be categorized into two types as screw-type TD and edge-type TD. The TDD of each type presented in the GaN layer can be identified from the FWHM of (002) and (102) X-ray rocking curves (i.e., β$_{(002)}$ and β$_{(102)}$) via the following formulas [11].

$$D_{screw} = \frac{\beta^2_{(002)}}{4.35 \times b^2_{screw}} \quad (1)$$

$$D_{edge} = \frac{\beta^2_{(102)} - \beta^2_{(002)}}{4.35 \times b^2_{edge}} \quad (2)$$

Table 1. Dependence of dislocation density on the SiH$_4$ flow rate by XRD and EPD analysis.

Sample ID	SiH$_4$ Source (sccm)	FWHM (arcsec)		TDD (cm^{-2})		EPD (cm^{-2})	Wafer Bowing (μm)
		(002)	(102)	Screw-Type	Edge-Type		
A	0	466 ± 5	945 ± 5	$(4.36 \pm 0.09) \times 10^8$	$(3.59 \pm 0.02) \times 10^9$	6.52×10^8	22.0
B	25	468 ± 5	920 ± 5	$(4.40 \pm 0.09) \times 10^8$	$(3.33 \pm 0.02) \times 10^9$	5.76×10^8	15.7
C	50	466 ± 5	908 ± 5	$(4.36 \pm 0.09) \times 10^8$	$(3.23 \pm 0.02) \times 10^9$	5.52×10^8	15.7
D	75	467 ± 5	872 ± 5	$(4.38 \pm 0.09) \times 10^8$	$(2.88 \pm 0.02) \times 10^9$	5.34×10^8	14.8
E	100	441 ± 5	786 ± 5	$(3.91 \pm 0.09) \times 10^8$	$(2.25 \pm 0.02) \times 10^9$	3.24×10^8	9.6

The screw-type TDD (D_{screw}) and edge-type TDD (D_{edge}) are computed from $\beta_{(002)}$, $\beta_{(102)}$, and burger vector length (b_{screw} = 0.5185 nm and b_{edge} = 0.3189 nm). The extracted TDDs of all samples are displayed in Table 1. Interestingly, the edge-type TDD decreased considerably with the increasing SiH$_4$ flow rate, whereas the screw-type TDD showed a slight drop as SiH$_4$ flow is more than 75 sccm. Thus, the increasing SiH$_4$ flow rate during SiN$_x$ nano-mask forming has a dominant effect on the decrease in edge-type TDD rather than screw-type TDD. We notice that this result plays an important role in improving the 2DEG properties of the AlGaN/GaN HEMT structure.

To further understand the mechanisms of the decreasing edge-type TDD, the cross-sectional bright-field scanning transmission electron microscopy (STEM) micrograph of sample E was taken as can be seen in Figure 4. The edge dislocations, screw dislocations, and mixed dislocations (edge and screw dislocations) are labeled as E, S, and M, respectively. By the comparison between two STEM images observed along $g = [0002]$ and $g = [\bar{1}\bar{1}20]$ zone axis, the edge dislocation density is distinctly reduced. Apparently, the propagation of most of TDs was terminated or bent at the SiN$_x$ interlayer and further annihilated during the 2D lateral overgrowth on 3D island-like LT-GaN overlayer, which is marked as white-dash lines in Figure 4. Three kinds of interaction between edge-type TDs are visible: (1) Type I is parallel, two TDs passed through the SiN$_x$ nano-mask with the same Burgers vector. (2) Type II is fusion, two TDs combined to form a new TD where its Burgers vector equals to the sum of two componential Burgers vectors. (3) Type III is annihilation, two TDs with the opposite Burgers vectors react against each other [33,35]. We notice that only type II and III interactions benefit the annihilation of TDs.

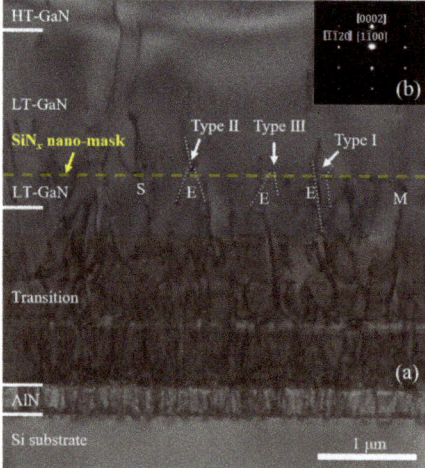

Figure 4. (a) Bright-field cross-sectional STEM micrograph of GaN-based HEMT structure on Si (111) with the SiN$_x$ nano-mask grown under 100 sccm SiH$_4$ flow rate and (b) Selective area electron diffraction pattern of the HT-GaN top layer taken along the $[1\bar{1}00]$ zone axis.

The effect of the SiH$_4$ flow rate on the transport properties of 2DEG HEMT structures was carried out by van der Pauw–Hall measurements at room temperature (RT) and the results are shown in Figure 5. Below a SiH$_4$ flow rate of 50 sccm, both the mobility and the sheet carrier concentration (N_S) of 2DEG structure enhanced simultaneously with increasing the SiH$_4$ flow rate. This behavior could be explained by the decrease in the edge-type TDs in the structure as observed from XRD results [32,36]. The edge-type TDs are usually accompanied by the dangling bonds. These dangling bonds generate an acceptor-like trap level in the band structure, capturing free charges and then forming negatively charged Coulombic scattering centers. Thus, any reduction in edge-type TDD is related to an improvement of the 2DEG mobility [37]. Meanwhile, the N_S was also influenced by the edge-type TDD. The increase of N_S with decreasing edge-type TDs is a result of the restrained charges at the acceptor-like traps [36,38]. As the SiH$_4$ flow rate was increased further to 75 sccm (sample D), both the mobility and N_S significantly degraded. It could be due to the formation of the nonuniform distribution of SMB defects as can be seen in Figure 3d. The appearance of SMB defects reflects the interface roughness and electrical field fluctuation at the 2DEG structure, leading to a decrease in mobility and N_S [39–41]. A further decrease in 2DEG mobility of sample E is understandable, ascribed to the formation of not only SMB but also HMP defects as can be observed in Figure 3e. Interestingly, the N_S of this sample increased as compared to sample D. This may be explained by the active diffusion of Si adatoms from the SiN$_x$ nanostructures. Under a SiH$_4$ flow rate as high as 100 sccm, the excess Si adatoms from the SiN$_x$ nano-mask could diffuse and precipitate on the sidewalls of the HMP defects via the TDs [26,33]. In addition, we notice that this diffusion process could be even boosted under a 3D growth mode of GaN channel layer as observed in our case (see sample E in Figure 2a). The Si adatoms preferentially sticking on the sidewalls of the HMP defects play a role as donors and releases electrons to the GaN channel layer. As a result, it causes an increase in the N_S of the 2DEG structure. Besides that, the Si diffusion and incorporation into TDs could be a pathway causing unexpected leakage currents. Similarly, this effect was also observed in the Mg-doped GaN film. The segregated Mg propagates through TDs and incorporates at the boundaries of pyramidal inversion domains (PIDs) to form an Mg-rich area, degrading device performance [42–44]. Up to now, the Si diffusion originated from the SiN$_x$ nano-mask, and how it involves in the 2DEG characteristics as well as GaN-based HEMT device performance have not been clearly understood yet, and needs to be explored further. Herein, we for the first time demonstrate that the V-defects existing in 2DEG structure can cause the reversion of its N_S as a SiH$_4$ overflow rate (>75 sccm) used during the SiN$_x$ nano-mask growth.

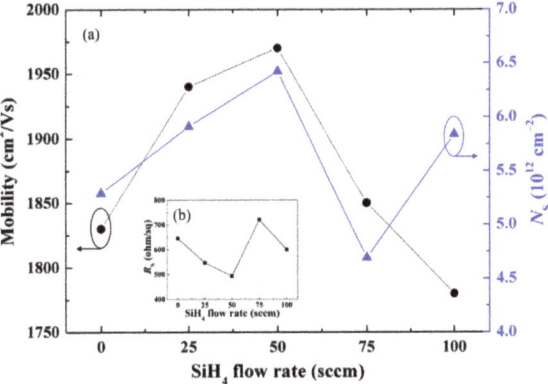

Figure 5. (a) Mobility and NS of the 2DEG structure as a function of the SiH$_4$ flow rate. The inset (b) shows the dependence of sheet resistance (R_S) on the SiH$_4$ flow rate.

4. Conclusions

In conclusion, the fabrication of AlGaN/AlN/GaN HEMT structures with high mobility and N_S has been demonstrated via modulating the SiH$_4$ flow rate of the SiN$_x$ nano-mask. The surface roughness of the samples was maintained at 0.22 nm, proving an effective elimination of surface desorption issue from the LT-GaN underlayer during inserting the SiN$_x$ nano-mask. More importantly, the SiN$_x$ nano-mask effectively contributed stress relaxation via promoting 3D-to-2D growth mode transformation of the LT-GaN overlayer and bending or annihilating TDs. Thus, it helped to reduce the edge-type TDD and EPD of the GaN-based HEMT structure as low as 2.25×10^9 and 3.24×10^8 cm^{-2}, respectively, as observed in the sample grown using 100 sccm SiH$_4$ flow rate. However, when the SiH$_4$ flow rate was larger than 50 sccm, an oversized SiN$_x$ nano-mask could be presented, resulting in the development of V-defects as SMB and HMP defects. These defects would be located at or near the 2DEG structure, causing degradation of the 2DEG mobility and sheet carrier concentration. Consequently, the highest mobility and N_S of the sample can be achieved to be ~1970 cm^2/V·s and 6.42×10^{12} cm^{-2}, respectively, under an optimized SiH$_4$ flow rate of 50 sccm. Interestingly, the N_S of sample E exhibited a reverse trend, which may be attributed to the accumulation of the diffused Si adatoms from the SiN$_x$ nano-mask on the sidewalls of HMP defects as the SiH$_4$ flux was used as high as 100 sccm.

Supplementary Materials: The following are available online at https://www.mdpi.com/2079-6412/11/1/16/s1, Figure S1: Normalized XRD rocking curves of GaN, (**a**) (002) and (**b**) (102) planes. (**c**) FWHM as a function of SiH$_4$ flow rate of (002) and (102) planes.

Author Contributions: J.-J.D. conceived and designed the experiments, performed the epitaxial growth, and wrote the manuscript. C.-W.H. and R.X. supported the epitaxial growth. C.-W.L., S.-K.W., J.-G.J., and S.-A.Y. supported structural, optical, and electrical characterizations. S.-H.H. performed revising the manuscript and analyzing data. T.T.M. and H.-C.W. provided suggestions and discussion. W.-C.C. is the advisor who supervised the experiments. All authors contributed to discuss the results and comment on the manuscript. All authors have read and agreed to the published version of the manuscript.

Funding: This research was funded by the National Science Council, Taiwan, R.O.C., grant No. MOST 108-2112-M-009-005.

Data Availability Statement: Data is contained within the article or supplementary material.

Conflicts of Interest: The authors declare no conflict of interest.

References

1. Amano, H.; Baines, Y.; Beam, E.; Borga, M.; Bouchet, T.; Chalker, P.R.; Charles, M.; Chen, K.J.; Chowdhury, N.; Chu, R.; et al. The 2018 GaN power electronics roadmap. *J. Phys. D Appl. Phys.* **2018**, *51*, 163001. [CrossRef]
2. Chen, K.J.; Haberlen, O.; Lidow, A.; Tsai, C.L.; Ueda, T.; Uemoto, Y.; Wu, Y. GaN-on-Si power technology: Devices and applications. *IEEE Trans. Electron Devices* **2017**, *64*, 779–795. [CrossRef]
3. Axelsson, O.; Gustafsson, S.; Hjelmgren, H.; Rorsman, N.; Blanck, H.; Splettstoesser, J.; Thorpe, J.; Roedle, T.; Thorsell, M. Application relevant evaluation of trapping effects in AlGaN/GaN HEMTs with Fe-doped buffer. *IEEE Trans. Electron Devices* **2016**, *63*, 326–332. [CrossRef]
4. Hu, A.; Yang, X.; Cheng, J.; Song, C.; Zhang, J.; Feng, Y.; Ji, P.; Xu, F.; Zhang, Y.; Yang, Z.; et al. Vertical leakage induced current degradation and relevant traps with large lattice relaxation in AlGaN/GaN heterostructures on Si. *Appl. Phys. Lett.* **2018**, *112*, 032104. [CrossRef]
5. Wespel, M.; Polyakov, V.M.; Dammann, M.; Reiner, R.; Waltereit, P.; Quay, R.; Mikulla, M.; Ambacher, O. Trapping effects at the drain edge in 600 V GaN-on-Si HEMTs. *IEEE Trans. Electron Devices* **2015**, *63*, 598–605. [CrossRef]
6. Yang, S.; Zhou, C.; Han, S.; Wei, J.; Sheng, K.; Chen, K.J. Impact of substrate bias polarity on buffer-related current collapse in AlGaN/GaN-on-Si power devices. *IEEE Trans. Electron Devices* **2017**, *64*, 5048–5056. [CrossRef]
7. Meneghini, M.; Rossetto, I.; Bisi, D.; Stocco, A.; Chini, A.; Pantellini, A.; Lanzieri, C.; Nanni, A.; Meneghesso, G.; Zanoni, E. Buffer traps in Fe-doped AlGaN/GaN HEMTs: Investigation of the physical properties based on pulsed and transient measurements. *IEEE Trans. Electron Devices* **2014**, *61*, 4070–4077. [CrossRef]
8. Liang, Y.; Jia, L.; He, Z.; Fan, Z.; Zhang, Y.; Yang, F. The study of the contribution of the surface and bulk traps to the dynamic Rdson in AlGaN/GaN HEMT by light illumination. *Appl. Phys. Lett.* **2016**, *109*, 182103. [CrossRef]

9. Liu, S.; Yang, S.; Tang, Z.; Jiang, Q.; Liu, C.; Wang, M.; Shen, B.; Chen, K.J. Interface/border trap characterization of Al_2O_3/AlN/GaN metal-oxide-semiconductor structures with an AlN interfacial layer. *Appl. Phys. Lett.* **2015**, *106*, 051605. [CrossRef]
10. Hua, M.; Lu, Y.; Liu, S.; Liu, C.; Fu, K.; Cai, Y.; Zhang, B.; Chen, K.J. Compatibility of AlN/SiN_x passivation with LPCVD-SiN_x gate dielectric in GaN-based MIS-HEMT. *IEEE Electron Device Lett.* **2016**, *37*, 265–268. [CrossRef]
11. Lee, H.-P.; Perozek, J.; Rosario, L.D.; Bayram, C. Investigation of AlGaN/GaN high electron mobility transistor structures on 200-mm silicon (111) substrates employing different buffer layer configurations. *Sci. Rep.* **2016**, *6*, 37588. [CrossRef] [PubMed]
12. Kim, M.-H.; Do, Y.-G.; Kang, H.C.; Noh, D.Y.; Park, S.-J. Effects of step-graded $Al_xGa_{1-x}N$ interlayer on properties of GaN grown on Si(111) using ultrahigh vacuum chemical vapor deposition. *Appl. Phys. Lett.* **2001**, *79*, 2713–2715. [CrossRef]
13. Cheng, J.; Yang, X.; Sang, L.; Guo, L.; Hu, A.; Xu, F.; Tang, N.; Wang, X.; Shen, B. High mobility AlGaN/GaN heterostructures grown on Si substrates using a large lattice-mismatch induced stress control technology. *Appl. Phys. Lett.* **2015**, *106*, 142106. [CrossRef]
14. Feltin, E.; Beaumont, B.; Laügt, M.; De Mierry, P.; Vennéguès, P.; Lahrèche, H.; Leroux, M.; Gibart, P. Stress control in GaN grown on silicon (111) by metalorganic vapor phase epitaxy. *Appl. Phys. Lett.* **2001**, *79*, 3230–3232. [CrossRef]
15. Selvaraj, S.L.; Suzue, T.; Egawa, T. Breakdown enhancement of AlGaN/GaN HEMTs on 4-in silicon by improving the GaN quality on thick buffer layers. *IEEE Electron Device Lett.* **2009**, *30*, 587–589. [CrossRef]
16. Sugawara, Y.; Ishikawa, Y.; Watanabe, A.; Miyoshi, M.; Egawa, T. Characterization of dislocations in GaN layer grown on 4-inch Si (111) with AlGaN/AlN strained layer superlattices. *Jpn. J. Appl. Phys.* **2016**, *55*, 05FB08. [CrossRef]
17. Tanaka, S.; Honda, Y.; Sawaki, N.; Hibino, M. Structural characterization of GaN laterally overgrown on a (111)Si substrate. *Appl. Phys. Lett.* **2001**, *79*, 955–957. [CrossRef]
18. Chang, S.; Wei, L.L.; Luong, T.T.; Chang, C.; Chang, L. Threading dislocation reduction in three-dimensionally grown GaN islands on Si (111) substrate with AlN/AlGaN buffer layers. *J. Appl. Phys.* **2017**, *122*, 105306. [CrossRef]
19. Lee, K.J.; Shin, E.H.; Lim, K.Y. Reduction of dislocations in GaN epilayers grown on Si(111) substrate using Si_xN_y inserting layer. *Appl. Phys. Lett.* **2004**, *85*, 1502. [CrossRef]
20. Hertkorn, J.; Lipski, F.; Brückner, P.; Wunderer, T.; Thapa, S.B.; Scholz, F.; Chuvilin, A.; Kaiser, U.; Beer, M.; Zweck, J. Process optimization for the effective reduction of threading dislocations in MOVPE grown GaN using in situ deposited masks. *J. Cryst. Growth* **2008**, *310*, 4867–4870. [CrossRef]
21. Dadgar, A.; Poschenrieder, M.; Reiher, A.; Bläsing, J.; Christen, J.; Krtschil, A.; Finger, T.; Hempel, T.; Diez, A.; Krost, A. Reduction of stress at the initial stages of GaN growth on Si(111). *Appl. Phys. Lett.* **2003**, *82*, 28–30. [CrossRef]
22. Cheng, K.; Leys, M.; DeGroote, S.; Germain, M.; Borghs, G. High quality GaN grown on silicon(111) using a Si_xN_y interlayer by metal-organic vapor phase epitaxy. *Appl. Phys. Lett.* **2008**, *92*, 192111. [CrossRef]
23. Scholz, F.; Forghani, K.; Klein, M.; Klein, O.; Kaiser, U.; Neuschl, B.; Tischer, I.; Feneberg, M.; Thonke, K.; Lazarev, S.; et al. Studies on defect reduction in AlGaN heterostructures by integrating an in-situ SiN interlayer. *Jpn. J. Appl. Phys.* **2013**, *52*, 08JJ07. [CrossRef]
24. Forghani, K.; Klein, M.; Lipski, F.; Schwaiger, S.; Hertkorn, J.; Leute, R.A.R.; Scholz, F.; Feneberg, M.; Neuschl, B.; Thonke, K.; et al. High quality AlGaN epilayers grown on sapphire using SiN_x interlayers. *J. Cryst. Growth* **2011**, *315*, 216–219. [CrossRef]
25. Neuschl, B.; Fujan, K.J.; Feneberg, M.; Tischer, I.; Thonke, K.; Forghani, K.; Klein, M.; Scholz, F. Cathodoluminescence and photoluminescence study on AlGaN layers grown with SiN_x interlayers. *Appl. Phys. Lett.* **2010**, *97*, 192108. [CrossRef]
26. Zang, K.Y.; Wang, Y.; Wang, L.S.; Chow, S.Y.; Chua, S.J. Defect reduction by periodic SiN_x interlayers in gallium nitride grown on Si (111). *J. Appl. Phys.* **2007**, *101*, 093502. [CrossRef]
27. Wang, T.-Y.; Ou, S.-L.; Horng, R.-H.; Wuu, D.-S. Improved GaN-on-Si epitaxial quality by incorporating various Si_xN_y interlayer structures. *J. Cryst. Growth* **2014**, *399*, 27–32. [CrossRef]
28. Xie, J.; Chevtchenko, S.A.; Özgür, Ü.; Morkoç, H. Defect reduction in GaN epilayers grown by metal-organic chemical vapor deposition with in situ SiN_x nanonetwork. *Appl. Phys. Lett.* **2007**, *90*, 262112. [CrossRef]
29. Klein, O.; Biskupek, J.; Forghani, K.; Scholz, F.; Kaiser, U. TEM investigations on growth interrupted samples for the correlation of the dislocation propagation and growth mode variations in AlGaN deposited on SiN_x interlayers. *J. Cryst. Growth* **2011**, *324*, 63–72. [CrossRef]
30. Kappers, M.J.; Datta, R.; Oliver, R.A.; Rayment, F.D.G.; Vickers, M.E.; Humphreys, C.J. Threading dislocation reduction in (0001) GaN thin films using SiN_x interlayers. *J. Cryst. Growth* **2007**, *300*, 70–74. [CrossRef]
31. Riemann, T.; Hempel, T.; Christen, J.; Veit, P.; Clos, R.; Dadgar, A.; Krost, A.; Haboeck, U.; Hoffmann, A. Optical and structural microanalysis of GaN grown on SiN submonolayers. *J. Appl. Phys.* **2006**, *99*, 123518. [CrossRef]
32. Lee, Y.S.; Chung, S.J.; Suh, E.-K. Effects of dislocations on the carrier transport and optical properties of GaN films grown with an in-situ SiN_x insertion layer. *Electron. Mater. Lett.* **2012**, *8*, 141–146. [CrossRef]
33. Wang, T.Y.; Ou, S.L.; Horng, R.-H.; Wuu, D.-S. Growth evolution of Si_xN_y on the GaN underlayer and its effects on GaN-on-Si (111) heteroepitaxial quality. *CrystEngComm* **2014**, *16*, 5724–5731. [CrossRef]
34. Szymański, T.; Wośko, M.; Wzorek, M.; Paszkiewicz, B.; Paszkiewicz, R. Origin of surface defects and influence of an in situ deposited SiN nanomask on the properties of strained AlGaN/GaN heterostructures grown on Si (111) using metal–organic vapour phase epitaxy. *CrystEngComm* **2016**, *18*, 8747–8755. [CrossRef]

35. Contreras, O.; Ponce, F.A.; Christen, J.; Dadgar, A.; Krost, A. Dislocation annihilation by silicon delta-doping in GaN epitaxy on Si. *Appl. Phys. Lett.* **2002**, *81*, 4712–4714. [CrossRef]
36. Weimann, N.G.; Eastman, L.F.; Doppalapudi, D.; Ng, H.M.; Moustakas, T.D. Scattering of electrons at threading dislocations in GaN. *J. Appl. Phys.* **1998**, *83*, 3656–3659. [CrossRef]
37. Chen, K.X.; Dai, Q.; Lee, W.; Kim, J.K.; Schubert, E.F.; Grandusky, J.; Mendrick, M.; Li, X.; Smart, J.A. Effect of dislocations on electrical and optical properties of n-type Al0.34Ga0.66N. *Appl. Phys. Lett.* **2008**, *93*, 192108. [CrossRef]
38. Krtschil, A.; Dadgar, A.; Krost, A. Decoration effects as origin of dislocation-related charges in gallium nitride layers investigated by scanning surface potential microscopy. *Appl. Phys. Lett.* **2003**, *82*, 2263–2265. [CrossRef]
39. Asgari, A.; Babanejad, S.; Faraone, L. Electron mobility, Hall scattering factor, and sheet conductivity in AlGaN/AlN/GaN heterostructures. *J. Appl. Phys.* **2011**, *110*, 113713. [CrossRef]
40. Xu, X.; Liu, X.; Han, X.; Yuan, H.; Wang, J.; Guo, Y.; Song, H.; Zheng, G.; Wei, H.; Yang, S.; et al. Dislocation scattering in $Al_xGa_{1-x}N$/GaN heterostructures. *Appl. Phys. Lett.* **2008**, *93*, 182111. [CrossRef]
41. Li, H.; Liu, G.; Wei, H.; Jiao, C.; Wang, J.; Zhang, H.; Jin, D.-D.; Feng, Y.; Yang, S.; Wang, L.; et al. Scattering due to Schottky barrier height spatial fluctuation on two dimensional electron gas in AlGaN/GaN high electron mobility transistors. *Appl. Phys. Lett.* **2013**, *103*, 232109. [CrossRef]
42. Iwata, K.; Narita, T.; Nagao, M.; Tomita, K.; Kataoka, K.; Kachi, T.; Ikarashi, N. Atomic resolution structural analysis of magnesium segregation at a pyramidal inversion domain in a GaN epitaxial layer. *Appl. Phys. Express* **2019**, *12*, 031004. [CrossRef]
43. Duguay, S.; Echeverri, A.; Castro, C.; Latry, O. Evidence of Mg segregation to threading dislocation in normally-off GaN-HEMT. *IEEE Trans. Nanotechnol.* **2019**, *18*, 995–998. [CrossRef]
44. Usami, S.; Mayama, N.; Toda, K.; Tanaka, A.; Deki, M.; Nitta, S.; Honda, Y.; Amano, H. Direct evidence of Mg diffusion through threading mixed dislocations in GaN p–n diodes and its effect on reverse leakage current. *Appl. Phys. Lett.* **2019**, *114*, 232105. [CrossRef]

Article

TiO$_2$ Nanowires on TiO$_2$ Nanotubes Arrays (TNWs/TNAs) Decorated with Au Nanoparticles and Au Nanorods for Efficient Photoelectrochemical Water Splitting and Photocatalytic Degradation of Methylene Blue

Ngo Ngoc Uyen [1,2], Le Thi Cam Tuyen [3], Le Trung Hieu [4], Thi Thu Tram Nguyen [5], Huynh Phuong Thao [6,7], Tho Chau Minh Vinh Do [6], Kien Trung Nguyen [8], Nguyen Thi Nhat Hang [9], Sheng-Rui Jian [10], Ly Anh Tu [1,*], Phuoc Huu Le [2,*] and Chih-Wei Luo [11,12,13,14]

^(citation block)

Citation: Uyen, N.N.; Tuyen, L.T.C.; Hieu, L.T.; Nguyen, T.T.T.; Thao, H.P.; Do, T.C.M.V.; Nguyen, K.T.; Hang, N.T.N.; Jian, S.-R.; Tu, L.A.; et al. TiO$_2$ Nanowires on TiO$_2$ Nanotubes Arrays (TNWs/TNAs) Decorated with Au Nanoparticles and Au Nanorods for Efficient Photoelectrochemical Water Splitting and Photocatalytic Degradation of Methylene Blue. *Coatings* **2022**, *12*, 1957. https://doi.org/10.3390/coatings12121957

Academic Editor: Alexandru Enesca

Received: 23 October 2022
Accepted: 8 December 2022
Published: 13 December 2022

Publisher's Note: MDPI stays neutral with regard to jurisdictional claims in published maps and institutional affiliations.

Copyright: © 2022 by the authors. Licensee MDPI, Basel, Switzerland. This article is an open access article distributed under the terms and conditions of the Creative Commons Attribution (CC BY) license (https:// creativecommons.org/licenses/by/ 4.0/).

1 Faculty of Applied Science, Ho Chi Minh City University of Technology-VNUHCM, 268 Ly Thuong Kiet Street, District 10, Ho Chi Minh City 70000, Vietnam
2 Department of Physics and Biophysics, Faculty of Basic Sciences, Can Tho University of Medicine and Pharmacy, 179 Nguyen Van Cu Street, Can Tho City 94000, Vietnam
3 Faculty of Chemical Engineering, Can Tho University, 3/2 Street, Ninh Kieu District, Can Tho City 94000, Vietnam
4 Department of Materials Science and Engineering, National Yang Ming Chiao Tung University, Hsinchu 30010, Taiwan
5 Department of Chemistry, Faculty of Basic Sciences, Can Tho University of Medicine and Pharmacy, 179 Nguyen Van Cu, Can Tho 94000, Vietnam
6 Department of Drug Quality Control-Analytical Chemistry-Toxicology, Faculty of Pharmacy, Can Tho University of Medicine and Pharmacy, 179 Nguyen Van Cu Street, Can Tho City 94000, Vietnam
7 Faculty of Pharmacy, Nam Can Tho University, 168 Nguyen Van Cu (Ext) Street, Can Tho City 94000, Vietnam
8 Department of Physiology, Faculty of Medicine, Can Tho University of Medicine and Pharmacy, 179 Nguyen Van Cu, Can Tho 94000, Vietnam
9 Institute of Applied Technology, Thu Dau Mot University, Thu Dau Mot City 820000, Vietnam
10 Department of Materials Science and Engineering, I-Shou University, Kaohsiung 84001, Taiwan
11 Department of Electrophysics, National Yang Ming Chiao Tung University, Hsinchu 30010, Taiwan
12 National Synchrotron Radiation Research Center, Hsinchu 30076, Taiwan
13 Institute of Physics and Center for Emergent Functional Matter Science, National Yang Ming Chiao Tung University, Hsinchu 30010, Taiwan
14 Department of Physics, University of Washington, Seattle, WA 98195, USA
* Correspondence: lyanhtu@hcmut.edu.vn (L.A.T.); lhuuphuoc@ctump.edu.vn (P.H.L.); Tel.: +84-29-2373-9730 (P.H.L.)

Abstract: In this study, TiO$_2$ nanowires on TiO$_2$ nanotubes arrays (TNWs/TNAs) and Au-decorated TNWs/TNAs nanostructures are designed and fabricated as a new type of photoanode for photoelectrochemical (PEC) water splitting. The TNWs/TNAs were fabricated on Ti foils by anodization using an aqueous NH$_4$F/ethylene glycol solution, while Au nanoparticles (NPs) and Au nanorods (NRs) were synthesized by Turkevich methods. We studied the crystal structure, morphology, and PEC activity of four types of nanomaterial photoanodes, including TNWs/TNAs, Au NPs- TNWs/TNAs, Au NRs-TNWs/TNAs, and Au NPs-NRs-TNWs/TNAs. The TiO$_2$ and Au-TiO$_2$ samples exhibited pure anatase phase of TiO$_2$ with (0 0 4), (1 0 1), and (1 0 5) preferred orientations, while Au-TiO$_2$ presented a tiny XRD peak of Au (111) due to a small Au decorated content of 0.7 ± 0.2 at.%. In addition, the samples obtained a well-defined and uniformed structure of TNAs/TNWs; Au NPs (size of 19.0 ± 1.9 nm) and Au NRs (width of 14.8 ± 1.3 nm and length of 99.8 ± 15.1 nm) were primarily deposited on TNWs top layer; sharp Au/TiO$_2$ interfaces were observed from HRTEM images. The photocurrent density (J) of the photoanode nanomaterials was in the range of 0.24–0.4 mA/cm^2. Specifically, Au NPs-NRs- decorated TNWs/TNAs attained the highest J value of 0.4 mA/cm^2 because the decoration of Au NPs and Au NRs mixture onto TNWs/TNAs improved the light harvesting capability and the light absorption in the visible-infrared region, enhanced photogenerated carriers' density, and increased electrons' injection efficiency via the localized surface plasmon resonance (LSPR) effect occurring at the Au nanostructures. Furthermore, amongst the investigated nanophotocatalysts, the Au NPs-NRs TNWs/TNAs exhibited the highest photocatalytic activity in the degradation of methylene blue with a high reaction rate constant of 0.7 ± 0.07 h^{-1}, which was 2.5 times higher than that of the pristine TNWs/TNAs.

Keywords: Au nanomaterials; anodic TiO$_2$; photoelectrochemical water splitting; localized surface plasmon effect; photocatalysts

1. Introduction

Since the discovery of photoelectrochemical (PEC) water splitting using a TiO$_2$ electrode in 1972 [1,2], this effect has widely become a promising route for hydrogen generation. In recent years, global environmental problems are becoming more and more concerned due to severe pollution, especially organic pollutants. During the dyeing process in the textile industry, a large amount (~15%) of total world dye is lost and released in the textile effluents [3]. When the colored wastewaters are discharged into the aquatic environment, it becomes a source of non-aesthetic pollution and causes eutrophication as well as perturbations to the aquatic life.

Semiconductor metal oxides offer a promising method for wastewater treatment and the environmentally friendly production of hydrogen [4,5]. Among different semiconductors, TiO$_2$ presents exceptional electrochemical and photocatalyst characteristics because of its intriguing electrical and optical properties [5,6]. TiO$_2$ also possesses excellent chemical- and photo-stability, cost-effectiveness, and nontoxicity, which is suitable for PEC and photocatalytic applications [5–7]. Furthermore, one-dimensional (1D) nanostructures such as TiO$_2$ nanotubes arrays (TNAs), TiO$_2$ nanowires on TiO$_2$ nanotubes arrays (TNWs/TNAs) and TiO$_2$ nanorods provide a direct conduction pathway for the photogenerated that can improve charge transport and reduce the recombination rate of electron–hole pairs [2,8–11]. However, TiO$_2$ had a wide band gap of ~3.2 eV for anatase phase, thus it only absorbs the ultraviolet (UV) light, which only accounts for 3%–5% of the total sunlight [2,11,12]. An approach for enhancing the visible-light (Vis) photoactivity of TiO$_2$ is the decoration of noble metal nanostructures with TiO$_2$ by utilizing the plasmonic effect [2,7,11,13–17]. The use of noble metal nanostructures as a decoration component for TiO$_2$ offers better photostability as compared with the use of semiconductor quantum dots with the anodic corrosion drawback [9,18,19]. For localized surface plasmon resonance (LSPR), the oscillation frequency is sensitively affected by the size, the shape of metal nanostructures and the dielectric constant of the surrounding environment [20–22]. The Au nanostructures with LSPR act as an antenna to localize the optical energy and sensitize TiO$_2$ by light with energy below the band gap. Consequently, they generate additional charge carriers for water oxidation or produce additional highly active free radicals such as hydroxyl ($^{\bullet}$OH) and superoxide ($^{\bullet}$O$_2{}^-$). It has been reported that several unique Au/TiO$_2$ composite systems achieved significant photoactivity enhancements for efficient solar water splitting owing to the LSPR effect [2,11,17,23] and for photocatalytic degradation pollutants [14,16,24]. For instance, the LSPR-induced electric field amplification near the TiO$_2$ surface allowed the enhancing of the 66-fold-PEC water splitting performance of Au NP-deposited TiO$_2$ films under Vis-light illumination [23].

The photoactivity of TiO$_2$ is enhanced by exploiting the LSPR absorption of Au NPs in the Vis region (typically ~550 nm) [25–28], while the infrared (IR) region of the sunlight spectrum was not utilized fully. Previous studies on Au nanorods (NRs) found that the LSPR absorption in the IR range of Au NRs can be tuned by controlling the aspect ratio and the medium dielectric constant [29–31]. Herein, we demonstrate that the photoactivity of TiO$_2$ with Au NPs and Au NRs enhances significantly via monitoring the PEC performance and degradation rate of methylene blue (MB) under UV-Vis irradiation. In this study, we are particularly interested in the TiO$_2$ nanowires on TiO$_2$ nanotube arrays (TNWs/TNAs) films fabricated by the anodic oxidation on immobilized titanium folds, since the material can provide a unidirectional electrical channel, large surface-to-volume ratio, and a higher photocatalytic performance than the well-known TNAs [10,14]. This study provides the detailed preparations and characterizations of various Au-TiO$_2$ nanostructured films, and the mechanism for the enhanced PEC and photocatalytic properties of the Au-TiO$_2$ systems.

2. Materials and Methods

2.1. Preparation of TiO$_2$ Nanowires on TiO$_2$ Nanotubes Arrays (TNWs/TNAs)

TNWs/TNAs were fabricated on titanium (Ti) foil substrates (10 mm × 25 mm × 0.4 mm, 99.9% purity,) by anodic oxidation. Before anodization, the substrate was first ultrasonically cleaned using acetone, methanol, and deionized (DI) water, and then dried by N$_2$ gas flow. The anodization was conducted using a two-electrode system with the Ti foil as an anode and a stainless-steel foil (SS304) as a cathode (Figure 1a). The electrolyte included ethylene glycol (97 vol%) with additions of 3 vol% DI water and 0.5 wt.% NH$_4$F (SHOWA, Tokyo, Japan). The anodizing voltage and time were 30 V and 5 h to grow TNWs/TNAs. To induce crystallization for TiO$_2$, the samples were annealed at 400 °C for 2 h in the air.

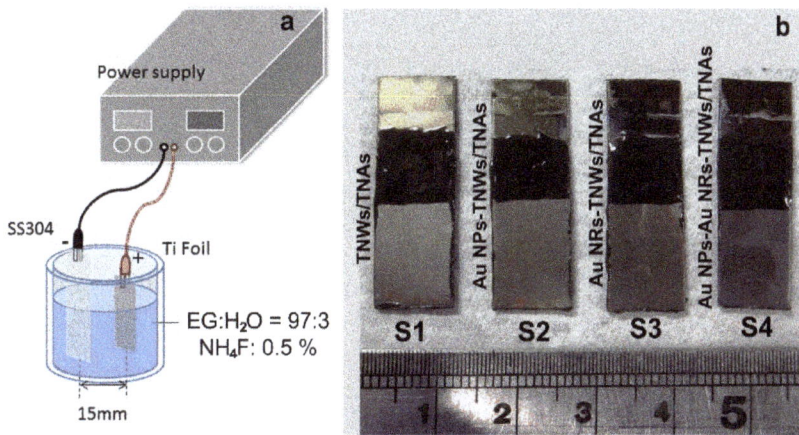

Figure 1. (**a**) A schematic of an electrochemical anodization process. (**b**) A picture of as-prepared TNWs/TNAs, Au NPs-TNWs/TNAs, Au NRs-TNWs/TNAs, and Au NPs-NRs-TNWs/TNAs.

2.2. Synthesis of Au Nanoparticles and Au Nanorods

Au nanoparticles (NPs) were synthesized by the Turkevick method [32,33]. In a typical experiment, 10 mL deionized (DI) water and 100 µL of 25 mM HAuCl$_4$·3H$_2$O (Merck) were placed in a conical flask. The solution was heated to boiling, and 300 µL of 1% trisodium citrate was added to the solution under vigorous stirring. The color of the solution immediately changed to light red, which indicates the formation of Au NPs. After continuing the vigorous stirring for 5 min, the solution was cooled to room temperature.

For synthesizing Au nanorods (NRs), a seed solution was first prepared by slowly mixing 100 µL of 0.025 M HAuCl$_4$·3H$_2$O with 10 mL of 0.1 M hexadecyltrimethylammonium bromide (CTAB) in a test tube. Then, 640 µL of 0.01 M ice-cold NaBH$_4$ solution, which was freshly prepared with 100 mL of 0.01 M NaOH and 3.78 mg NaBH$_4$, was added all at once under stirring for 30 min. The color of the mixture changed from light yellow to light brown. A growth solution of Au NRs was prepared by adding 1.75 µM AgNO$_3$ to a mixture of 1 mM CTAB, 2.5 µM HAuCl$_4$·3H$_2$O and 50 µM hydroquinone. Finally, 80 µL seed solution was added to the growth solution and left overnight before cleaning through two centrifugation cycles of 10,000 rpm for 20 min.

2.3. Preparation of Au NPs-, Au NRs-, and Au-NRs-NPs-Decorated TNWs/TNAs

Au NPs-decorated-TNWs/TNAs (S$_2$), Au NRs-decorated-TNWs/TNAs (S$_3$) and Au NPs-NRs-decorated-TNWs/TNAs (S$_4$) were prepared by drop casting technique using

1 mL Au NPs solution, 1 mL Au NRs solution, and a mixture of 0.5 mL Au NPs and 0.5 mL Au NRs solution, respectively (Figure 1b). The samples were then heated at 120 °C for 60 min for drying, removing the residual solvent, and improving the connectivity at Au/TiO$_2$ interfaces.

2.4. Characterization Methods

The orientation and crystallinity of the materials will be determined using X-ray diffraction (XRD) (XRD, Bruker D2, Billerica, MA, USA) and using Cu Kα radiation (λ = 1.5406 Å) in the θ–2θ configuration. The grain sizes can be estimated by the Scherrer formula. Morphologies and film thicknesses of the samples were characterized by scanning electron microscopy (SEM, JEOL JSM-6500, Pleasanton, CA, USA). The compositions of samples were analyzed using an energy-dispersive X-ray spectroscopy (EDS) equipped with the SEM instrument. The elemental atomic percentage of each sample was obtained by averaging the values measured at 5 distinct 10 μm × 12 μm areas on the film's surfaces. Structural characterization at atomic scale was performed in a JEOL JEM-ARM200F (Tokyo, Japan) high-resolution scanning transmission electron microscope (HRTEM), operated at 200 kV. A TEM specimen was prepared by scratching the film surface using a pointed diamond tip and transferring the fragments onto a Cu grid. To study the elemental composition and chemical state of the materials, X-ray photoelectron spectra of a selected Au-TiO$_2$ film were performed by an XPS instrument (ThermoVG 350, East Grinstead, UK) with an X-ray source of Mg Kα 1253.6 eV and 300 W. The C1s peak at 284.8 eV was used as an internal standard, and the freeware XPSPEAK 4.1 was employed for XPS curve fitting with the Shirley background subtraction and assuming a Gaussian–Lorentzian peak shape.

Photoelectrochemical measurements were conducted using a three-electrode cell with a reference electrode of Ag/AgCl, a counter electrode of Pt, and 0.5 M Na$_2$SO$_4$ electrolyte. The working electrodes were the TNWs/TNAs and Au-TNWs/TNAs films on Ti metal substrate. The substrate edges and the metal contact region were sealed with insulating epoxy resin to leave a working electrode area of 1.0 cm^2. Linear sweeps and J–t scans were measured by an electrochemical workstation (Jiehan 5000, Jiehan Technology Co., Taichung, Taiwan). Incident photon to current conversion efficiencies spectra were collected under illumination light from a 100 W xenon lamp. The photocatalytic activity of the selected films was determined by the decomposition of methylene blue (MB) under UV-VIS irradiation from a 100 W xenon lamp. Prior to illumination, investigated samples were immersed in a MB solution (10 mg/L or 3.13 × 10^{-5} M) in the dark for 20 min to achieve absorption-desorption equilibrium. All photocatalytic reactions were maintained at 32–34 °C. After a certain photocatalytic reaction time, 1 mL MB was withdrawn to determine the relative concentration by measuring absorption spectra in the wavelength range of 400–800 nm using a UV–Vis spectrophotometer (Hitachi U-2900, Hitachi, Tokyo, Japan). This spectrophotometer was also used to measure the absorption spectra of the TNWs/TNAs and Au-TNWs/TNAs samples.

3. Results and Discussion

Figure 2 presents the XRD patterns of TNWs/TNAs, Au NPs-TNWs/TNAs, Au NRs-TNWs/TNAs, and Au NPs-NRs-TNWs/TNAs. All the samples exhibited the anatase phase of TiO$_2$ with preferred lattice planes of (101) at 25.1°, (004) at 37.8°, and (105) at 53.8° (JCPDS No. 21-1272). In addition, there was no rutile TiO$_2$ phase peak, confirming that the TiO$_2$ nanomaterials in this study were pure anatase phases. These XRD results are similar to those reported in Refs. [14,34–36] for TiO$_2$ and Au-TiO$_2$ nanostructures. A closer inspection of the (004) peaks of Au-TiO$_2$ samples, Au (111) component was observed by the shoulder peak at 38.3°. Indeed, the fittings of Au-TiO$_2$ (004) peaks allowed us to extract Au (111) components, as demonstrated in Figure 2b for Au NPs-NRs-TNWs/TNAs (S$_4$). This confirms the presence of crystalline Au nanomaterials in S$_2$, S$_3$, S$_4$ samples. We employed the Scherrer equation to estimate the grain sizes (D) of the samples, $D = 0.9\lambda/\beta\cos\theta$, where λ, β, and θ are the X-ray wavelength, full width at half maximum of the TiO$_2$ (004) peak, and Bragg diffraction angle, respectively [14,37]. As a result, the D values varied in a narrow range of 22.2–24.0 nm (Figure 2c), suggesting a similar grain size and crystallinity level amongst the four nanomaterials.

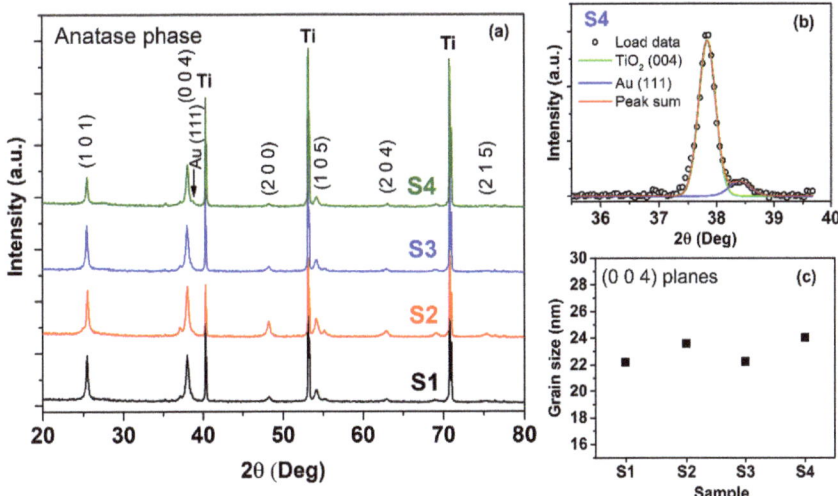

Figure 2. (a) The XRD patterns of TNWs/TNAs (S_1), Au NPs-TNWs/TNAs (S_2), Au NRs-TNWs/TNAs (S_3), and Au NPs-NRs-TNWs/TNAs (S_4). (b) The (004) peak of S_4 shows two components of TiO_2 (004) and Au (111). (c) The estimated grain size of the four nanomaterials.

Figure 3 shows the morphologies and absorption spectra of as-prepared Au NPs and Au NRs colloidal solutions. The Au NPs had a uniform spherical shape with a size of 19.0 ± 1.9 nm (Figure 3a), which induced an LSPR peak at 512 nm (Figure 3c), which was consistent with the LSPR-peaks of Au NPs in Refs. [15,38]. Meanwhile, Au NRs exhibited well-defined rods with a length of 99.8 ± 15.1 nm and a width of 14.8 ± 1.3 nm (Figure 3b); they also include a minor amount of Au NPs with an average size of 19.6 nm (Figure 3b). Consequently, the Au NRs had a strong broad LSPR absorption peak at 1192 nm in the infrared region, and a small absorption peak at 512 nm due to the presence of a small amount of Au NPs (Figure 3b,d). The insets in Figure 3c,d are the photographs of the Au NPs and Au NRs colloidal solutions, which exhibit as light red and light brown, respectively. The LSPR peak of Au NPs was consistent with those in refs. [15,38], but the present Au NRs peak at 1192 nm is longer than the Au NRs peaks (range of 740–840 nm) in ref. [30], owing to the differences in the length, width, and aspect ratio of the Au NRs. The interesting optical properties of the Au nanomaterials with LSPR peaks in Vis and/or IR regions are of great interest for their use for enhancing the PEC and photocatalytic activities of the Au-TiO_2 heterostructures (see later).

Figure 4 presents the morphologies of TNWs/TNAs, Au NPs-TNWs/TNAs, Au NRs-TNWs/TNAs, and Au NPs-NRs-TNWs/TNAs. Obviously, the TNWs/TNAs exhibited well-defined nanowires (length of ~6.0 µm) covering nanotube arrays (tube diameter of ~80 nm and length of ~6.0 µm, Figure 4a,e). For Au-TNWs/TNAs samples (S_2, S_3, S_4), Au NPs distributed relatively uniformly on TNWs/TNAs surfaces (Figure 4b). Similarly, S_3 exhibited Au NRs- decoration on TNWs/TNAs surface (Figure 4c), meanwhile S_4 had both Au NPs and Au NRs on the surface of TNWs/TNAs (Figure 4d). Typical EDS spectra collected from S_1 and S_4 shows Ti, O peaks for S_1, and Au, Ti, O peaks for S_4. Moreover, the elemental contents of S_1 were [Ti] = 31.6 ± 0.3 at.% and [O] = 68.4 ± 0.3 at.%, while S_4 had [Au] = 0.7 ± 0.2 at.% [Ti] = 31.2 ± 1.1 at.% and [O] = 68.1 ± 1.3 at.%. The Ti and O contents had a relatively close stoichiometry of TiO_2. These EDS results suggest the successful fabrications of TiO_2 and Au-TiO_2 films in this study (Figure 4f).

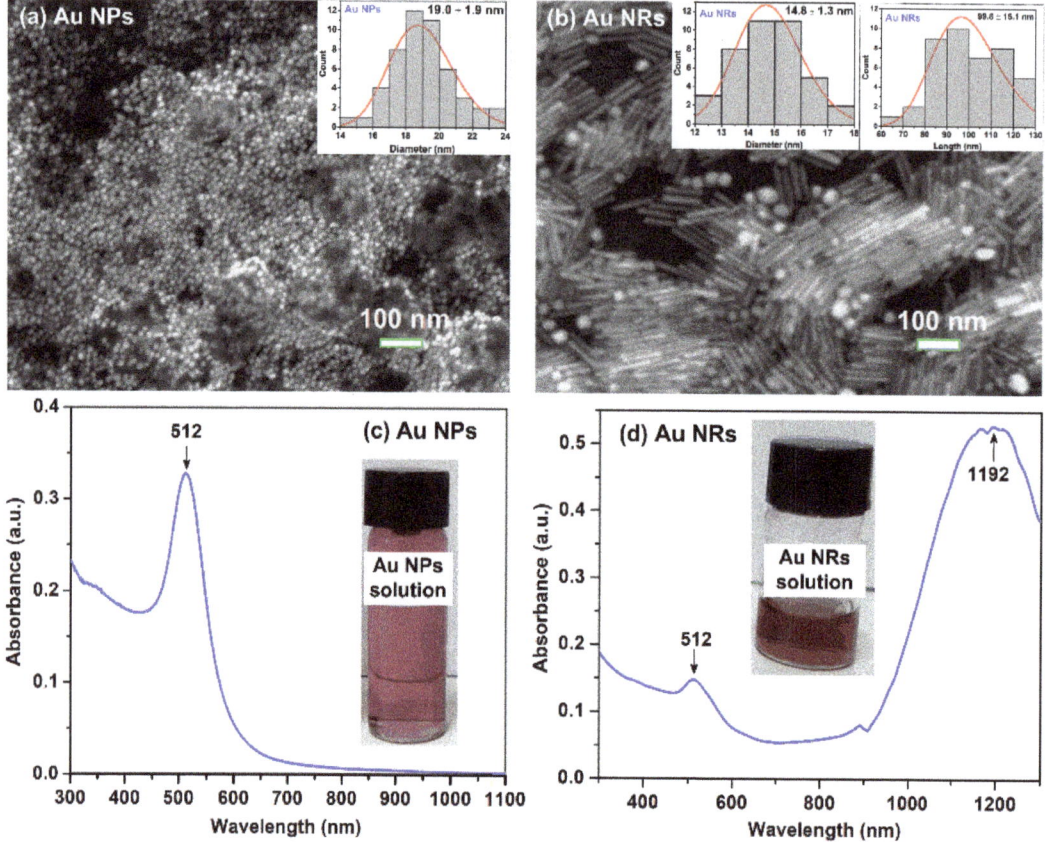

Figure 3. (**a,b**) SEM images of as-prepared Au nanoparticles and Au nanorods colloidal solutions. The insets in (**a,b**) are the size distribution histograms of Au NPs and Au NRs. (**c,d**) Absorption spectra of Au nanoparticles and Au nanorods solutions, showing a localized surface plasmon resonance (LSPR) peak at 512 nm for Au NPs, and two peaks at 512 nm and 1192 nm for Au NRs; the insets in (**c,d**) are the corresponding photographs of the Au NPs and Au NRs solutions.

The HRTEM image of S_4 in Figure 5a shows Au NP and Au NR decorated onto TiO_2. To reveal the structural quality at the interface between Au/TiO_2, an HRTEM image obtained from the dashed square area in Figure 5a presents a sharp interface between Au NP-, Au NR- and TiO_2 that facilitates the charge transfer to reduce the charge recombination (Figure 5b). In addition, the three crystallite domains have different orientations and lattice spacings of approximately 0.23 nm, 0.20 nm, and 0.35 nm, which correspond to the Au (111), Au (200) (AMCSD–0011140), and anatase TiO_2 (101) planes (AMCSD–0019093), respectively. Thus, it is evidenced that the close contact metal-semiconductor (Au-TiO_2) heterostructure is successfully formed.

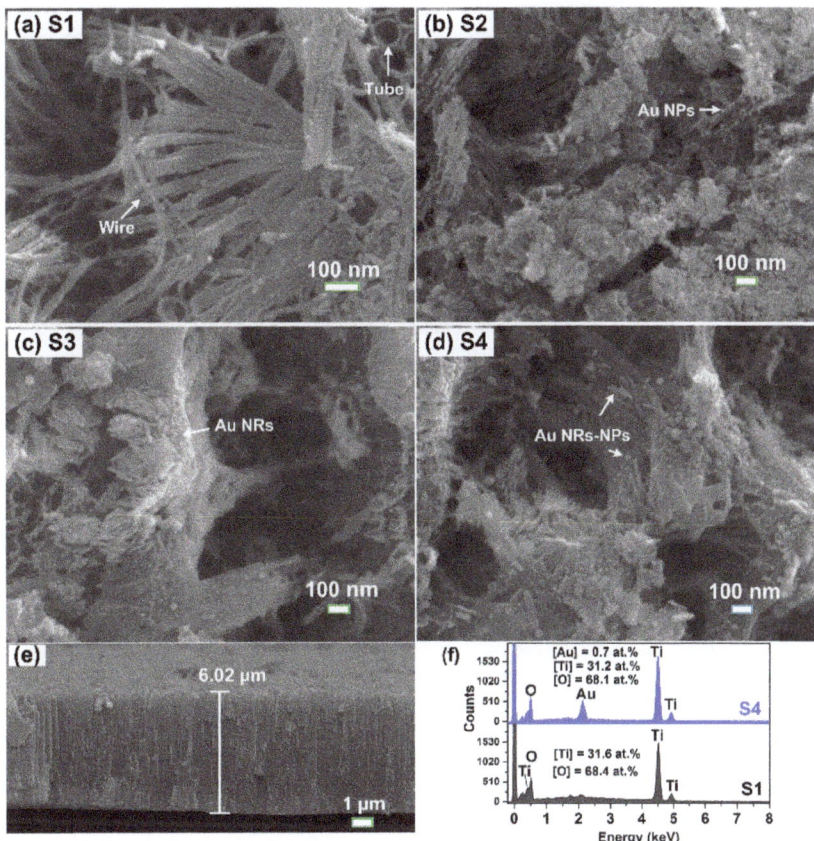

Figure 4. SEM images of (**a**) TNWs/TNAs (S_1), (**b**) Au NPs-TNWs/TNAs (S_2), (**c**) Au NRs-TNWs/TNAs (S_3), and (**d**) Au NPs-NRs-TNWs/TNAs (S_4). (**e**) A typical cross-sectional SEM image of the TiO_2 films in this study. (**f**) The typical EDS spectra of S_1 and S_4 in this study.

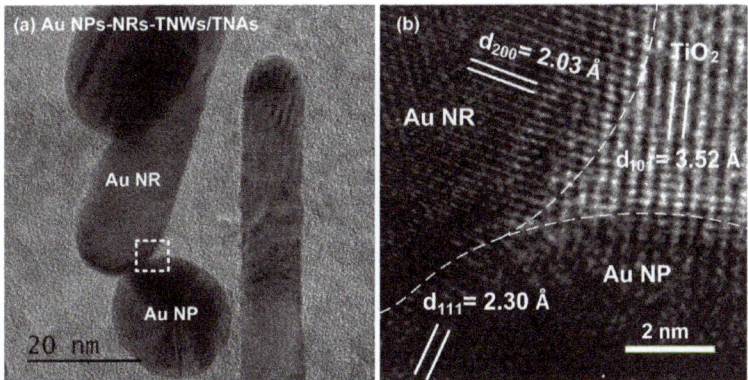

Figure 5. (**a**) An HRTEM image of Au-NPs-NRs-TNWs/TNAs (S_4). (**b**) An HRTEM image obtained from the dashed square area in panel (**a**), showing the crystal structure at the interface of Au NP–Au NR and TiO_2.

To elucidate the chemical states, compositions, and functional groups on the surfaces of the studied nanomaterials, the XPS spectrum of a representative Au NPs-NRs-TNWs/TNAs film (S_4) was measured. Figure 6a is a wide-scan XPS spectrum of S_4, which clearly shows the spectra of Au 4f, Ti 2p, and O 1s. In Figure 6b, Au $4f_{7/2}$ peaks at 83.2 eV, which was lower than the Au $4f_{7/2}$ at 84.0 eV of Au in the metallic state (see Figure 6a inset for confirming the Au binding energy) [2,39]. The Au $4f_{7/2}$ peak exhibited a negative shift by 0.8 eV, which could be due to the electron transfer from oxygen vacancies of TiO_2 to Au, suggesting the strong interaction between TiO_2 and Au. This result agreed well with the results for the dendritic Au/TiO_2 nanorod arrays [2] and Au/TiO_2 nanotubes [40]. The C 1s presented a main peak at 184.8 eV of C=C and a small peak of O=C– at 188.3 eV, suggesting the formation of bond type O=C–O–Ti and C–O group (Figure 6c) [41]. In addition, according to the deconvolution results, Ti $2p_{1/2}$ and Ti $2p_{3/2}$ peaks were located at 464.4 eV and 458.7 eV, respectively (Figure 6d), indicating the Ti^{4+} oxidation state [2,40,41]. As shown in Figure 6e, the O 1s spectrum is asymmetrical and it has an extending tail towards the higher energy. The O 1s was deconvoluted into three component peaks at 530.0 eV, 531.8 eV, and 533.7 eV, which could be assigned to (i) O^{2-} in the TiO_2 lattice, (ii) Ti–OH or C=O groups bound to two titanium atoms, (iii) OH groups bound to C and/or Ti, respectively [42,43].

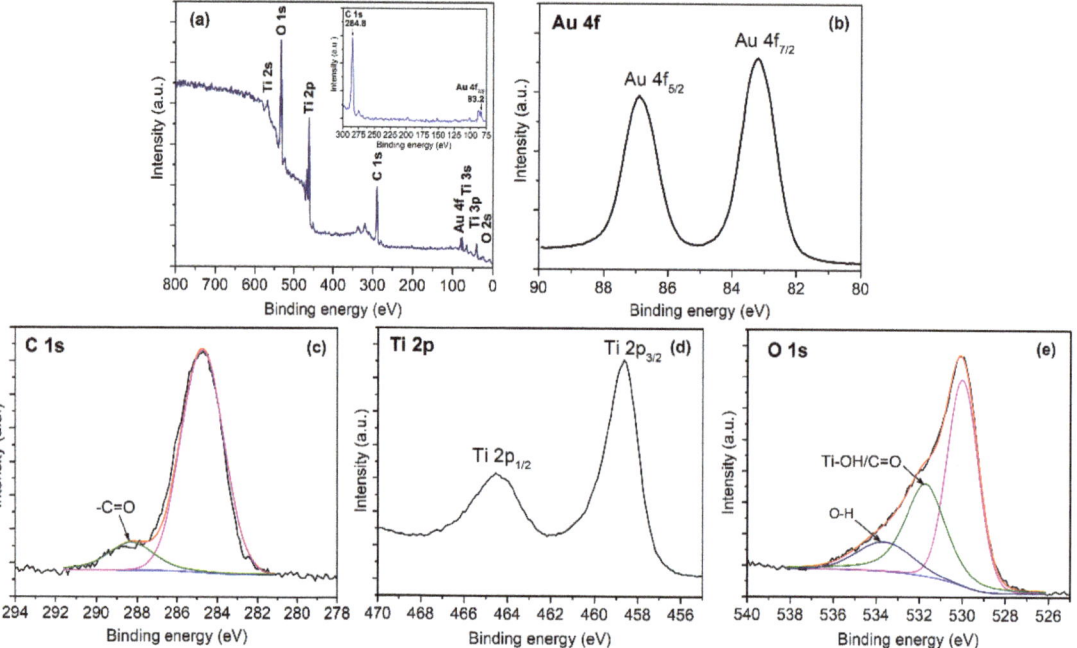

Figure 6. (**a**) Wide-scan XPS spectrum of Au NPs-NRs-TNWs/TNAs (S_4); Inset shows Au 4f and C1s calibrated at 284.8 eV. High-resolution spectrum for (**b**) Au 4f; (**c**) C 1s; (**d**) Ti 2p; (**e**) O 1s.

To evaluate the PEC activity of TNWs/TNAs and Au-TNWs/TNAs, linear sweeping voltammetry (LSV) curves were performed using the sample as photoanode, Pt as a counter electrode, and Ag/AgCl as a reference electrode and under the UV-Vis irradiation. As shown in Figure 7a, without illumination, all the TNWs/TNAs and Au-TNWs/TNAs photoanodes exhibited negligible photocurrent density (J) in the whole potential range [44]. The J increased remarkably for all the photoanodes with illumination, indicating typical properties of the semiconductor. Moreover, the J of Au-decorated TNWs/TNAs was remarkably higher than that of TNWs/TNAs photoanode. At 0.15 V, the J values of S_1, S_2, S_3, and S_4 were 0.24, 0.32, 0.31, and 0.40 mA/cm^2, respectively (Figure 7a). This means that Au NPs-NRs-TNWs/TNAs (S_4) possessed the highest PEC activity among all the investigated photoanode nanomaterials.

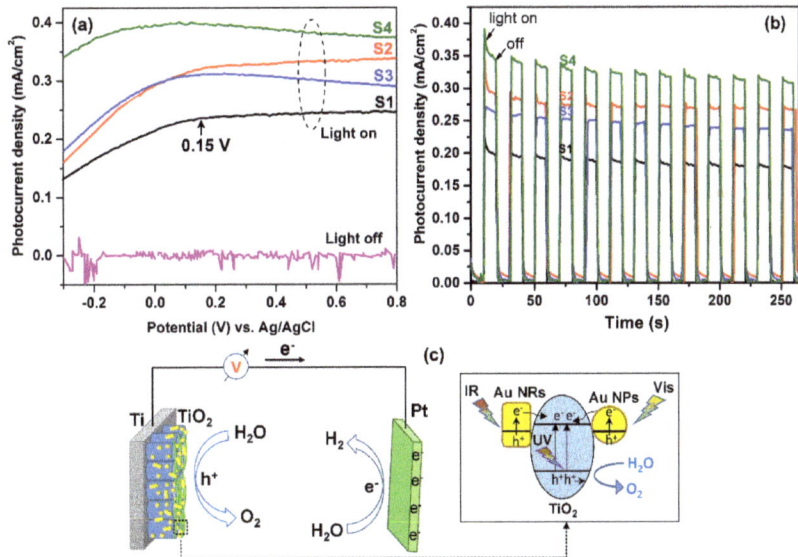

Figure 7. (a) Linear sweep voltammetry (LSV) of TNWs/TNAs and Au-TNWs/TNAs (S$_1$–S$_4$) recorded in a 0.5 M Na$_2$SO$_4$ solution under the illumination of 100 W xenon lamp. (b) Chronoamperometric I−t curves for the samples collected at 0.15 V versus RHE under the UV-Vis irradiation of 100 W xenon lamp. (c) A schematic illustration for PEC water splitting using Au NPs-NRs-TNWs/TNAs array as a photoanode.

Table 1 summarizes the photocurrent densities of the optimal Au-TiO$_2$-based photoanodes developed in this study and in the literature [45–49]. The J of the Au NPs-NRs-TiO$_2$ photoanode (0.4 mA/cm^2) was higher than that of Au/reduced graphene oxide/hydrogenated TiO$_2$ nanotube arrays (Au/RGO/H-TNTs, 0.22 mA/cm^2) under the visible light irradiation [45], but it was 2.4–6.8 times lower than the optimum J values of Au/TiO$_2$/Au heterostructure (0.94 mA/cm^2) [46], Au/TiO$_2$ nanorod arrays (1.1 mA/cm^2) [47], Au NPs-NRs/TiO$_2$ nanowires (1.49 mA/cm^2) [11], Au NPs- branched TiO$_2$ (2.5 mA/cm^2) [48], and Au NPs/3D TiO$_2$ nanorods (2.7 mA/cm^2) [49]. The different J values from different studies are attributed to both the intrinsic different PEC properties of the photoanodes and the different PEC experimental conditions (e.g., light source, potential, and electrolyte).

Figure 7c presents a schematic diagram of a possible PEC water-splitting process for Au NPs-NRs-TNWs/TNAs. The photoanode exhibited an enhanced PEC activity owing to the following factors: (1) the LSPR of Au NPs and Au NRs improve the light-harvesting capability and the light absorption in both Vis and IR regions (Figure 3c,d), (2) the enhancement of photogenerated carriers' density due to LSPR at Au NPs and Au NRs, (3) the LSPR hot electrons (e$^−$) in Au NPs and Au NRs can inject into the conduction band of TiO$_2$ to improve the electrons' injection efficiency and reduce electron-hole recombination rate [48]. Therefore, these synergistic effects of Au NPs-NRs-TNWs/TNAs resulted in the highest PEC performance over either the single-type Au-decorated TNWs/TNAs or the pristine TNWs/TNAs.

The photocurrent response graphs at 0.15 V of the photoanodes with controlling of the on–off irradiation cycle are shown in Figure 7b. The measured J values were exactly consistent with those recorded in the LSV experiments at 0.15 V. Based on the J values, the PEC performance of the four photoanode nanomaterials was in the order of S$_4$ > S$_2$ > S$_3$ > S$_1$. Furthermore, though witnessing a considerable photocurrent drop in the first 25 s, the photocurrent remained relatively stable during the test lasting 270 s, with only a 13% J decrease with respect to the maximum J value.

Table 1. Photocurrent density and photoelectrochemical measurement conditions of the optimal Au/TiO$_2$-based photoanodes developed in this study and in the literature.

Photoanode Nanomaterial	Electrolyte/Potential (V) vs. RHE	Illumination	Photocurrent Density (mA/cm^2)	Ref.
Au NPs-NRs/TNWs/TNAs	0.5 M Na$_2$SO$_4$/0.15 V	100 W Xe lamp	0.40	This study
Au/RGO/H-TNTs	1 M KOH/1.23 V	Xe lamp, AM 1.5G filter, λ > 400 nm	0.22	[45]
Au/TiO$_2$/Au heterostructure	1 M KOH/0.2 V	300 W Xenon Lamp, 100 mW/cm^2 UV–100 mW/cm^2 Vis	0.94	[46]
Au/TiO$_2$ nanorod arrays	0.5 M Na$_2$SO$_4$/1.0 V	500 W Xe lamp, 100 mW/cm^2	1.1	[47]
Au NPs-NRs/TiO$_2$ nanowires	1 M KOH/0 V	AM 1.5G, 100 mW/cm^2	1.49	[11]
Au NPs-branched TiO$_2$	0.5 M Na$_2$SO$_4$/1.23 V	500 W Xe lamp, 100 mW/cm^2	2.5	[48]
Au NPs/3D TiO$_2$ nanorods	1 M NaOH/0 V	150 W Xe lamp, AM 1.5G, 100 mW/cm^2	2.7	[49]

Since S$_4$ possessed the highest PEC activity, it was of interest to further study its photocatalytic activity in the degradation of MB. Figure 8a,b show MB degradation by photolysis and photocatalysis using S$_1$ and S$_4$ under the UV-Vis irradiation of a 100 W xenon lamp. It is seen that the main MB absorbance at λ$_{max}$ ~659 nm decreases substantially with the increase in the irradiation time (Figure 8c). Clearly, both photolysis and photocatalysis induced the decrease of MB concentration with the exponential decay, $C_t = C_0 \times e^{-kt}$, where C_0 is the initial concentration, C_t is the concentration of MB at time t, and k is the reaction rate constant (h^{-1}). In Figure 8d, the k is obtained by performing the linear fitting on the plot of $-\ln(C_t/C_0)$ vs. t. Specifically, the k of the photolysis process was a low value of 0.13 h^{-1}, indicating that MB is quite stable under UV-Vis irradiation (Figure 8d inset). Meanwhile, under the photocatalytic reactions, the k values of S$_1$ and S$_4$ were 0.28 ± 0.02 h^{-1} and 0.70 ± 0.07 h^{-1}, respectively (Figure 8d inset). This means that the k value of Au NPs-NRs- TNWs/TNAs was 2.5 times higher than that of TNWs/TNAs, which indicates a dramatic enhancement in the photocatalytic activity of TiO$_2$ by introducing a mixture of Au NPs and Au NRs.

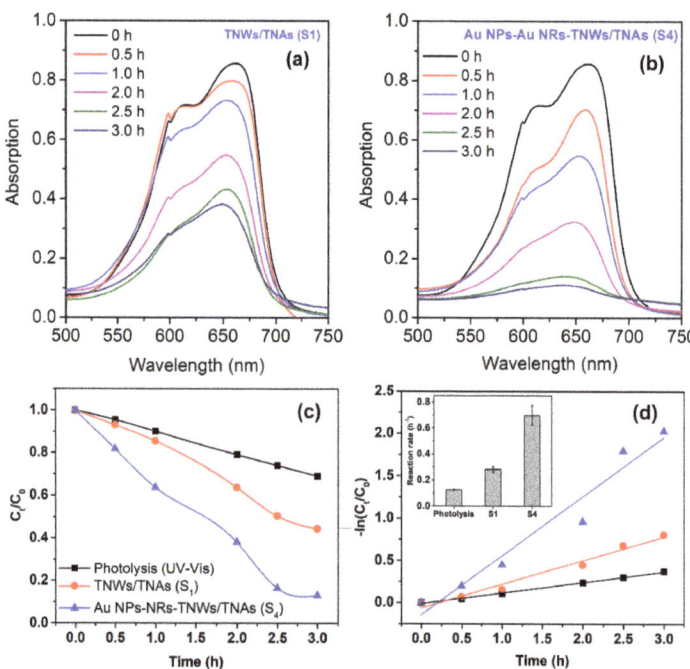

Figure 8. (a,b) Changes in UV-visible absorption spectra of MB by TNWs/TNAs (S$_1$) and Au NPs-NRs- TNWs/TNAs (S$_4$) as a function of irradiation time. (c,d) Variations in (C_t/C_0) and $-\ln(C_t/C_0)$ as a function of irradiation time. Inset in (d) is the reaction rate constant (k) of S$_1$ and S$_4$.

For reference, Table 2 summarizes the k values of TiO$_2$-based nanophotocatalysts prepared by some methods in this study and the relevant literature [10,50–53]. It is very hard to make the right comparison when each developed TiO$_2$-based nanophotocatalyst has its intrinsic properties (e.g., morphology-surface area, crystal structure, composition, decoration, or doping content) and different photocatalytic reaction conditions (e.g., catalyst dosage, light source, initial concentration of MB). Generally, the k value of the present TNWs/TNAs (0.28 h^{-1}) was twice higher than those of the 40 nm-TNAs/20 nm-TNWs (0.13 h^{-1}) and the TiO$_2$ nanoparticles film (0.14 h^{-1}) in ref. [10], which could be due to the larger surface area associated with the thicker TNAs and longer TNWs film in this study. Meanwhile, the k value of Au NPs-NRs- TNWs/TNAs (0.7 h^{-1}) was comparable with those of the optimal Ag/TiO$_2$ nanoparticles (0.65 h^{-1}) [50] and the Bi-Fe doped TiO$_2$ (0.78 h^{-1}) [51]. However, the k value of Au NPs-NRs- TNWs/TNAs was 2.1–2.7 times lower than the k values of the brookite phase TNAs (1.45 h^{-1}) synthesized by anodization and annealed at 500 °C [53], and of the SnO$_2$ NPs-decorated TNAs (1.86 h^{-1}) prepared by anodization and solvothermal process [52].

Table 2. The synthesis methods and photocatalytic reaction rate constants in the degradation of methylene of the selected TiO$_2$-based nanophotocatalysts in this study and the literature.

Photocatalyst	Synthesis Methods	Reaction Rate (h^{-1})	Ref.
TNWs/TNAs	Anodization	0.28 ± 0.02	This study
Au NPs-NRs/TNWs/TNAs	Turkevick method–anodization–drop casting method	0.70 ± 0.07	This study
40 nm-TNWs/20 nm-TNAs	Anodic oxidation	0.13	[10]
TiO$_2$ nanoparticles P25 film	-	0.14	[10]
TNAs	Anodization	1.45	[53]
Ag/TiO$_2$ P25	Photo-reduction method	0.65	[50]
Bi-Fe doped TiO$_2$	Wet impregnation technique	0.78	[51]
SnO$_2$ NPs-decorated TNAs	Anodization–solvothermal process	1.86	[52]

To elucidate the mechanism of the photocatalytic activity enhancement in Au NPs-NRs-TNWs/TNAs, the absorption spectra of S$_1$ and S$_4$ were measured and shown in Figure 9a. Obviously, the TNWs/TNAs exhibited a typical absorption spectrum of TiO$_2$, characterized by a gradually increased absorbance in the IR-Vis region and a sharply increased absorbance in the UV region. Meanwhile, Au NPs-NRs-TNWs/TNAs had an LSPR peak of Au NPs at approximately 512 nm, and a broad intense absorption peak at 1192 nm owing to the LSPR peak of Au NRs.

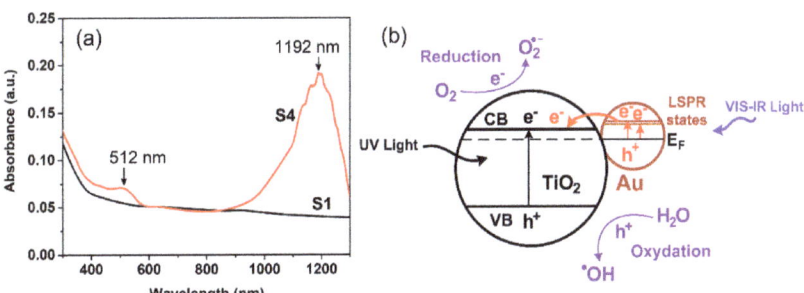

Figure 9. (a) Absorption spectra of the solution of S$_1$ and S$_4$ prepared by scratching 1 mg powder of TNWs/TNAs and Au-TNWs/TNAs. (b) A proposed mechanism for the photocatalytic activity of Au-TiO$_2$ upon the excitation of the Au surface plasmon band.

As shown in Figure 9b, a proposed mechanism of the significant k enhancement by decorating TNWs/TNAs with Au NPs and Au NRs is due to the LSPR effect [11,25,28,54,55]. In fact, the LSPR peaks at 112 nm for Au NPs and 1192 nm for Au NRs were observed, as shown in Figure 3c,d and Figure 7a. For describing LSPR, when the electromagnetic field of the incident light becomes associated with the oscillations of the conduction electrons of Au NPs and Au NRs, the local electromagnetic fields near the surface of Au NPs and Au NRs enhanced strongly [11,23]. Indeed, by 3D finite-difference time-domain simulation, the electrical field amplification at the Au/TiO$_2$ interfaces upon SPR excitation was observed clearly [11,23,49]. Therefore, the enhanced photocatalytic activity of Au NPs-NRs-TiO$_2$ in this study is attributed to the LSPR absorption of Au NPs and Au NRs under Vis-IR illumination to generate photoexcited electrons in LSPR states and holes. Then, the energetic electrons can inject into the conduction band of TiO$_2$ that leads to the enhanced separation of photo-excited electron-hole pairs, and generate a larger amount of reactive oxygenated free radicals (e.g., $O_2^{\bullet-}$ and $^\bullet OH$) to trigger photocatalytic reactions (Figure 9b) [15,28,38,56,57]. Since Au NPs-NRs-TNWs/TNAs obtained the LSPR-absorption in both Vis and IR regions, they achieved the best PEC activity among the four nanomaterials and presented a high photocatalytic performance. It is worth mentioning that a further study on Au NPs-NRs-TNWs/TNAs with different Au NPs- and Au NRs- sizes, amounts, and mixing ratios is of great interest to further enhance the PEC and photocatalytic properties of the nanomaterial.

4. Conclusions

TNWs/TNAs and Au-TNWs/TNAs nanomaterials were successfully synthesized by the anodic oxidation method combined with a chemical reduction method, and their PEC and photocatalytic activities were studied. The TNWs/TNAs and Au-TNWs/TNAs exhibited pure anatase phase of TiO$_2$ with (0 0 4), (1 0 1), and (1 0 5) preferred orientations, and componential Au (111) peak was observed for Au-TNWs/TNAs. The samples presented well-defined and uniform structure of TNWs/TNAs (i.e., TiO$_2$ nanowire length of ~6.0 μm covering on TiO$_2$ nanotubes arrays with tube diameter of ~80 nm and tube length of ~6.0 μm). Additionally, sharp Au/TiO$_2$ interfaces were observed via HRTEM images, and XPS results confirm for the strong interaction Au-TiO$_2$, the Ti^{4+} oxidation state, and the typical functional groups on the material surfaces. In addition, Au NPs (size of 19.0 ± 1.9 nm) and Au NRs (width of 14.8 ± 1.3 nm and length of 99.8 ± 15.1 nm) were relatively even decoration on TNWs/TNAs. The EDS results show that Au-decorated content onto TiO$_2$ was 0.7 ± 0.2 at.%. For PEC properties, the photocurrent density (J) at 0.15 V was 0.24 mA/cm^2 for TNWs/TNAs, 0.32 mA/cm^2 for Au NRs-TNWs/TNAs, 0.31 mA/cm^2 for Au NPs-TNWs/TNAs, and 0.40 mA/cm^2 for Au NRs-NRs-TNWs/TNAs. This means that Au NRs-NRs-TNWs/TNAs achieved the best PEC activity owing to the LSPR-absorption of Au NPs and Au NRs in the Vis and IR regions. Furthermore, Au NPs-NRs-TNWs/TNAs possessed a high photocatalytic performance in MB degradation with k = 0.7 h^{-1}, which was 2.5 times higher than that of the pristine TNWs/TNAs. These study results demonstrate that the PEC and photocatalytic properties of semiconductors can be enhanced by combining various nanostructures of plasmonic noble metals.

Author Contributions: N.N.U. performed the experiments, analyzed the data, and wrote the first paper-draft; P.H.L. and L.A.T. revised and edited the paper and supervised the study; P.H.L. was project administrator and funding acquisition. L.T.C.T., L.T.H. and H.P.T. contributed to the material characterizations. T.T.T.N., T.C.M.V.D., K.T.N., N.T.N.H. and S.-R.J. contributed to the useful discussions and reviewed the paper. S.-R.J. and C.-W.L. supported for the advanced characterizations and review and made minor revisions to the paper. All authors have read and agreed to the published version of the manuscript.

Funding: This research is funded by Vietnam National Foundation for Science and Technology Development (NAFOSTED) under grant number 103.02-2019.374.

Institutional Review Board Statement: Not applicable.

Informed Consent Statement: Not applicable.

Data Availability Statement: Not applicable.

Acknowledgments: The authors acknowledge the support of time and facilities from Ho Chi Minh City University of Technology-VNU-HCM, National Yang Ming Chiao Tung University, and Can Tho University of Medicine and Pharmacy for this study.

Conflicts of Interest: The authors declare no conflict of interest.

References

1. Fujishima, A.; Honda, K. Electrochemical photolysis of water at a semiconductor electrode. *Nature* **1972**, *238*, 37–38. [CrossRef] [PubMed]
2. Su, F.; Wang, T.; Lv, R.; Zhang, J.; Zhang, P.; Lu, J.; Gong, J. Dendritic Au/TiO$_2$ nanorod arrays for visible-light driven photoelectrochemical water splitting. *Nanoscale* **2013**, *5*, 9001. [CrossRef] [PubMed]
3. Houas, A.; Lachheb, H.; Ksibi, M.; Elaloui, E.; Guillard, C.; Herrmann, J.-M. Photocatalytic degradation pathway of methylene blue in water. *Appl. Catal. B* **2001**, *31*, 145–157. [CrossRef]
4. Hoffmann, M.R.; Martin, S.T.; Choi, W.Y.; Bahnemann, D.W. Environmental Applications of Semiconductor Photocatalysis. *Chem. Rev.* **1995**, *95*, 69–96. [CrossRef]
5. Asahi, R.; Morikawa, T.; Irie, H.; Ohwaki, T. Nitrogen-doped titanium dioxide as visible-light-sensitive photocatalyst: Designs, developments, and prospects. *Chem. Rev.* **2014**, *114*, 9824–9852. [CrossRef] [PubMed]
6. Chen, X.; Mao, S.S. Titanium Dioxide Nanomaterials: Synthesis, Properties, Modifications, and Applications. *Chem. Rev.* **2007**, *107*, 2891–2959. [CrossRef] [PubMed]
7. Singh, J.; Manna, A.K.; Soni, R.K. Bifunctional Au—TiO$_2$ thin films with enhanced photocatalytic activity and SERS based multiplexed detection of organic pollutant. *J. Mater. Sci. Mater. Electron.* **2019**, *30*, 16478–16493. [CrossRef]
8. Hoang, S.; Guo, S.; Hahn, N.T.; Bard, A.J.; Mullins, C.B. Visible Light Driven Photoelectrochemical Water Oxidation on Nitrogen-Modified TiO$_2$ Nanowires. *Nano Lett.* **2012**, *12*, 26–32. [CrossRef]
9. Baker, D.R.; Kamat, P.V. Photosensitization of TiO$_2$ Nanostructures with CdS Quantum Dots: Particulate versus Tubular Support Architectures. *Adv. Funct. Mater.* **2009**, *46556*, 805–811. [CrossRef]
10. Hsu, M.-Y.; Hsu, H.-L.; Leu, J. TiO$_2$ Nanowires on Anodic TiO$_2$ Nanotube Arrays (TNWs/TNAs): Formation Mechanism and Photocatalytic Performance. *J. Electrochem. Soc.* **2012**, *159*, H722–H727. [CrossRef]
11. Pu, Y.C.; Wang, G.; Chang, K.-D.; Ling, Y.; Lin, Y.K.; Fitzmorris, B.C.; Liu, C.M.; Lu, X.; Tong, Y.; Zhang, J.Z.; et al. Au nanostructure-decorated TiO$_2$ nanowires exhibiting photoactivity across entire UV-visible region for photoelectrochemical water splitting. *Nano Lett.* **2013**, *13*, 3817–3823. [CrossRef] [PubMed]
12. Mo, S.-D.; Ching, W.Y. Electronic and optical properties of three phases of titanium dioxide: Rutile, anatase, and brookite. *Phys. Rev. B* **1995**, *51*, 13023. [CrossRef] [PubMed]
13. Khan, M.A.M.; Siwach, R.; Kumar, S.; Alhazaa, A.N. Role of Fe doping in tuning photocatalytic and photoelectrochemical properties of TiO$_2$ for photodegradation of methylene blue. *Opt. Laser Technol.* **2019**, *118*, 170–178. [CrossRef]
14. Do, T.C.M.V.; Nguyen, D.Q.; Nguyen, K.T.; Le, P.H. TiO$_2$ and Au-TiO$_2$ Nanomaterials for Rapid Photocatalytic Degradation of Antibiotic Residues in Aquaculture Wastewater. *Materials* **2019**, *12*, 2434. [CrossRef] [PubMed]
15. Linic, S.; Christopher, P.; Ingram, D.B. Plasmonic-metal nanostructures for efficient conversion of solar to chemical energy. *Nat. Mater.* **2011**, *10*, 911–921. [CrossRef] [PubMed]
16. Yu, Y.; Wen, W.; Qian, X.-Y.; Liu, J.-B.; Wu, J.-M. UV and visible light photocatalytic activity of Au/TiO$_2$ nanoforests with Anatase/Rutile phase junctions and controlled Au locations. *Sci. Rep.* **2017**, *7*, 41253. [CrossRef] [PubMed]
17. Zhang, J.; Jin, X.; Morales-Guzman, P.I.; Yu, X.; Liu, H.; Zhang, H.; Razzari, L.; Claverie, J.P. Engineering the Absorption and Field Enhancement Properties of Au−TiO$_2$ Nanohybrids via Whispering Gallery Mode Resonances for Photocatalytic Water Splitting. *ACS Nano* **2016**, *10*, 4496–4503. [CrossRef]
18. Luo, J.; Ma, L.; He, T.; Ng, C.F.; Wang, S.; Sun, H.; Fan, H.J. TiO$_2$/(CdS, CdSe, CdSeS) Nanorod Heterostructures and Photoelectrochemical Properties. *J. Phys. Chem. C* **2012**, *116*, 11956–11963. [CrossRef]
19. Lin, H.; Mao, Z.; Zhou, N.; Wang, M.; Li, L.; Li, Q. Fabrication of CdS quantum dots sensitized TiO$_2$ nanowires/nanotubes arrays and their photoelectrochemical properties. *SN Appl. Sci.* **2019**, *1*, 391. [CrossRef]
20. Kelly, K.L.; Coronado, E.; Zhao, L.L.; Schatz, G.C. The Optical Properties of Metal Nanoparticles: The Influence of Size, Shape, and Dielectric Environment. *J. Phys. Chem. B* **2003**, *107*, 668–677. [CrossRef]
21. Chou, C.-H.; Chen, C.-D.; Wang, C.R.C. Highly Efficient, Wavelength-Tunable, Gold Nanoparticle Based Optothermal Nanoconvertors. *J. Phys. Chem. B* **2005**, *109*, 11135–11138. [CrossRef]
22. Li, C.; Shuford, K.L.; Chen, M.; Lee, E.J.; Cho, S.O. A Facile Polyol Route to Uniform Gold Octahedra with Tailorable Size and Their Optical Properties. *ACS Nano* **2008**, *2*, 1760. [CrossRef] [PubMed]
23. Liu, Z.; Hou, W.; Pavaskar, P.; Aykol, M.; Cronin, S.B. Plasmon Resonant Enhancement of Photocatalytic Water Splitting Under Visible Illumination. *Nano Lett.* **2011**, *11*, 1111–1116. [CrossRef] [PubMed]
24. Paul, K.K.; Giri, P.K. Role of Surface Plasmons and Hot Electrons on the Multi-Step Photocatalytic Decay by Defect Enriched Ag@TiO$_2$ Nanorods Under Visible Light Illumination. *J. Phys. Chem. C* **2017**, *121*, 20016–20030. [CrossRef]
25. Wang, C.; Astruc, D. Nanogold plasmonic photocatalysis for organic synthesis and clean energy conversion. *Chem. Soc. Rev.* **2014**, *43*, 7188. [CrossRef] [PubMed]

26. Singh, J.; Sahu, K.; Satpati, B.; Shah, J.; Kotnala, R.K.; Mohapatra, S. Facile synthesis, structural and optical properties of Au-TiO$_2$ plasmonic nanohybrids for photocatalytic applications. *J. Phys. Chem. Solids* **2019**, *135*, 109100. [CrossRef]
27. Fu, F.; Zhang, Y.; Zhang, Z.; Zhang, X.; Chen, Y.; Zhang, Y. The preparation and performance of Au loads TiO$_2$ nanomaterials. *Mater. Res. Express* **2019**, *6*, 095041. [CrossRef]
28. Veziroglu, S.; Ullrich, M.; Hussain, M.; Drewes, J.; Shondo, J.; Strunskus, T.; Adam, J.; Faupel, F.; Aktas, O.C. Plasmonic and non-plasmonic contributions on photocatalytic activity of Au-TiO$_2$ thin film under mixed UV—Visible light. *Surf. Coat. Technol.* **2020**, *389*, 125613. [CrossRef]
29. Link, S.; Mohamed, M.B.; El-Sayed, M.A. Simulation of the Optical Absorption Spectra of Gold Nanorods as a Function of Their Aspect Ratio and the Effect of the Medium Dielectric Constant. *J. Phys. Chem. B* **1999**, *103*, 3073–3077. [CrossRef]
30. Sau, T.K.; Murphy, C.J. Seeded High Yield Synthesis of Short Au Nanorods in Aqueous Solution. *Langmuir* **2004**, *20*, 6414–6420. [CrossRef]
31. Kumar, R.; Binetti, L.; Nguyen, T.H.; Alwis, L.S.M.; Agrawal, A.; Sun, T.; Grattan, K.T.V. Determination of the Aspect-ratio Distribution of Gold Nanorods in a Colloidal Solution using UV-visible absorption spectroscopy. *Sci. Rep.* **2019**, *9*, 17469. [CrossRef] [PubMed]
32. Enustun, B.V.; Turkevich, J. Coagulation of Colloidal Gold. *J. Am. Chem. Soc.* **1963**, *85*, 3317–3328. [CrossRef]
33. Kimling, J.; Maier, M.; Okenve, B.; Kotaidis, V.; Ballot, H.; Plech, A. Turkevich Method for Gold Nanoparticle Synthesis Revisited. *J. Phys. Chem. B* **2006**, *110*, 15700–15707. [CrossRef]
34. Sun, L.; Cai, J.; Wu, Q.; Huang, P.; Su, Y.; Lin, C. N-doped TiO$_2$ nanotube array photoelectrode for visible-light-induced photoelectrochemical and photoelectrocatalytic activities. *Electrochim. Acta* **2013**, *108*, 525–531. [CrossRef]
35. Preethi, L.K.; Antony, R.P.; Mathews, T.; Walczak, K.; Gopinath, C.S. A Study on Doped Heterojunctions in TiO$_2$ Nanotubes: An Efficient Photocatalyst for Solar Water Splitting. *Sci. Rep.* **2017**, *7*, 14314. [CrossRef] [PubMed]
36. Do, T.C.M.V.; Nguyen, D.Q.; Nguyen, T.D.; Le, P.H. Development and Validation of a LC-MS/MS Method for Determination of Multi-Class Antibiotic Residues in Aquaculture and River Waters, and Photocatalytic Degradation of Antibiotics by TiO$_2$ Nanomaterials. *Catalysts* **2020**, *10*, 356. [CrossRef]
37. Tuyen, L.T.C.; Jian, S.-R.; Tien, N.T.; Le, P.H. Nanomechanical and Material Properties of Fluorine-Doped Tin Oxide Thin Films Prepared by Ultrasonic Spray Pyrolysis: Effects of F-Doping. *Materials* **2019**, *12*, 1665. [CrossRef]
38. Chen, Y.; Bian, J.; Qi, L.; Liu, E.; Fan, J. Efficient Degradation of Methylene Blue over Two-Dimensional Au/TiO$_2$ Nanosheet Films with Overlapped Light Harvesting Nanostructures. *J. Nanomater.* **2015**, *2015*, 905259. [CrossRef]
39. Padikkaparambil, S.; Narayanan, B.; Yaakob, Z.; Viswanathan, S.; Tasirin, S.M. Au/TiO$_2$ Reusable Photocatalysts for Dye Degradation. *Inter. J. Photoenergy* **2013**, *2013*, 752605. [CrossRef]
40. Liu, W.; Duan, W.; Jia, L.; Wang, S.; Guo, Y.; Zhang, G.; Zhu, B.; Huang, W.; Zhang, S. Surface Plasmon-Enhanced Photoelectrochemical Sensor Based on Au Modified TiO$_2$ Nanotubes. *Nanomaterials* **2022**, *12*, 2058. [CrossRef]
41. Kruse, N.; Chenakin, S. XPS characterization of Au/TiO$_2$ catalysts: Binding energy assessment and irradiation effects. *Appl. Catal. A.* **2011**, *391*, 367–376. [CrossRef]
42. Yoshiiri, K.; Wang, K.; Kowalska, E. TiO$_2$/Au/TiO$_2$ Plasmonic Photocatalysts: The Influence of Titania Matrix and Gold Properties. *Inventions* **2022**, *7*, 54. [CrossRef]
43. Yu, J.; Zhao, X.; Zhao, Q. Effect of surface structure on photocatalytic activity of TiO$_2$ thin films prepared by sol-gel method. *Thin Solid Film.* **2000**, *379*, 7–14. [CrossRef]
44. Xu, F.; Mei, J.; Li, X.; Sun, Y.; Wu, D.; Gao, Z.; Zhang, Q.; Jiang, K. Heterogeneous three-dimensional TiO$_2$/ZnO nanorod array for enhanced photoelectrochemical water splitting properties. *J. Nanopart Res.* **2017**, *19*, 297. [CrossRef]
45. Luo, J.; Li, D.; Yang, Y.; Liu, H.; Chen, J.; Wang, H. Preparation of Au/reduced graphene oxide/hydrogenated TiO$_2$ nanotube arrays ternary composites for visible-light-driven photoelectrochemical water splitting. *J. Alloys Compd.* **2016**, *661*, 380–388. [CrossRef]
46. Li, Y.; Yu, H.; Zhang, C.; Fu, L.; Li, G.; Shao, Z.; Yi, B. Enhancement of photoelectrochemical response by Au modified in TiO$_2$ nanorods. *Int. J. Hydrogen Energy* **2013**, *38*, 13023–13030. [CrossRef]
47. Xu, F.; Bai, D.; Mei, J.; Wu, D.; Gao, Z.; Jiang, K.; Liu, B. Enhanced photoelectrochemical performance with in-situ Au modified TiO$_2$ nanorod arrays as photoanode. *J. Alloys Compd.* **2016**, *688*, 914–920. [CrossRef]
48. Xu, F.; Mei, J.; Zheng, M.; Bai, D.; Wu, D.; Gao, Z.; Jiang, K. Au nanoparticles modified branched TiO$_2$ nanorod array arranged with ultrathin nanorods for enhanced photoelectrochemical water splitting. *J. Alloys Compd.* **2017**, *693*, 1124–1132. [CrossRef]
49. Li, H.; Li, Z.; Yu, Y.; Ma, Y.; Yang, W.; Wang, F.; Yin, X.; Wang, X. Surface-Plasmon-Resonance-Enhanced Photoelectrochemical Water Splitting from Au-Nanoparticle-Decorated 3D TiO$_2$ Nanorod Architectures. *J. Phys. Chem. C* **2017**, *121*, 12071–12079. [CrossRef]
50. Tseng, H.-C.; Chen, Y.-W. Facile Synthesis of Ag/TiO$_2$ by Photoreduction Method and Its Degradation Activity of Methylene Blue under UV and Visible Light Irradiation. *Mod. Res. Catal.* **2020**, *9*, 1–19. [CrossRef]
51. Mishra, S.; Chakinala, N.; Chakinala, A.G.; Surolia, P.K. Photocatalytic degradation of methylene blue using monometallic and bimetallic Bi-Fe doped TiO$_2$. *Catal. Commun.* **2022**, *171*, 106518. [CrossRef]
52. Li, Y.; Zhang, X.; Hu, X.; Li, Z.; Fan, J.; Liu, E. Facile fabrication of SnO$_2$/TiO$_2$ nanotube arrays for efficient degradation of pollutants. *Opt. Mater.* **2022**, *127*, 112252. [CrossRef]

53. Kang, X.; Chen, S. Photocatalytic reduction of methylene blue by TiO$_2$ nanotube arrays: Effects of TiO$_2$ crystalline phase. *J. Mater. Sci.* **2010**, *45*, 2696–2702. [CrossRef]
54. Chen, Y.; Tian, G.; Pan, K.; Tian, C.; Zhou, J.; Zhou, W.; Ren, Z.; Fu, H. In situ controlled growth of well-dispersed gold nanoparticles in TiO$_2$ nanotube arrays as recyclable substrates for surface-enhanced Raman scattering. *Dalt. Trans.* **2012**, *41*, 1020–1026. [CrossRef] [PubMed]
55. Zhang, Z.; Zhang, L.; Hedhili, M.N.; Zhang, H.; Wang, P. Plasmonic gold nanocrystals coupled with photonic crystal seamlessly on TiO$_2$ nanotube photoelectrodes for efficient visible light photoelectrochemical water splitting. *Nano Lett.* **2013**, *13*, 14–20. [CrossRef] [PubMed]
56. Kowalska, E.; Mahaney, O.O.P.; Abe, R.; Ohtani, B. Visible-light-induced photocatalysis through surface plasmon excitation of gold on titania surfaces. *Phys. Chem. Chem. Phys.* **2010**, *12*, 2344–2355. [CrossRef]
57. Nosaka, Y.; Nosaka, A. Understanding Hydroxyl Radical (•OH) Generation Processes in Photocatalysis. *ACS Energy Lett.* **2010**, *1*, 356–359. [CrossRef]

Design and Characterization of Nanostructured Titanium Monoxide Films Decorated with Polyaniline Species

Tomas Sabirovas [1], Simonas Ramanavicius [1,*], Arnas Naujokaitis [2], Gediminas Niaura [3] and Arunas Jagminas [1,*]

1. State Research Institute Centre for Physical Sciences and Technology, Department of Electrochemical Material Science, Sauletekio Ave. 3, LT-10257 Vilnius, Lithuania
2. State Research Institute Centre for Physical Sciences and Technology, Department of Characterization of Materials Structure, Sauletekio Ave. 3, LT-10257 Vilnius, Lithuania
3. State Research Institute Centre for Physical Sciences and Technology, Department of Organic Chemistry, Sauletekio Ave. 3, LT-10257 Vilnius, Lithuania
* Correspondence: simonas.ramanavicius@ftmc.lt (S.R.); arunas.jagminas@ftmc.lt (A.J.)

Abstract: The fabrication of nanostructured composite materials is an active field of materials chemistry. However, the ensembles of nanostructured titanium monoxide and suboxide species decorated with polyaniline (PANI) species have not been deeply investigated up to now. In this study, such composites were formed on both hydrothermally oxidized and anodized Ti substrates via oxidative polymerization of aniline. In this way, highly porous nanotube-shaped titanium dioxide (TiO_2) and nano leaflet-shaped titanium monoxide (TiO_x) species films loaded with electrically conductive PANI in an emeraldine salt form were designed. Apart from compositional and structural characterization with Field Emission Scanning Electron Microscopy (FESEM) and Raman techniques, the electrochemical properties were identified for each layer using cyclic voltammetry and electrochemical impedance spectroscopy (EIS). Based on the experimentally determined EIS parameters, it is envisaged that TiO-based nanomaterials decorated with PANI could find prospective applications in supercapacitors and biosensing.

Keywords: polyaniline; titanium; monoxide; titanium dioxide; oxidative polymerization; anodizing; electrochemical impedance spectroscopy

1. Introduction

Electrically conducting polymers are promising alternative materials for technological applications in many areas, including chemistry, material sciences, and engineering [1,2]. Conducting polymers-based nanocomposites also have outstanding potential applications in biomedical fields [3], such as antimicrobial therapy [4], drug delivery [5], wearable energy harvesting devices [6], nerve regeneration [7], and tissue engineering [8]. Conducting polymers are also frequently applied in the design of supercapacitors [9]. Modification of metal oxides by conducting polymers significantly improves electrochemical capacitance [10]. Capacity variations of conducting polymer-based structures can be well applied in various conducting–polymer-based sensors [11] and biosensors [12]. Due to its high electrical conductivity, biocompatibility, low toxicity, and good environmental stability, polyaniline (PANI) is one of the most studied conducting polymers [13]. Three redox forms of polyaniline are usually distinguished: leucoemeraldine (reduced), emeraldine (half-oxidized), and pernigraniline (fully oxidized), differing in color and electrical conductivity [14,15]. Traditionally, PANI layers are designed onto various substrates by electrochemical treatment in acidic solutions containing aniline. PANI can also be synthesized by chemical oxidative polymerization (COP) of aniline monomer using ammonium persulfate as a redox initiator [16]. However, to the extent of the authors' knowledge, only a few reports have covered titanium monoxide and suboxide structures with PANI at the time of this writing [17,18], whereas none of them investigated nanostructures or nanostructured surfaces as it is done in this study.

There is a variety of applications for titania [19], but for titanium suboxides, despite their advantageous properties such as high electrical conductivity and low bandgap [20], application areas are still not clear. Up to now, several publications have reported on titanium–suboxides-based gas sensors [21,22], microbial fuel cells [18], substrates for electrocatalysts [23], solar cells [24], optoelectronic devices [25], batteries [26], etc. While many common synthesis methods exist for TiO_2, such as anodization [27], microwave synthesis [28], laser beam treating [29] and chemical oxidation [30], these methods are not suitable for the formation of nonstoichiometric titanium oxide structures. One of the well-known approaches for the formation of titanium suboxide materials is the reduction of TiO_2 by heating in an H_2 atmosphere, with TiH_4, $NaBH_4$, Al, Mg, either with Zn powder [31–33]. Note that the presence of oxygen vacancy, Ti^{3+} and Ti-OH in titanium monoxides and suboxides contributed to their high photocatalytic efficiency and significantly lower band gaps [20]. However, these processes require lengthy processing at high temperatures in a vacuum, and both structure size and surface morphology are difficult to control [34]. Therefore, in this study, we designed the chemical oxidation of the Ti surface into titanium suboxide thin films by hydrothermal treatment under appropriate oxidation/dissolution reactivity with a controllable thickness and surface morphology. For this, an alkaline solution of 0.4 mol L^{-1} H_2SeO_3, already reported by us [35], has been slightly modified and used in this study. In this way, highly porous nanostructured films composed of titanium monoxide, TiO_x, species possessing a low band gap (1.29 eV) value and hierarchical morphology have been prepared.

The aim of this study is the formation of nanoplatelet titanium monoxide films decorated with polyaniline (PANI) for application as a supercapacitor. Decorated titanium monoxide film surface morphology was investigated by FESEM, the formation of PANI in Emeraldine form was proved by Raman spectroscopy, whereas cyclic voltammetry and electrochemical impedance spectroscopy were applied for the investigation of capacitive properties. Moreover, for purposes of comparison, PANI deposited onto anodized Ti substrates was investigated in this study.

2. Materials and Methods
2.1. Synthesis and Characterization

Analytical grade ammonium fluoride, sodium hydroxide, potassium peroxydisulfate, $K_2S_2O_8$, acids: HCl, H_3PO_4, $HClO_4$, H_2SeO_3, and aniline purchased from Sigma-Aldrich were used for the preparation of aqueous solutions. The 99.7% purity Ti working samples in dimensions of 10 × 10 mm^2 were cut from a Ti foil, 0.127 mm thick (Sigma-Aldrich, Taufkirchen, Germany). The samples were ultrasonically cleaned by sonication in acetone, ethanol and water baths, 6 min each. An aqueous solution containing 2 mol L^{-1} H_3PO_4 and 0.3 mol L^{-1} NH_4F was applied for Ti nanoporous anodizing at 20 V for 2 h., whereas chemical oxidation of Ti foil was performed by hydrothermal treatment in an alkaline solution of 0.3 mol L^{-1} H_2SeO_3 and NaOH at 150–200 °C for 15 h., as previously reported by us [35], followed by rigorous rinsing. To obtain a denser and thicker layer compared to the previous study, the pH of the solution was increased to 10.0 and the temperature of autoclaving to 200 °C. Prior to further depositions, all samples were annealed in argon at 350 °C for 2 h. Similar to the methodology reported by Rahman et al. [36], an aqueous solution of HCl, aniline and potassium peroxydisulfate ($K_2S_2O_8$) was used for the chemical deposition of polyaniline. For this purpose, 0.02 mL aniline was dissolved under vigorous stirring in 22 mL 1.0 mol L^{-1} HCl solution at the ice temperature. Then, 2 mL of a solution containing 25 mg mL^{-1} $K_2S_2O_8$ salt also kept at ice temperature was added dropwise under intense stirring, then the Ti sample was inserted in the mixture and kept from 12 to 48 h. Finally, the sample was washed with water several times, dried at 60 °C, and stored in a desiccator for characterization. SEM images were obtained using the Helios Nanolab 650 field emission scanning electron microscope (FEI Eindhoven, The Netherlands).

2.2. Raman Spectroscopy

A Renishaw InVia spectrometer (Wotton-under-Edge, UK) equipped with a thermoelectrically cooled (−70 °C) CCD camera and a microscope was used for Raman spectroscopy measurements. Raman spectra were excited with 632.8 nm radiation from a He-Ne gas laser. The 20×/0.40 NA objective lens and 1800 lines/mm grating was used, where the accumulation time of Raman spectra was 400 s. The power of the laser was decreased to 0.3 mW to avoid possible destruction of the samples. The polystyrene standard spectrum was used to calibrate the Raman scattering wavenumber axis. Experimental data were fitted with Gaussian-Lorentzian shape components using GRAMS/A1 8.0 (Thermo Scientific, Waltham, MA, USA) software to determine the parameters of the bands.

2.3. Electrochemical Measurements

Electrochemical measurements of cyclic voltammetry and electrochemical impedance spectroscopy were carried out using an electrochemical workstation Zahner Zennium (Kronach, Germany) in a standard three-electrode cell with the working electrode, Ag/AgCl$_{sat}$ reference electrode, and platinum (99.99% purity, Aldrich) wire as a counter electrode. As a working electrode, pristine substrates and PANI-decorated substrates (12 h incubation time) were selected. Electrochemical impedance spectroscopy (EIS) was conducted in a frequency range of 0.1 Hz–10^5 Hz with 10 mV amplitude. The data were normalized to the surface area of the electrodes (0.16 cm^2). Multiple experiments (up to 5) were carried out to account for data discrepancy.

3. Results

3.1. Structural Characterization

Hydrothermal treatment of titanium in strongly alkaline solutions at 150–200 °C results in its surface etching and oxidation covering by densely packed crystallite grains (Figure 1A). The average size of titanium hydrate species varied from several to several tens of nanometers, depending on the processing conditions, such as the solution pH, temperature, and treatment duration. The subsequent calcination of this ensemble at 300–350 °C in air results in titanium hydrate crystallization to TiO$_2$ [37,38].

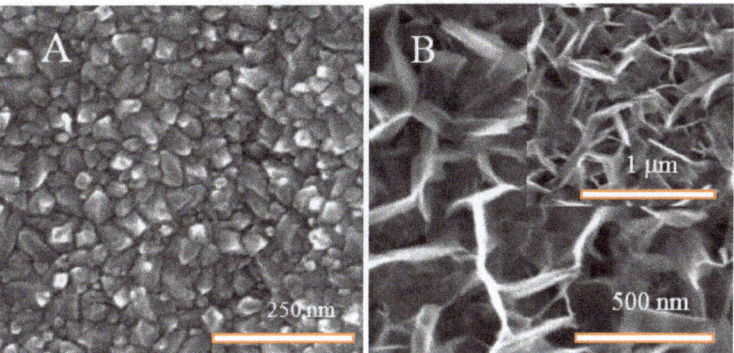

Figure 1. Top-side SEM images of the Ti surface after hydrothermal treatment in the solution of NaOH (pH = 11.0) at 150 °C for 15 h (**A**) and 0.3 mol L^{-1} H$_2$SeO$_3$ + NaOH solution (pH = 10.0) at 150 °C (**B**).

However, variations in the hydrothermal treatment conditions and pH did not allow us to form nanostructured titanium monoxide or suboxide films. Therefore, in this study, we applied the hydrothermal processing in an aqueous alkaline solution containing SeO$_3^{2-}$, which upon a slow hydrolysis reaction at high temperature and pressure produces reactive OH$^-$ species and TiOH [35]. Long-time hydrothermal treatment of a Ti substrate in this solution even at 150 °C resulted in the formation of a thin ($\cong$1.0 μm) layer comprising a leaflet-type species array (Figure 1B). The rise in the autoclave temperature to 200 °C results in the formation of thicker films of the more densely packed and needle-shaped leaflets with lateral dimensions of 8–13 nm × 0.5–1.1 μm (Figure 2A). The SEM images shown in the Figure 2B–D panels imply that the entire surface of all TiO_x needle-shaped leaflets is decorated with frost-type precipitates after prolonged sonication of the TiO_x in the cold aniline solution with potassium persulfate. Upon vigorous mixing for 12, 24, and 48 h, the initial thickness of TiO_x needles increased to 50–60 nm, 150–170 nm, and 200–250 nm, respectively, although the porous morphology of the film was preserved even after 48 h of processing. From the previous research, after calcination in an oxygen-free atmosphere, these leaflets comprise titanium monoxide [35].

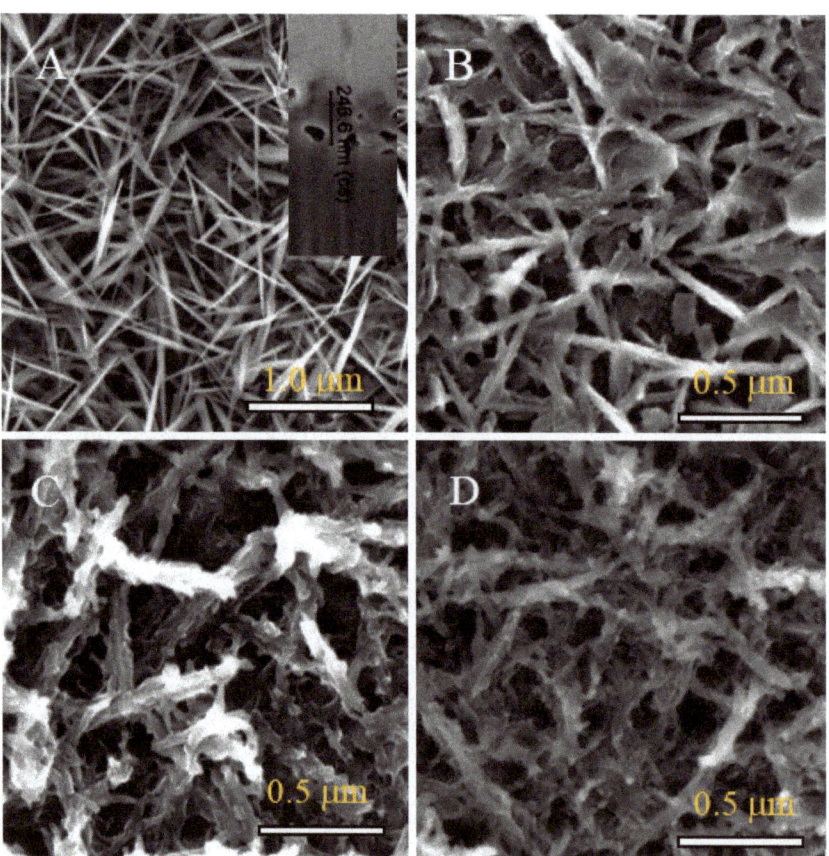

Figure 2. Top-side FESEM images of TiO_x film before (**A**) and after (**B**–**D**) covering with polyaniline by chemical oxidation of aniline with potassium persulfate at 2–4 °C for 12 (**B**), 24 (**C**), and 48 h. (**D**). The inset shows the film thickness of film grown at 200 °C for 15 h: pH = 10; 0.3 M SeO_3^{2-}.

For purposes of comparison, we also prepared Ti samples covered with porous anodic oxide film sandwiched with PANI, focusing on the relation between structure and electrochemical properties between these layers. The Ti surface was prepared by anodizing in an aqueous solution containing 2.0 mol L^{-1} H_3PO_4 and 0.2 mol L^{-1} NH_4F at 20 V for 3 h and subsequent oxidation of aniline with $K_2S_2O_8$ at ice temperature for 12 h. The aniline-$K_2S_2O_8$ solution was found to greatly influence the microstructure of PANI species deposited on the nanotubed anatase TiO_2 surface (Figure 3). We established that formation of polymer species on the TiO_2 substrate by chemical oxidation of aniline proceeds easily and required no surface pretreatment, as reported previously [39]. Furthermore, the morphology of as-formed species onto the TiO$_2$ surface differs significantly from the frost-type precipitates decorated on the TiO$_x$ leaflets (Figure 3).

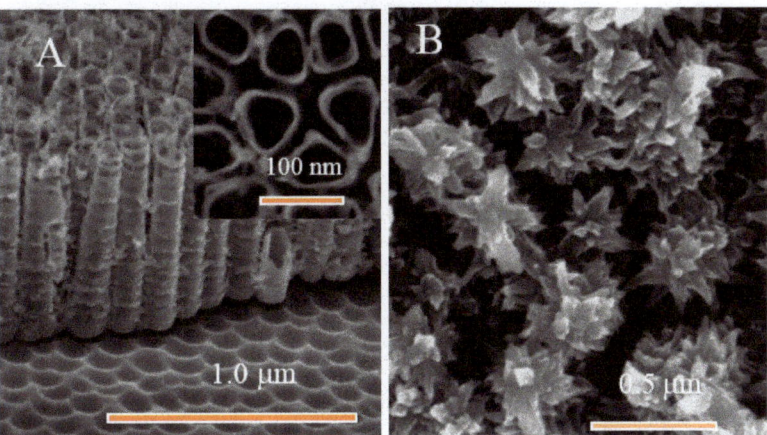

Figure 3. Panoramic (**A**) and top-side [inset in (**A**)] SEM image of TiO$_2$ film before and after polyaniline chemical deposition (**B**) from an aqueous solution of 1 mol L^{-1} HCl containing 0.9 mL L^{-1} aniline and 90 mL L^{-1} K$_2$S$_2$O$_8$ solution (2.5 wt.%) at ice temperature for 12 h.

3.2. Raman Spectroscopy

Resonance Raman spectroscopy provides rich molecular level information on the structure, oxidation and protonation states of polyaniline (PANI) [14,40–44]. Figure 4 compares 632.8-nm excited Raman spectra of Ti/TiO$_x$/PANI and Ti/TiO$_2$/PANI samples. Two strong bands at 1169 and 1589 cm^{-1} belong to C−H bending and C=C stretching vibrations of quinone rings, respectively, whereas the broad feature at 1487 cm^{-1} is associated with C=N stretching vibration of an emeraldine base (imine sites) [14,40]. The shoulder near 1618 cm^{-1} is related to the stretching vibration of the benzene ring. Both samples exhibit a strong band near 1339 cm^{-1} which was assigned to stretching vibration of polaronic structures C~N$^+$ possessing a bond intermediate between the single and double bonds [43]. These structures are responsible for the conductivity of the film. The middle-intensity feature near 812 cm^{-1} is related to C−N−C bending motion [14], whereas well-defined bands at 521 and 428 cm^{-1} are associated with amine group in-plane deformation and ring deformation modes [14,40]. The sharp band at 579 cm^{-1} was previously assigned to the presence of phenazine-like crosslinked structures in the film [44,45]. Such structures are usually synthesized during the annealing procedure at higher temperatures. The amount of such structures is higher in the case of TiO$_x$/PANI film. The low-intensity band at 1223 cm^{-1} is characteristic of the C−N stretching mode of emeraldine (amine sites) [14].

In general, the spectral pattern shown in Figure 4 indicates that the polyaniline films are in the emeraldine structure. A relatively intense polaron band at 1339 cm^{-1} evidences the presence of conductive emeraldine salt form in the films. However, because the relative intensity of the polaronic band is higher for the sample Ti/TiO$_x$/PANI, this sample should possess higher conductivity. In addition, it should be noted that sample Ti/TiO$_x$/PANI contains a higher number of phenazine-like structures.

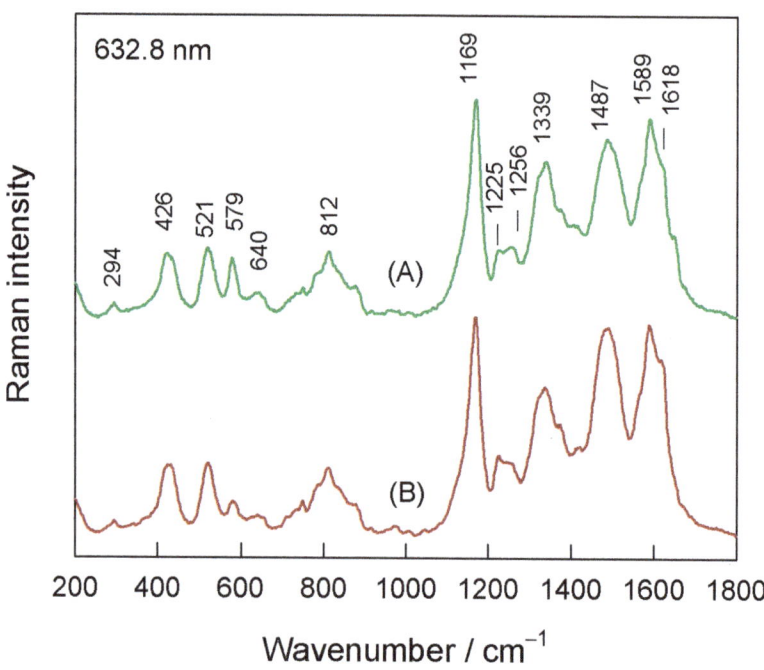

Figure 4. Raman spectrum of (**A**)–Ti/TiO$_x$/PANI sample and (**B**) Ti/TiO$_2$-/PANI sample. Excitation wavelength 632.8 nm (0.1 mW).

3.3. Electrochemical Analysis

Cyclic voltammetry and electrochemical impedance spectroscopy techniques were used to elucidate the electrochemical properties of Ti/TiO$_x$ and Ti/TiO$_2$ electrodes before and after samples were decorated with PANI. Both Ti/TiO$_2$ and Ti/TiO$_2$/PANI samples exhibited oxidation and reduction peaks upon injection of a redox probe (Figure 5A,C). However, different redox reaction responses were observed among Ti/TiO$_2$ and Ti/TiO$_2$/PANI samples. First, the peak-to-peak potential separation value (ΔE_p) of the [Fe(CN)$_6$]$^{3-}$/[Fe(CN)$_6$]$^{4-}$ redox pair (Figure 5A,C) on the Ti/TiO$_2$ sample decreased from 1020 mV to 100 mV when nanotubed TiO$_2$ film was decorated with PANI, a tenfold difference, implying a significant acceleration of the electron-transfer process. Secondly, in the presence of a redox probe, the PANI-decorated Ti/TiO$_2$ specimens also exhibited significantly higher anodic and cathodic peak currents.

Contrary to Ti/TiO$_2$ samples, no distinct oxidation and reduction peaks were observed in the presence of a redox probe for the Ti/TiO$_x$ sample (Figure 5B). Furthermore, the peak-to-peak potential separation value for Ti/TiO$_x$ was estimated to be $\Delta E_p \sim 1500$ mV, which is significantly higher than the value observed for the Ti/TiO$_2$ electrode, indicating that electron-transfer kinetics are much slower due to the presence of a titanium monoxide barrier. After decoration of TiO$_x$ leaflets with PANI, an obvious enhancement of redox reaction was observed (Figure 5D). The Ti/TiO$_x$ electrode decorated with PANI displayed significantly lower ΔE_p value of 150 mV compared to ΔE_p of 1500 mV obtained for Ti/TiO$_x$ electrodes. Notably, both electrodes, i.e., Ti/TiO$_x$/PANI and Ti/TiO$_2$/PANI, display redox reaction peaks, which are located at the potential associated with the transition of PANI forms between leucoemeraldine and emeraldine (A$_1$ and A$_2$) and of by-products, namely, hydroquinine/benzoquinone redox reaction (B$_1$ and B$_2$) [46,47].

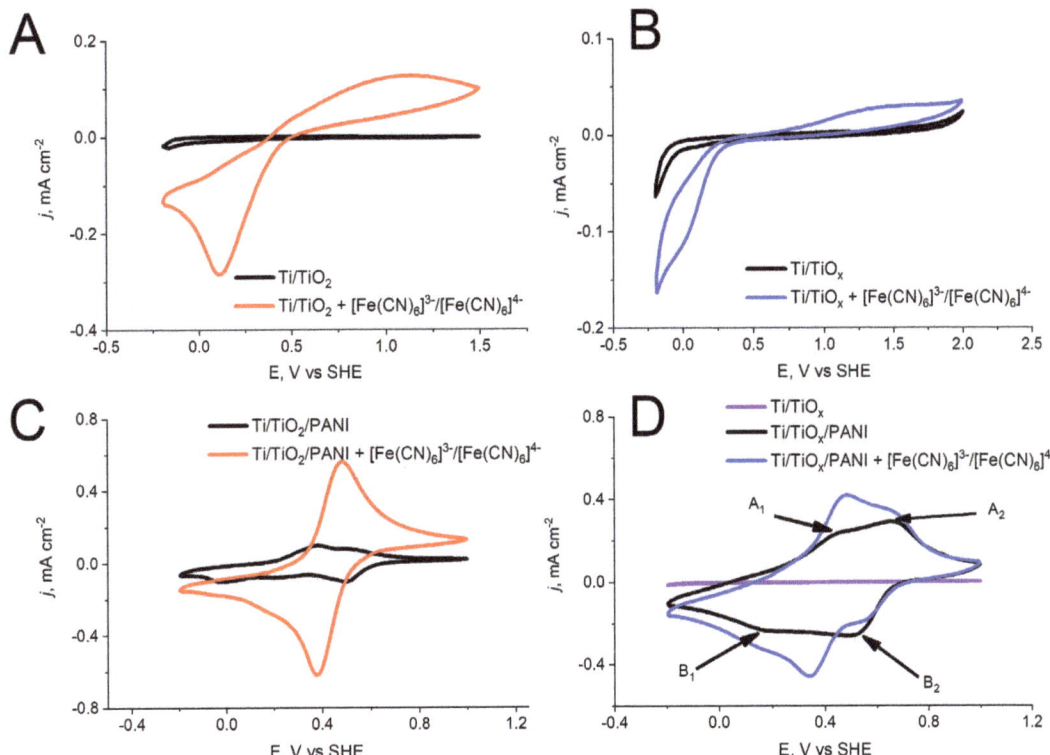

Figure 5. Nyquist plots of Ti/TiO$_2$, Ti/TiO$_2$/PANI (**A**) and TiO$_x$, TiO$_x$/PANI (**B**) with and without (**C**,**D**) 5 mM [Fe(CN)$_6$]$^{3-}$/[Fe(CN)$_6$]$^{4-}$ in 0.5 M HCl (pH = 3) solution. Scan rate: 10 mV s^{-1}.

Electrochemical properties of PANI-decorated Ti/TiO$_x$ and Ti/TiO$_2$ electrodes were also investigated by electrochemical impedance spectroscopy. From the cyclic voltammetry data (Figure 5), it was determined that [Fe(CN)$_6$]$^{3-}$/[Fe(CN)$_6$]$^{4-}$ redox reaction potential depends on the film nature. For this reason, for EIS measurements, cathodic peak potential obtained for the film in the presence of a redox probe was chosen and compared with the same film without a redox probe. The obtained data are presented in Nyquist plots (Figure 6). In most cases, two distinct parts can be observed in Nyquist plots—a semicircle in the high-frequency region and the tail in a lower-frequency region. The high-frequency region is related to the overall charge transfer resistance, R_{ct}, between the electrode-electrolyte interface. It was observed that the Nyquist plots of unmodified Ti/TiO$_2$ and Ti/TiO$_x$ samples do not display a semicircular shape (Figure 6A—filled rhombus and open rhombus; Figure 6B—blue circles), producing a straight line approaching 90 degrees, meaning that impedance at the high and low frequency region is dominated by capacitive behavior of the films. However, TiO$_x$ films in the presence of a redox probe in a high-frequency region exhibit a semicircular shape (Figure 6B orange circles), indicating that the process is limited to charge transfer. Unlike the TiO$_x$ films, an inclined line at 45° in a lower-frequency region was observed on Ti/TiO$_2$ films in the presence of a redox probe (Figure 6A, open rhombus), showing that the redox reaction displays diffusion limitations, which is referred as a Warburg impedance.

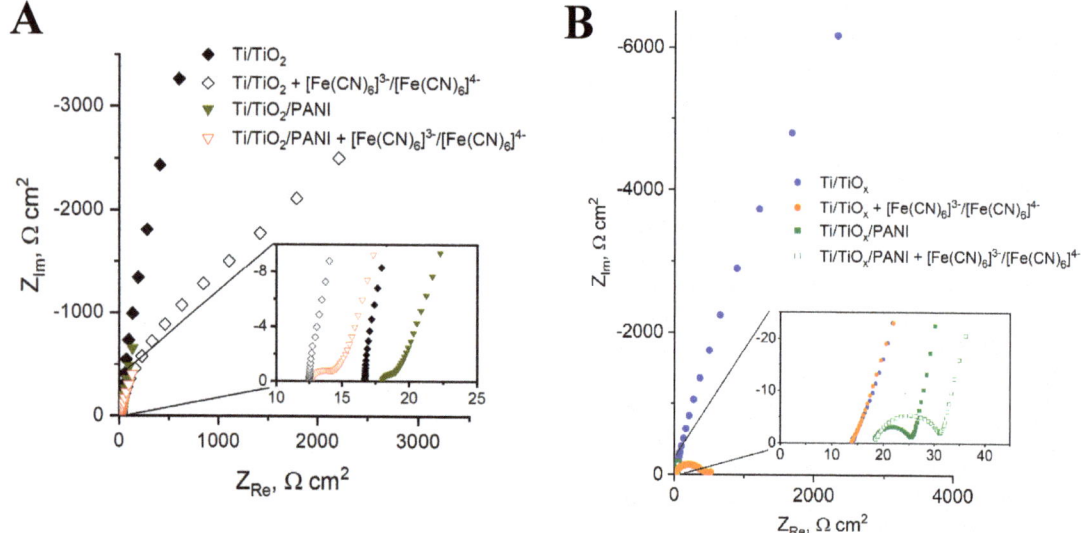

Figure 6. Nyquist plots of Ti/TiO$_2$, Ti/TiO$_2$/PANI (**A**) and TiO$_x$, TiO$_x$/PANI (**B**) with and without 5 mM [Fe(CN)$_6$]$^{3-}$/[Fe(CN)$_6$]$^{4-}$. Insets: magnification of the spectra.

PANI-decorated films exhibit semi-circular shape in the high frequency range, displaying charge transfer resistance (Figure 6A—open red triangles and filled dark yellow triangles, Figure 6B—red open and filled squares) and different conductive properties. Interestingly, smaller diameters of semicircle were obtained on both films in an absence of a redox probe (Figure 6A—filled dark yellow triangles and Figure 6B—filled green squares), exhibiting smaller charge transfer resistance. Another important characteristic of PANI-decorated film spectra with or without a redox probe is a tail in a lower-frequency region that is vertical and almost parallel to the imaginary impedance axis. Usually, the inclined line in the lower frequency region is related to the porous structure of the electrode [48]. However, the morphology of flowered PANI species deposited onto the nanotubed TiO$_2$ (Figure 3) differs significantly from the frost-type PANI species (Figure 2) deposited onto the leaflets of TiO$_x$ (Figure 2). Clearly, observed similarities of the inclined line in a lower frequency region of Ti/TiO$_x$/PANI and Ti/TiO$_2$/PANI derive from the PANI itself, which indicates capacitive behavior. Such capacitive-like behavior represents an accumulation of diffusion ions in the PANI structure [49,50].

All in all, PANI-decorated substrates exhibited significant difference from unmodified films in electrochemical properties: enhanced conductivity together with electron-transfer kinetics and near-ideal supercapacitive properties.

4. Conclusions

In this research, nanotubed TiO$_2$ and nano leaflet-shaped TiO$_x$ films possessing different surface morphology were designed and decorated with PANI species by chemical deposition. Based on SEM images, it was established that the surface morphology of titania films greatly affects the morphology of PANI species. The frost-type PANI precipitates were observed on TiO$_x$ nanoleaflets, and flower-like PANI fragments on a nanotubed TiO$_2$ substrate. Raman spectroscopy suggested that in both cases, PANI films are in the emaraldine base state. We also determined the electrochemical properties of nanotubed TiO$_2$ and TiO$_x$ films, based on cyclic voltammetry and electrochemical impedance spectroscopy. As expected, both films displayed capacitive behavior and impaired electron-transfer kinetics due to the presence of the oxide layer on the Ti surface. However, samples decorated with

PANI in all cases displayed different electrochemical properties: enhanced conductivity and electron-transfer kinetics, and near-ideal supercapacitive properties.

We suppose that these substrates have an indispensable potential for the development of novel supercapacitors. Nanotubed TiO_2, TiO_x, and PANI are considered to be biocompatible; therefore, the findings in this study could find potential use in bio-organic semiconductors for the detection of relevant biological material. Moreover, based on the low-band gap of TiO_x film, together with PANI decoration, possible applications for supercapacitors or biosensors may be considered.

Author Contributions: T.S.: Methodology, Investigation, Formal analysis, Data curation, Software, Validation, Writing—Review and Editing. S.R.: Investigation, Visualization, Writing—Review and Editing. A.N.: Investigation, Visualization. G.N.: Investigation, Formal analysis, Writing—Original Draft. A.J.: Supervision, Formal analysis, Conceptualization, Writing—Original Draft, Writing—Review and Editing. All authors have read and agreed to the published version of the manuscript.

Funding: This research received no external funding.

Institutional Review Board Statement: Not applicable.

Informed Consent Statement: Not applicable.

Data Availability Statement: Not applicable.

Conflicts of Interest: The authors declare no conflict of interest.

References

1. Geniès, E.; Boyle, A.; Lapkowski, M.; Tsintavis, C. Polyaniline: A historical survey. *Synth. Met.* **1990**, *36*, 139–182. [CrossRef]
2. Mozafari, M.; Chauhan, N.P.S. *Fundamentals and Emerging Applications of Polyaniline*; Elsevier: Amsterdam, The Netherlands, 2019.
3. Zare, E.N.; Makvandi, P.; Ashtari, B.; Rossi, F.; Motahari, A.; Perale, G. Progress in Conductive Polyaniline-Based Nanocomposites for Biomedical Applications: A Review. *J. Med. Chem.* **2020**, *63*, 1–22. [CrossRef] [PubMed]
4. Laourari, I.; Lakhdari, N.; Belgherbi, O.; Medjili, C.; Berkani, M.; Vasseghian, Y.; Golzadeh, N.; Lakhdari, D. Antimicrobial and antifungal properties of NiCu-PANI/PVA quaternary nanocomposite synthesized by chemical oxidative polymerization of polyaniline. *Chemosphere* **2021**, *291*, 132696. [CrossRef]
5. You, C.; Wu, H.; Wang, M.; Wang, S.; Shi, T.; Luo, Y.; Sun, B.; Zhang, X.; Zhu, J. A strategy for photothermal conversion of polymeric nanoparticles by polyaniline for smart control of targeted drug delivery. *Nanotechnology* **2017**, *28*, 165102. [CrossRef]
6. Nandihalli, N.; Liu, C.-J.; Mori, T. Polymer based thermoelectric nanocomposite materials and devices: Fabrication and characteristics. *Nano Energy* **2020**, *78*, 105186. [CrossRef]
7. Xu, D.; Fan, L.; Gao, L.; Xiong, Y.; Wang, Y.; Ye, Q.; Yu, A.; Dai, H.; Yin, Y.; Cai, J.; et al. Micro-Nanostructured Polyaniline Assembled in Cellulose Matrix via Interfacial Polymerization for Applications in Nerve Regeneration. *ACS Appl. Mater. Interfaces* **2016**, *8*, 17090–17097. [CrossRef]
8. Li, M.; Guo, Y.; Wei, Y.; MacDiarmid, A.G.; Lelkes, P.I. Electrospinning polyaniline-contained gelatin nanofibers for tissue engineering applications. *Biomaterials* **2006**, *27*, 2705–2715. [CrossRef]
9. Samukaite-Bubniene, U.; Valiūnienė, A.; Bucinskas, V.; Genys, P.; Ratautaite, V.; Ramanaviciene, A.; Aksun, E.; Tereshchenko, A.; Zeybek, B.; Ramanavicius, A. Towards supercapacitors: Cyclic voltammetry and fast Fourier transform electrochemical impedance spectroscopy based evaluation of polypyrrole electrochemically deposited on the pencil graphite electrode. *Colloids Surf. A Physicochem. Eng. Asp.* **2021**, *610*, 125750. [CrossRef]
10. Viter, R.; Kunene, K.; Genys, P.; Jevdokimovs, D.; Erts, D.; Sutka, A.; Bisetty, K.; Viksna, A.; Ramanaviciene, A.; Ramanavicius, A. Photoelectrochemical Bisphenol S Sensor Based on ZnO-Nanoroads Modified by Molecularly Imprinted Polypyrrole. *Macromol. Chem. Phys.* **2019**, *221*, 1900232. [CrossRef]
11. Ratautaite, V.; Brazys, E.; Ramanaviciene, A.; Ramanavicius, A. Electrochemical sensors based on l-tryptophan molecularly imprinted polypyrrole and polyaniline. *J. Electroanal. Chem.* **2022**, *917*, 116389. [CrossRef]
12. Ratautaite, V.; Boguzaite, R.; Brazys, E.; Ramanaviciene, A.; Ciplys, E.; Juozapaitis, M.; Slibinskas, R.; Bechelany, M.; Ramanavicius, A. Molecularly imprinted polypyrrole based sensor for the detection of SARS-CoV-2 spike glycoprotein. *Electrochim. Acta* **2022**, *403*, 139581. [CrossRef] [PubMed]
13. Bhadra, S.; Khastgir, D.; Singha, N.K.; Lee, J.H. Progress in preparation, processing and applications of polyaniline. *Prog. Polym. Sci.* **2009**, *34*, 783–810. [CrossRef]
14. Mažeikienė, R.; Niaura, G.; Malinauskas, A. A comparative multiwavelength Raman spectroelectrochemical study of polyaniline: A review. *J. Solid State Electrochem.* **2019**, *23*, 1631–1640. [CrossRef]
15. Gicevicius, M.; Kucinski, J.; Ramanaviciene, A.; Ramanavicius, A. Tuning the optical pH sensing properties of polyaniline-based layer by electrochemical copolymerization of aniline with o-phenylenediamine. *Dye. Pigment.* **2019**, *170*, 107457. [CrossRef]

16. Abu-Thabit, N.Y. Chemical Oxidative Polymerization of Polyaniline: A Practical Approach for Preparation of Smart Conductive Textiles. *J. Chem. Educ.* **2016**, *93*, 1606–1611. [CrossRef]
17. Li, Z.; Yang, S.; Song, Y.; Xu, H.; Wang, Z.; Wang, W.; Zhao, Y. Performance evaluation of treating oil-containing restaurant wastewater in microbial fuel cell using in situ graphene/polyaniline modified titanium oxide anode. *Environ. Technol.* **2018**, *41*, 420–429. [CrossRef]
18. Li, Z.; Yang, S.; Song, Y.; Xu, H.; Wang, Z.; Wang, W.; Dang, Z.; Zhao, Y. In-situ modified titanium suboxides with polyaniline/graphene as anode to enhance biovoltage production of microbial fuel cell. *Int. J. Hydrog. Energy* **2019**, *44*, 6862–6870. [CrossRef]
19. Dahl, M.; Liu, Y.; Yin, Y. Composite Titanium Dioxide Nanomaterials. *Chem. Rev.* **2014**, *114*, 9853–9889. [CrossRef]
20. Xu, B.; Sohn, H.Y.; Mohassab, Y.; Lan, Y. Structures, preparation and applications of titanium suboxides. *RSC Adv.* **2016**, *6*, 79706–79722. [CrossRef]
21. Ramanavicius, S.; Tereshchenko, A.; Karpicz, R.; Ratautaite, V.; Bubniene, U.; Maneikis, A.; Jagminas, A.; Ramanavicius, A. TiO2-x/TiO2-Structure Based 'Self-Heated' Sensor for the Determination of Some Reducing Gases. *Sensors* **2019**, *20*, 74. [CrossRef]
22. Ramanavicius, S.; Ramanavicius, A. Insights in the Application of Stoichiometric and Non-Stoichiometric Titanium Oxides for the Design of Sensors for the Determination of Gases and VOCs (TiO_{2-x} and Ti_nO_{2n-1} vs. TiO_2). *Sensors* **2020**, *20*, 6833. [CrossRef] [PubMed]
23. Jagminas, A.; Naujokaitis, A.; Gaigalas, P.; Ramanavičius, S.; Kurtinaitienė, M.; Trusovas, R. Substrate Impact on the Structure and Electrocatalyst Properties of Molybdenum Disulfide for HER from Water. *Metals* **2020**, *10*, 1251. [CrossRef]
24. Kim, G.; Kong, J.; Kim, J.; Kang, H.; Back, H.; Kim, H.; Lee, K. Overcoming the Light-Soaking Problem in Inverted Polymer Solar Cells by Introducing a Heavily Doped Titanium Sub-Oxide Functional Layer. *Adv. Energy Mater.* **2014**, *5*, 1401298. [CrossRef]
25. Verrelli, E.; Tsoukalas, D. Cluster beam synthesis of metal and metal-oxide nanoparticles for emerging memories. *Solid-State Electron.* **2014**, *101*, 95–105. [CrossRef]
26. Singh, A.; Kalra, V. TiO Phase Stabilized into Freestanding Nanofibers as Strong Polysulfide Immobilizer in Li–S Batteries: Evidence for Lewis Acid–Base Interactions. *ACS Appl. Mater. Interfaces* **2018**, *10*, 37937–37947. [CrossRef] [PubMed]
27. Dronov, A.; Gavrilin, I.; Kirilenko, E.; Dronova, D.; Gavrilov, S. Investigation of anodic TiO_2 nanotube composition with high spatial resolution AES and ToF SIMS. *Appl. Surf. Sci.* **2018**, *434*, 148–154. [CrossRef]
28. Mahmood, P.H.; Amiri, O.; Ahmed, S.S.; Hama, J.R. Simple microwave synthesis of TiO_2/NiS_2 nanocomposite and $TiO_2/NiS_2/Cu$ nanocomposite as an efficient visible driven photocatalyst. *Ceram. Int.* **2019**, *45*, 14167–14172. [CrossRef]
29. Wiener, J.; Shahidi, S.; Goba, M. Laser deposition of TiO_2 nanoparticles on glass fabric. *Opt. Laser Technol.* **2012**, *45*, 147–153. [CrossRef]
30. Zhou, B.; Jiang, X.; Liu, Z.; Shen, R.; Rogachev, A.V. Preparation and characterization of TiO_2 thin film by thermal oxidation of sputtered Ti film. *Mater. Sci. Semicond. Process.* **2012**, *16*, 513–519. [CrossRef]
31. Chen, B.T.D.; Dammann, J.F.; Boback, J.L. Nanomaterials. *Chem. Soc. Rev.* **2017**, *44*, 1861. [CrossRef]
32. Chen, X.; Liu, L.; Yu, P.Y.; Mao, S.S. Increasing Solar Absorption for Photocatalysis with Black Hydrogenated Titanium Dioxide Nanocrystals. *Science* **2011**, *331*, 746–750. [CrossRef] [PubMed]
33. Zhu, G.; Shan, Y.; Lin, T.; Zhao, W.; Xu, J.; Tian, Z.; Zhang, H.; Zheng, C.; Huang, F. Hydrogenated blue titania with high solar absorption and greatly improved photocatalysis. *Nanoscale* **2016**, *8*, 4705–4712. [CrossRef] [PubMed]
34. Xu, J.; Tian, Z.; Yin, G.; Lin, T.; Huang, F. Controllable reduced black titania with enhanced photoelectrochemical water splitting performance. *Dalton Trans.* **2016**, *46*, 1047–1051. [CrossRef] [PubMed]
35. Jagminas, A.; Ramanavičius, S.; Jasulaitiene, V.; Šimėnas, M. Hydrothermal synthesis and characterization of nanostructured titanium monoxide films. *RSC Adv.* **2019**, *9*, 40727–40735. [CrossRef] [PubMed]
36. Rahman, K.H.; Kar, A.K. Effect of band gap variation and sensitization process of polyaniline (PANI)-TiO_2 p-n heterojunction photocatalysts on the enhancement of photocatalytic degradation of toxic methylene blue with UV irradiation. *J. Environ. Chem. Eng.* **2020**, *8*, 104181. [CrossRef]
37. Zhao, Y.-N.; Lee, U.; Suh, M.-K.; Kwon, Y.-U. Synthesis and Characterization of Highly Crystalline Anatase Nanowire Arrays. *Bull. Korean Chem. Soc.* **2004**, *25*, 1341–1345.
38. Zárate, R.; Fuentes, S.; Wiff, J.; Fuenzalida, V.; Cabrera, A. Chemical composition and phase identification of sodium titanate nanostructures grown from titania by hydrothermal processing. *J. Phys. Chem. Solids* **2007**, *68*, 628–637. [CrossRef]
39. Jagminas, A.; Balčiūnaitė, A.; Niaura, G.; Tamašauskaitė-Tamašiūnaitė, L. Electrochemical synthesis and characterisation of polyaniline in TiO_2 nanotubes. *Trans. IMF* **2012**, *90*, 311–315. [CrossRef]
40. Morávková, Z.; Dmitrieva, E. Structural changes in polyaniline near the middle oxidation peak studied by in situ Raman spectroelectrochemistry. *J. Raman Spectrosc.* **2017**, *48*, 1229–1234. [CrossRef]
41. Mažeikienė, R.; Niaura, G.; Malinauskas, A. Raman spectroelectrochemical study of polyaniline at UV, blue, and green laser line excitation in solutions of different pH. *Synth. Met.* **2018**, *243*, 97–106. [CrossRef]
42. Mažeikienė, R.; Niaura, G.; Malinauskas, A. Study of deprotonation processes of polyaniline by differential multiwavelength Raman spectroscopy in an electrochemical system. *Chemija* **2019**, *30*, 219–226. [CrossRef]
43. Niaura, G.; Mažeikienė, R.; Malinauskas, A. Structural changes in conducting form of polyaniline upon ring sulfonation as deduced by near infrared resonance Raman spectroscopy. *Synth. Met.* **2004**, *145*, 105–112. [CrossRef]

44. Cochet, M.; Louarn, G.; Quillard, S.; Buisson, J.P.; Lefrant, S. Theoretical and experimental vibrational study of emeraldine in salt form. Part II. *J. Raman Spectrosc.* **2000**, *31*, 1041–1049. [CrossRef]
45. Šeděnková, I.; Prokeš, J.; Trchová, M.; Stejskal, J. Conformational transition in polyaniline films—Spectroscopic and conductivity studies of ageing. *Polym. Degrad. Stab.* **2008**, *93*, 428–435. [CrossRef]
46. Yoon, S.-B.; Yoon, E.-H.; Kim, K.-B. Electrochemical properties of leucoemeraldine, emeraldine, and pernigraniline forms of polyaniline/multi-wall carbon nanotube nanocomposites for supercapacitor applications. *J. Power Sources* **2011**, *196*, 10791–10797. [CrossRef]
47. Li, X.; Rafie, A.; Smolin, Y.Y.; Simotwo, S.; Kalra, V.; Lau, K.K. Engineering conformal nanoporous polyaniline via oxidative chemical vapor deposition and its potential application in supercapacitors. *Chem. Eng. Sci.* **2019**, *194*, 156–164. [CrossRef]
48. Lasia, A. Modeling of Impedance of Porous Electrodes. In *Modeling and Numerical Simulations*; Springer: Berlin, Germany, 2008; pp. 67–137.
49. Bieńkowski, K.; Strawski, M.; Szklarczyk, M. The determination of the thickness of electrodeposited polymeric films by AFM and electrochemical techniques. *J. Electroanal. Chem.* **2011**, *662*, 196–203. [CrossRef]
50. Gholivand, M.B.; Heydari, H.; Abdolmaleki, A.; Hosseini, H. Nanostructured CuO/PANI composite as supercapacitor electrode material. *Mater. Sci. Semicond. Process.* **2015**, *30*, 157–161. [CrossRef]

Article

Van der Waals Epitaxial Growth of ZnO Films on Mica Substrates in Low-Temperature Aqueous Solution

Hou-Guang Chen *, Yung-Hui Shih, Huei-Sen Wang *, Sheng-Rui Jian, Tzu-Yi Yang and Shu-Chien Chuang

Department of Materials Science and Engineering, I-Shou University, Kaohsiung City 84001, Taiwan; yhshi@isu.edu.tw (Y.-H.S.); srjian@isu.edu.tw (S.-R.J.); ziyiiyun119@gmail.com (T.-Y.Y.); r262524526@gmail.com (S.-C.C.)
* Correspondence: houguang@isu.edu.tw (H.-G.C.); huei@isu.edu.tw (H.-S.W.)

Abstract: In this article, we demonstrate the van der Waals (vdW) epitaxial growth of ZnO layers on mica substrates through a low-temperature hydrothermal process. The thermal pretreatment of mica substrates prior to the hydrothermal growth of ZnO is essential for growing ZnO crystals in epitaxy with the mica substrates. The addition of sodium citrate into the growth solution significantly promotes the growth of ZnO crystallites in a lateral direction to achieve fully coalesced, continuous ZnO epitaxial layers. As confirmed through transmission electron microscopy, the epitaxial paradigm of the ZnO layer on the mica substrate was regarded as an incommensurate van der Waals epitaxy. Furthermore, through the association of the Mist-CVD process, the high-density and uniform distribution of ZnO seeds preferentially occurred on mica substrates, leading to greatly improving the epitaxial qualities of the hydrothermally grown ZnO layers and obtaining flat surface morphologies. The electrical and optoelectrical properties of the vdW epitaxial ZnO layer grown on mica substrates were comparable with those grown on sapphire substrates through conventional solution-based epitaxy techniques.

Keywords: zinc oxide; van der Waals epitaxy; hydrothermal growth; Mist-CVD; mica

Citation: Chen, H.-G.; Shih, Y.-H.; Wang, H.-S.; Jian, S.-R.; Yang, T.-Y.; Chuang, S.-C. Van der Waals Epitaxial Growth of ZnO Films on Mica Substrates in Low-Temperature Aqueous Solution. *Coatings* **2022**, *12*, 706. https://doi.org/10.3390/coatings12050706

Academic Editor: Alexandru Enesca

Received: 9 May 2022
Accepted: 19 May 2022
Published: 20 May 2022

Publisher's Note: MDPI stays neutral with regard to jurisdictional claims in published maps and institutional affiliations.

Copyright: © 2022 by the authors. Licensee MDPI, Basel, Switzerland. This article is an open access article distributed under the terms and conditions of the Creative Commons Attribution (CC BY) license (https://creativecommons.org/licenses/by/4.0/).

1. Introduction

Zinc oxide (ZnO) has attracted intensive research efforts for its versatile applications in transparent electronics, solar cells, gas sensors, light-emitting diodes, laser diodes, and photodetectors due to its excellent optoelectronic properties including wide band gap (3.37 eV), high exciton binding energy at room temperature (60 meV), and high optical transparency within the visible spectrum [1–8]. The epitaxial growth of ZnO layers is essential for the development of advanced ZnO-based optoelectronic devices. For the conventional epitaxy paradigm, due to the strong covalent interactions at the heterointerface, the heteroepitaxial growth of semiconductors or oxides on single-crystal substrates requires a lattice-matched or small lattice-mismatched system, leading to only limited combinations of materials suitable for heteroepitaxial growth. Recently, the van der Waals (vdW) epitaxy, regarded as an incommensurate epitaxy, has been considered to enable an heteroepitaxy with large lattice mismatching [9,10]. In contrast to a conventional covalent epitaxy, the van der Waals epitaxy, mediated by the weak van der Waals force, can overcome the limitation of the lattice mismatch between the overlayer and substrate. To date, a considerable number of experimental and computational studies have been made on the van der Waals epitaxy, such as the epitaxial growth of the two-dimensional layered (2D) materials on either 2D or 3D (e.g., sapphire) single-crystal substrates [11–14]. In addition, the epitaxial growth of 3D materials on van der Waals crystals whose surface is chemically inert, and the lack of dangling bonds, such as muscovite mica, is also considered a category of the vdW epitaxy paradigm [15]. By taking advantage of the weak-interlayer van der Waals force, the ultra-thin mica sheet, with a thickness in the range of several micrometers, can be readily obtained by cleaving along the (001) plane; the resulting muscovite mica sheet is highly transparent and flexible, making it a suitable substrate for application in flexible optoelectronics [15–17]. Thus, to realize an advanced functional oxide heteroepitaxy, there is considerable interest in the vdW epitaxial growth of various functional-oxide nanostructures or layers on the muscovite mica substrates [16–22].

In the last few years, many articles have been devoted to the study of the vdW epitaxial growth of ZnO nanostructures or layers on muscovite mica substrates by using the vapor-phase deposition process [23]. The Al-doped ZnO epitaxial layers, grown on mica substrates, exhibited good flexibility and superior durability [17]. Li et al. reported the epitaxial growth of quasi-2D ZnO single crystal plates on mica substrates by applying the technique of pulse-laser-deposition (PLD)-assisted vdW epitaxy. Moreover, due to weak interfacial interaction, the resulting ZnO plates could be easily transferred from the mica substrate onto the SiO_2/Si substrate to demonstrate the applications of vdW epitaxial ZnO crystals in optoelectronic devices, including the self-powered ultraviolet (UV) photodetector and UV light-emitting diode, respectively [24]. Thus far, the vdW epitaxial growth of ZnO films on mica substrates has been mainly accomplished through the high-vacuum vapor phase epitaxy technique. Although the solution-phase vdW epitaxy of ZnO nanowires on mica substrates has been proven feasible [25], little attention has been given to the point of vdW epitaxial growth of ZnO layers on mica substrates by the solution-based or atmospheric-pressure process. In this study, we employed hydrothermal growth and atmospheric pressure solution-processed mist chemical vapor deposition (Mist-CVD) to implement the vdW epitaxial growth of ZnO films on mica substrates, because of their low-cost equipment and non-vacuum system [26]. Furthermore, the study also proposed a seed-assisted hydrothermal growth of vdW epitaxial ZnO films by combining the Mist-CVD process with low-temperature hydrothermal growth, and investigated the effect of growth conditions on the vdW epitaxial growth of ZnO films on mica substrates.

2. Materials and Methods

A freshly cleaved muscovite mica (V-1 grade) was used as the substrate for the van der Waals epitaxial growth of ZnO crystals. For hydrothermal growth of ZnO, the samples were immersed in an aqueous solution of 0.1 M zinc nitrate ($Zn(NO_3)_2 \cdot 6H_2O$, 99%, Showa) and 0.1 M hexamethylenetetramine (HMT) (($CH_2)_6N_4$, 99%, Showa) at 90 °C for 5 h. A different concentration of tri-sodium citrate dihydrate ($Na_3C_6H_5O_7 \cdot 2H_2O$, 99%, Aencore) was added to the solution. The surface pretreatment of mica substrates was performed by thermal annealing in air at various temperatures for 3 h. For seed-assisted hydrothermal growth, the atmospheric-pressure Mist-CVD process was employed to implement the van der Waals epitaxial growth of ZnO seeds on mica substrates, followed by subsequent hydrothermal growth of ZnO layers. The configuration of the Mist-CVD apparatus used in this work is illustrated in Figure 1. The apparatus consisted of a mist generator and a horizontal quartz-tube furnace. In this study, the zinc source was zinc acetic dehydrate ($Zn(CH_3COO)_2 \cdot 2H_2O$, 99%, Showa). An amount of 0.1 M zinc acetic was dissolved in a solution mixture of deionized water and acetic acid (70:30). The precursor solution was atomized into liquid aerosol particles by a 2.4 MHz ultrasonic transducer, and the aerosols formed were transferred into a tube furnace with the carrier gas of either nitrogen (N_2, 99.99% purity) or oxygen (O_2, 99.99% purity), respectively, at a flow rate of 2 L/min. The temperature and growth time for the vdW epitaxial growth of ZnO seeds on mica were 600 °C and 30 min, respectively.

To start with, surface morphologies of the samples were observed by field-emission scanning electron microscopy (SEM, Hitachi, S4700). The structural properties of ZnO films grown on mica were examined using an X-ray diffractometer (XRD, PANalytical, X'Pert PRO-MPD). A cross-sectional sample was prepared with the focused ion beam technique (FIB, FEI, Nova-200 NanoLab). The ZnO–mica interface was also observed by high-resolution transmission electron microscopy (HRTEM, JEOL, JEM-2100F). The HRTEM image of the ZnO–mica interface was further analyzed by Gatan Digital Micrograph™ software. The geometric phase analysis (GPA), developed by Hüytch et al., was performed on the HRTEM image to produce the strain map around the interface [27]. The Strain++ program was used for GPA [28–30]. Next, the room-temperature photoluminescence (PL) spectrum was recorded by a Horiba Jobin Yvon HR800 system, with a 325 nm He-Cd laser as an excitation source. The transmission spectrum was obtained using a UV-Vis spectrophotometer (UV-Vis, Thermal Scientific, Evolution 201). Finally, the electrical resistivity, carrier concentration, and mobility of the ZnO films were measured by a Hall-effect measurement analyzer (MarChannel, AHM-800B).

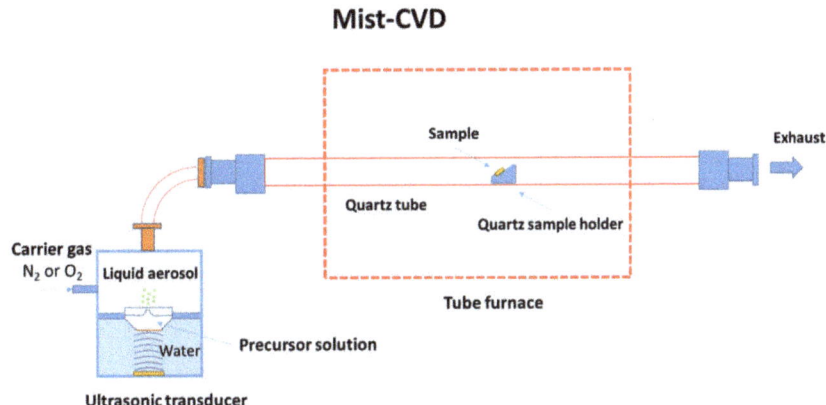

Figure 1. The schematic illustration of Mist-CVD system.

3. Results

3.1. Van der Waals Epitaxial Growth of ZnO Films on Mica Substrates by Low-Temperature Hydrothermal Growth Process

Figure 2a demonstrates the SEM image of ZnO micro-rods hydrothermally grown on the as-cleaved mica substrate with 0.25 mM sodium citrate. The image shows that the majority of ZnO micro-rods were oriented in random directions on the mica substrates. Despite a few ZnO micro-rods with in-plane alignment existing on the sample (Figure 2b), the epitaxial growth of ZnO crystals on as-cleaved mica, without other pretreatments in the low-temperature aqueous solution, was indeed very difficult. Although a freshly cleaved mica surface was usually used as substrates for the van der Waals epitaxy of various oxide systems, the mica surface pretreatment was also the critical for the van der Waals epitaxy of other semiconductor materials [31,32]. Thermal annealing is a common means to modify the substrate surface for epitaxial growth [33,34]. Hence, we performed the thermal annealing treatment of the mica substrates at various temperatures. Figure 2c shows the hydrothermal growth of ZnO micro-rods performed on the 400 °C annealed mica substrate; notably, several epitaxial ZnO micro-rods, exhibiting in-plane alignment, can be observed on such substrates. Furthermore, with the increase in annealing temperature to 600 °C, a high density of vertical aligned ZnO micro-rods grew on mica substrates (Figure 2d). The magnified SEM image (Figure 2e) shows an in-plane alignment of most of the vertically aligned ZnO micro-rods, indicating that the high-temperature annealing treatment (~600 °C) of mica substrates is beneficial to the van der Waals epitaxial growth of ZnO on mica in the aqueous solution at low temperature (~90 °C). In addition, highly dense in-plane aligned ZnO rods, partially coalescing into continuous layers, can be found in the other regions of the same sample, as shown in Figure 2f. For conventional hydrothermal growth of ZnO micro-rods, the addition of tri-sodium citrate into the growth solution can inhibit the growth along the *c*-axis of ZnO and promote the growth of ZnO along the lateral direction. To demonstrate the addition effect of tri-sodium citrate in the zinc nitrate/HMT precursor solution for the growth of ZnO crystals on mica substrates, the hydrothermal growth of ZnO crystals in the solutions containing various citrate concentrations was implemented on 600 °C annealed mica substrates. Figure 3a shows the SEM image of the ZnO micro-rods epitaxially grown on mica substrates in the precursor solution without citrate. The diameters of the ZnO micro-rods were, obviously, much smaller (~1.3 µm) than those grown (3 µm) in the solution with 0.25 mM citrate (Figure 2d). Remarkably, a fully continuous ZnO layer was formed on the mica substrate in the solution containing the citrate concentration of 0.5 mM, as shown in Figure 3b, indicating that the addition of tri-sodium citrate in the precursor solution is a key factor for growing fully coalesced ZnO films on mica substrates.

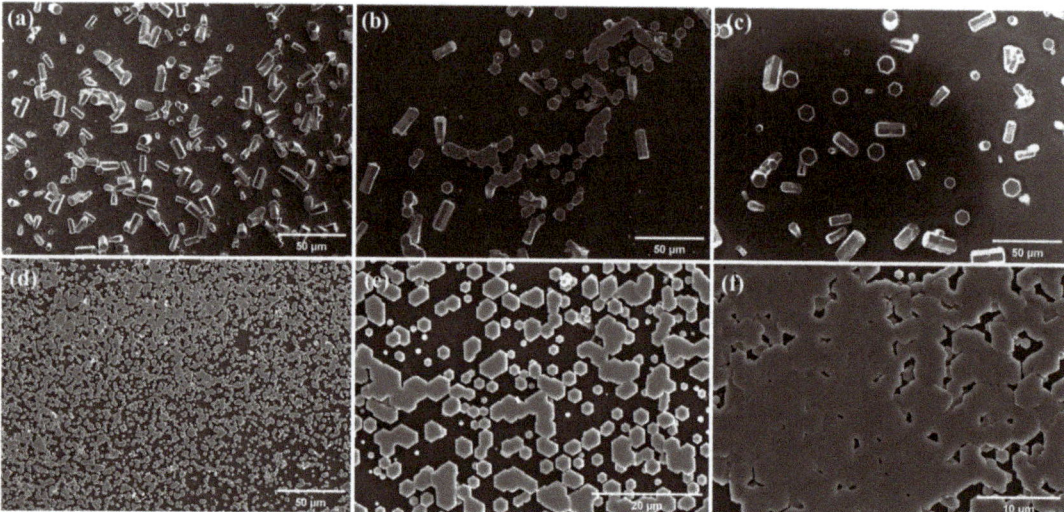

Figure 2. (**a**) SEM images of randomly oriented hydrothermally grown ZnO micro-rods, from the solution containing 0.25 mM sodium citrate, on as-cleaved mica substrate and (**b**) a few ZnO micro-rods with in-plane alignment existing on the other region of the same sample. SEM images of ZnO micro-rods hydrothermally grown on (**c**) 400 °C and (**d**) 600 °C annealed mica substrates. (**e**) Magnified SEM image of (**d**). (**f**) Partially coalesced ZnO layers formed on the other region of 600 °C annealed mica substrate.

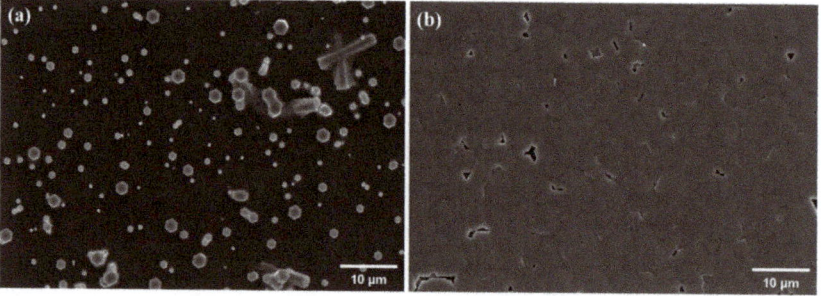

Figure 3. SEM images of ZnO micro-rods hydrothermally grown on 600 °C annealed mica substrates from the solution containing (**a**) 0 mM and (**b**) 0.5 mM sodium citrate.

The XRD measurement was carried out to gain insight into the microstructural properties of the epitaxial ZnO layer grown on the mica substrate in the solution with 0.5 mM citrate. In Figure 4a, the profile of the θ-2θ scan of the continuous ZnO film grown on the mica substrate revealed the preferred orientation of the c-axis, consisting of the results of SEM observation, as shown in Figure 3b. Figure 4b displays the XRD φ-scan profiles of ZnO {10Ī1} and mica {202} planes. Based on the results of XRD φ-scans, the ZnO film was in epitaxy with the mica substrate, with the in-plane orientation relationship of $(0001)_{ZnO} || (001)_{mica}$ and $[11\bar{2}0]_{ZnO} || [10]_{mica}$. To evaluate the mosaic spread and epitaxial quality of the epitaxial ZnO layer, hydrothermally grown on the mica substrate, we recorded the X-ray rocking curve (XRC) profile of the ZnO (0002) plane, as shown in Figure 4c. The full-width at half-maximum (FWHM) value of XRC, for the (0002) plane of ZnO, extracted by fitting a single pseudo-Voigt function was 0.51°, which was significantly broader than those of ZnO epitaxial layers grown on sapphire substrates under the similar solution conditions [35,36].

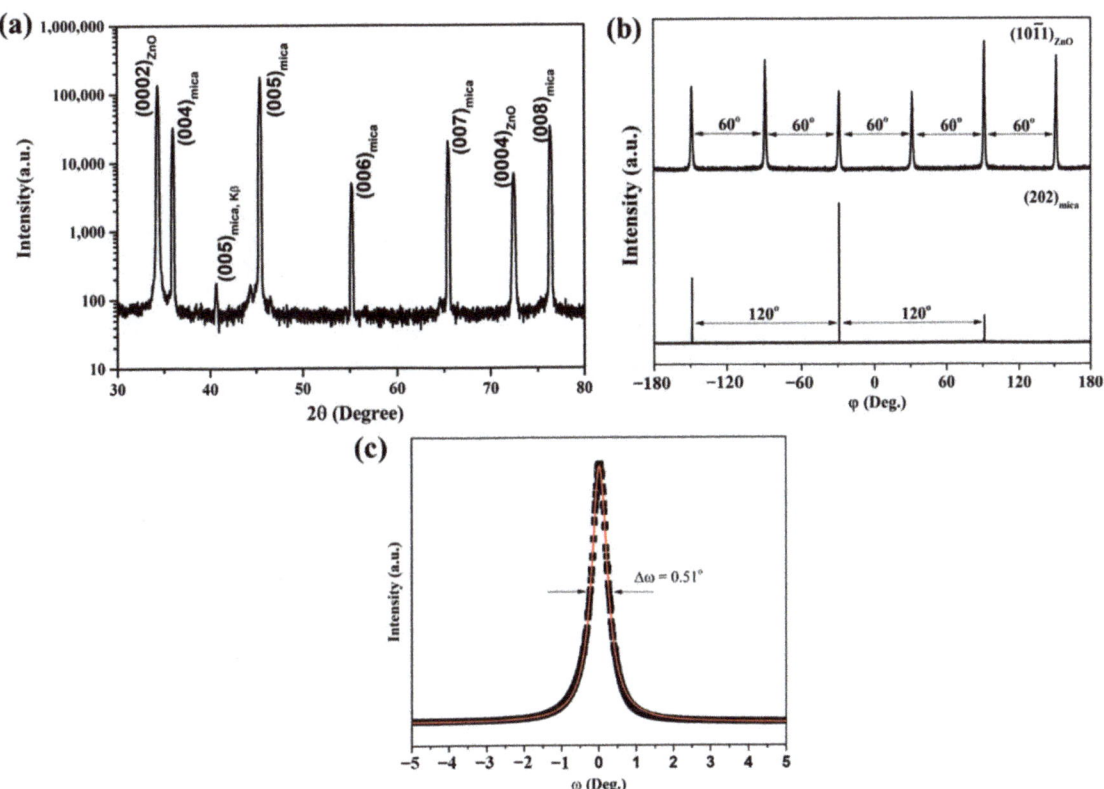

Figure 4. (a) XRD θ-2θ profile, (b) XRD φ-scan of the ZnO {10Ī1} and mica {202} reflections, and (c) Rocking curve profile of the ZnO (0002) reflection for the continuous ZnO film hydrothermally grown on 600 °C annealed mica substrate from a solution containing 0.5 mM sodium citrate.

We explored the interfacial characteristics of the vdW epitaxial ZnO micro-rods and film hydrothermally grown on mica substrates by cross-sectional transmission electron microscopy. Figure 5a,c show the cross-sectional TEM bright-field images of the ZnO micro-rods and film hydrothermally grown on the 600 °C annealed mica substrates from the solutions containing 0.25 mM and 0.5 mM sodium citrate, respectively. The corresponding selective-area-electron-diffraction (SAED) patterns were taken from their interfacial regions, as shown in Figure 5b,d, respectively, showing that the orientational relationship between ZnO and mica was $(0001)_{ZnO} || (001)_{mica}$ and $[11\bar{2}0]_{ZnO} || [310]_{mica}$ for the ZnO micro-rods grown from solution with 0.25 mM sodium citrate (Figure 5b) and $(0001)_{ZnO} || (001)_{mica}$ and $[11\bar{2}0]_{ZnO} || [10]_{mica}$ for the ZnO film grown from the solution with 0.5 mM sodium citrate (Figure 5d). Due to the (001) plane of mica possessing a quasi-hexagonal symmetry, the angle between directions [10] and [310] was 59.96°, which is very close to 60°. Therefore, while the ZnO <11$\bar{2}$0> directions were parallel to the mica <010> directions, the other ZnO <11$\bar{2}$0> directions were almost nearly parallel to mica <310> directions. Based on the SAED patterns, the epitaxial relationship between the ZnO layer and mica substrate can be estimated as $(0001)_{ZnO} || (001)_{mica}$ and $[11\bar{2}0]_{ZnO} || [010]_{mica}$, in agreement with the results of the XRD φ-scan. Based on the cross-sectional TEM images, the height of the resulting ZnO micro-rods (~4.2 µm) grown from the solution with 0.25 mM sodium citrate was higher than that from the solution with 0.5 mM sodium citrate (~2.3 µm); hence, the higher concentration of sodium citrate obviously reduced the growth rate along the *c*-axis of ZnO crystals, consistent with the results of other reported studies in the literature. The effect of sodium citrate as a crystal habit modifier on the morphology of ZnO has been widely studied in the previous works [37–40].

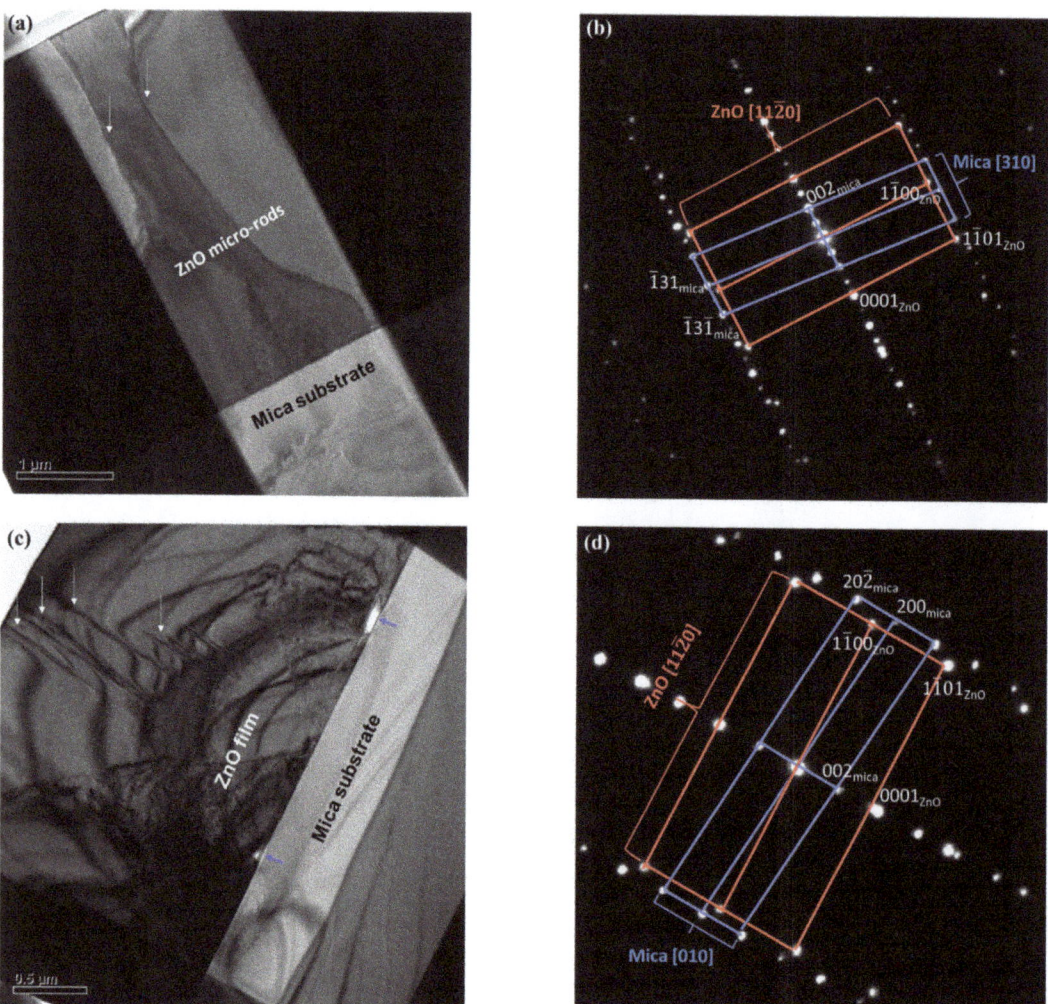

Figure 5. (**a**) Cross-sectional bright-field TEM images of ZnO micro-rods and film hydrothermally grown on mica substrates from solutions containing (**a**) 0.25 mM and (**c**) 0.5 mM sodium citrate. (**b**,**d**) Corresponding SAED patterns recorded at the interfacial regions.

In the previous report of the solution-phase vdW epitaxy of ZnO wires on the mica substrate, the initial growth behavior of ZnO on mica, following the three-dimensional island growth, was named the Volmer–Weber (VW) model. Two or more tiny wires grown from nucleation islands were merged to form a single ZnO wire with a larger diameter during the growing process. However, despite the coalesced ZnO wire being confirmed as a single crystalline, the resulting wires would suffer certain lattice distortions or strains due to the imperfect attachment of the different adjacent wires during the growth process [25]. In this work, remarkably, the cross-sectional TEM image revealed the threading dislocations present in either the micro-rods or the epitaxial layer, as indicated by white arrows in Figure 5a,c. It is considered that the threading dislocations formed at the merging boundaries accommodated the misorientations, originated from the imperfect attachment among adjacent crystallites, and resulted in the broadening of the XRC peak (also see Figure 4c) [41]. Furthermore, because of the less surface migration of growth species at low growth temperature, the spaces, among these nucleation islands, persisted during the growth process and left several tiny voids, as marked by blue arrows in Figure 5c, at the heterointerface between the ZnO film and mica substrate [25].

The interfacial structure between the ZnO film and mica substrate (Figure 5c) was visualized by using high-resolution transmission electron microscopy (HRTEM), viewed along the mica [10] zone-axis, as shown in Figure 6a. An atomically abrupt interface, without an obvious intermediate layer, can be observed at the ZnO/mica interface region. The corresponding fast Fourier transform (FFT) pattern (Figure 6b) of the HRTEM image was consistent with the SAED pattern (Figure 5d). The Fourier-filtered image reconstructed by utilizing ZnO $1\bar{1}00$ and mica 200 reflections is shown in Figure 6c. The measured average lattice spacings of $(1\bar{1}00)_{ZnO}$ and $(200)_{mica}$ were about 0.2814 nm and 0.260 nm, respectively; the values were nearly identical to the values of the standard JCPDS for ZnO (JCPDS card No. 36-1451) and muscovite mica (JCPDS card No. 06-0263), respectively. At the interface, the periodic misfit dislocations (indicated by red arrows) were evident in the Fourier-filtered image (Figure 6c), with every 12 planes of ZnO $(1\bar{1}00)$ matching with 13 planes of mica (200), which is the same as the results reported by Utama et al. [23]. The calculated lattice mismatch between $(1\bar{1}00)$ZnO and (200)mica was about 8.23%, which was accommodated by matching $12 \times d(1\bar{1}00)$ZnO with $13 \times d(200)$mica. Due to the near-strain relaxation of ZnO, the incommensurate van der Waals epitaxy was considered as the valid paradigm for epitaxial growth of the ZnO nanowire on the mica substrate.

For conventional covalent heteroepitaxy, the epitaxial films are strained to match the underlying substrate; hence, the coherently strained layers exist adjacent to the interface. To explore the lattice strain effect in the epitaxial ZnO layer, geometric phase analysis (GPA) based on the HRTEM image was conducted to produce nanoscale strain maps. Figure 7a shows the other HRTEM image of the ZnO layer/mica substrate interface, viewed along ZnO $[11\bar{2}0]$ direction. The x- and y-axis were parallel to the ZnO $[1\bar{1}00]$ and ZnO [1] directions, respectively. The $1\bar{1}00$ and 0001 reflections of the ZnO were chosen from the corresponding FFT pattern to implement geometric phase analysis (Figure 7b). The GPA strain maps of strain components ε_{xx} and ε_{yy} obtained from the region adjacent to the interface are illustrated in Figure 7c,d, respectively; notably, both strain maps of ε_{xx} and ε_{yy} exhibited a nearly uniform distribution of the strain with the value close to zero (0.0 ± 0.01) in the whole region of ZnO. The absence of a gradual variation in strain along the growth direction indicated that the strained layer was not present around the interface. Therefore, the epitaxial growth of ZnO on mica was nearly fully relaxed. In this work, the epitaxial growth paradigm of ZnO film on the mica substrate was regarded as an incommensurate van der Waals epitaxy.

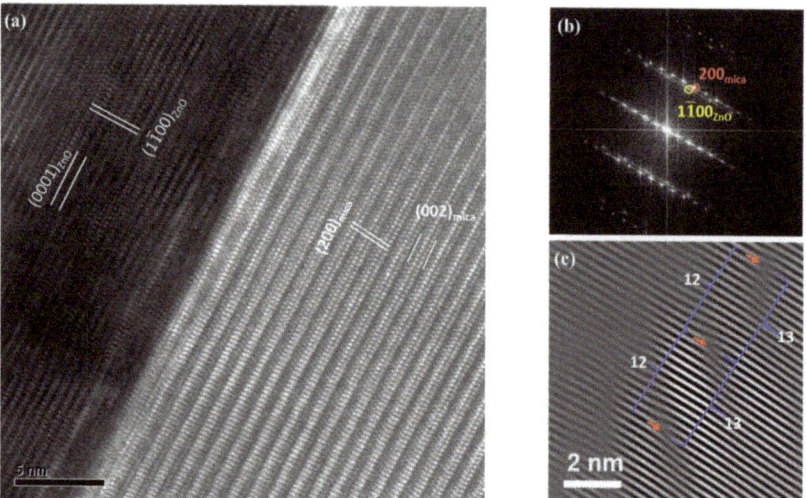

Figure 6. (**a**) HRTEM image of the ZnO layer–mica substrate interface viewed along the mica [10] zone axis. (**b**) Corresponding FFT pattern of (**a**). (**c**) Fourier-filtered image reconstructed utilizing ZnO $1\bar{1}00$ and mica 200 reflections. The black triangles in the Fourier-filtered image indicate the position of misfit dislocation.

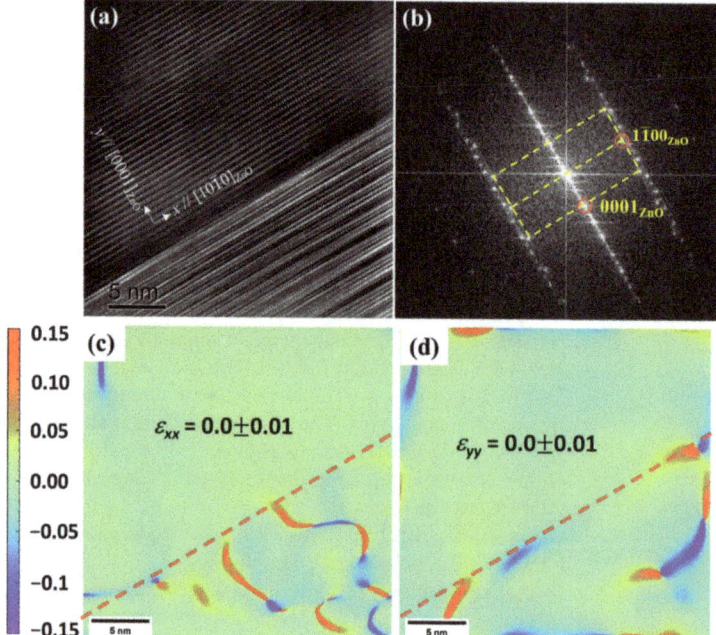

Figure 7. (**a**) HRTEM image of the region adjacent to the interface between the ZnO layer and the mica substrate viewed along the ZnO [$11\bar{2}0$] zone axis. (**b**) Corresponding FFT pattern of (**a**). The $1\bar{1}00$ and 0001 reflections of the ZnO were chosen to implement geometric phase analysis. The GPA strain maps of strain components (**c**) ε_{xx} and (**d**) ε_{yy} obtained from the region adjacent to the interface. The red-dashed lines indicate the interface.

3.2. Van der Waals Epitaxial Growth of ZnO Films on Mica Substrates through the High-Temperature Seed-Layer-Assisted Hydrothermal Growth

For solution-phase van der Waals epitaxial growth of ZnO films on mica, less surface migration of Zn-related species in the low-temperature environment led to the growth of ZnO islands with a sparse nucleation density at the initial growth stage. As a result, the coalesced epitaxial layer, following hydrothermal growth, featured a mosaic morphology with large misorientations. Namely, the increase in nucleation density of ZnO crystallites on mica at the initial growth stage is a straightforward strategy to ameliorate the growth quality of ZnO epitaxial layers on mica. In this study, the high-temperature seed-layer-assisted hydrothermal growth was performed. The seed layers can act as nucleation sites to facilitate the formation of epitaxial ZnO crystallites on mica substrates with high density. Prior to the hydrothermal growth of ZnO layers, the epitaxial ZnO seeds were grown on the mica substrate using an atmospheric-pressure Mist-CVD process for 30 min. Considering that the Mist-CVD process was carried out at high temperature (600 °C) under atmospheric pressure, the as-cleaved mica substrates without annealing treatments were used as the substrates for the vdW epitaxial growth of ZnO seeds. Taking advantage of the elevated temperature, the adatom received sufficient thermal energy to enhance surface migration, promoting the ZnO crystallite nucleation on mica at the initial growth stage and eliminating the defects or/and imperfect arrangement in lattices. In order to scrutinize the effect of ambient conditions on the growth behavior of ZnO seeds on mica during the Mist-CVD process, the nitrogen (N_2) and oxygen (O_2) were used as the carrier gases for the growth of ZnO seed layers on mica substrates, designated as Type A and Type B seeds, respectively. Figure 8a,b show the SEM images of two types of ZnO seeds grown on mica substrates. The surface morphologies revealed most of the ZnO crystallites with hexagonal facets and an in-plane alignment on both samples, and it suggested that the ZnO seeds were epitaxially grown on mica substrates by the Mist-CVD process under either nitrogen or oxygen ambient conditions. Two forms of ZnO crystallites were observed with different morphologies grown on the mica substrate under nitrogen ambient (Type A): One of the ZnO crystallites presented a hexagonal rod-like morphology with a sharp tip, and an average diameter of about 1.5 μm; the other form exhibited a flat top morphology with a larger diameter of 3~5 μm. Notably, several ZnO crystallites were rotated by 30° with respect to the mica substrate, as indicated by the white arrows. The 30°-rotated domain usually existed in the epitaxial ZnO layers grown on c-plane sapphire substrates and the (111) plane spinel substrates [42,43]. In contrast, the high density and uniform distribution of well-faceted ZnO crystallites preferentially occurred under oxygen ambient (Type B). Subsequently, both types of Mist-CVD-grown ZnO crystallites were used as a seed layer for the hydrothermal growth of continuous ZnO films. Figure 8c,d display the surface morphologies of the ZnO films grown on both types of ZnO seed layers through the hydrothermal process for 5 h under the same hydrothermal growth condition as that described before. Completely continuous ZnO films on either Type A or Type B seeds were evident. For the ZnO films grown on Type A seeds, they revealed a surface morphology featuring typical hexagonal mosaic structures. In addition, the 30°-rotated domains, as indicated by white arrows, were present in the ZnO layer (Figure 8c). On the other hand, the ZnO layer grown on Type B seeds exhibited a nearly flat and smooth surface morphology, as shown in Figure 8d.

XRD measurement was performed to evaluate the structural characteristic of the coalesced ZnO films grown by the seed-layer-assisted hydrothermal growth. Figure 9 shows the XRD measurements of ZnO films grown on both types of ZnO seed layers. The θ-2θ scan profiles of both samples revealed the ZnO films, featuring a c-axis-preferred orientation, as shown in Figure 9a. Interestingly, the obvious difference in the XRD φ-scan profiles of the two samples can be observed in Figure 9b,c. For the ZnO films grown on Type A seeds, two sets of peaks with six-fold symmetry, relative to each other through 30° rotation about the surface normal, were present in the ZnO {10$\bar{1}$1} φ-scan, indicating that two orientation variants of ZnO coexisted on the mica substrate, in agreement with the SEM observation (Figure 8c). We deduced the in-plane crystallographic relationship between the ZnO and mica to be $[11\bar{2}0]_{ZnO} || [10]_{mica}$ (original orientation variant) and $[11\bar{2}0]_{ZnO} || [100]_{mica}$ (30°-rotated domain), respectively. Instead of the original orientation variant, the 30°-rotated domain was the dominant variant in the ZnO layer grown on the seed layer (Type A) under nitrogen ambient conditions. In comparison, the XRD φ-scans for ZnO {$10\bar{1}1$} reflections corresponding to the ZnO film grown on Type B seeds featured merely one set of peaks with six-fold symmetry; moreover, the in-plane crystallographic relationship was confirmed to be $[11\bar{2}0]_{ZnO} || [10]_{mica}$. The ambience during Mist-CVD growth of ZnO seeds greatly influenced the epitaxial growth behavior of ZnO crystallites on the mica substrate at the initial stage. The reason for the epitaxial orientations of the ZnO seeds on the mica substrate as a function of growth ambient is still not explicit at the moment. Nevertheless, the Mist-CVD growth environment, as well as the mica surface conditions, might influence the adatom arrangement on the mica surface, and the detailed mechanism will be clarified in future work.

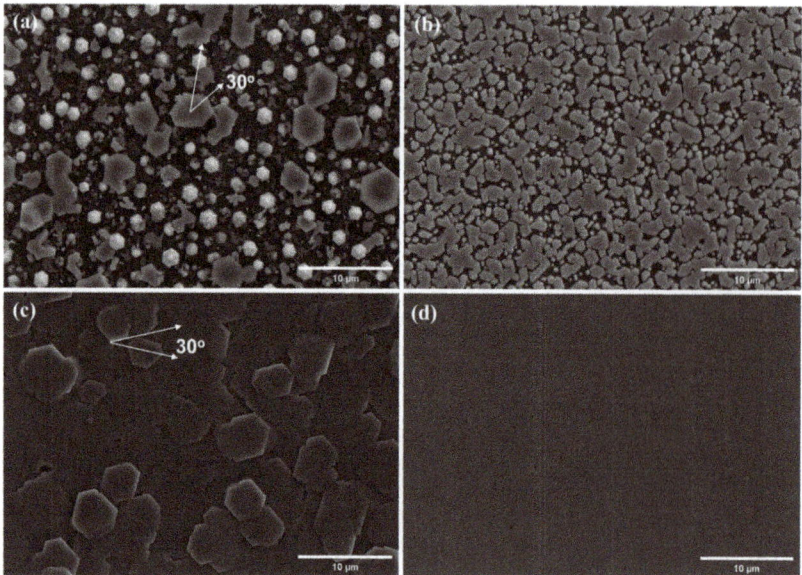

Figure 8. SEM images of ZnO seeds grown on as-cleaved mica substrates by Mist-CVD at 600 °C for 30 min under (**a**) nitrogen (Type A) and (**b**) oxygen (Type B) ambient. Subsequent hydrothermal growth of continuous ZnO layers on (**c**) Type A and (**d**) Type B seeds.

The epitaxial quality of the coalesced ZnO films grown through the seed-assisted hydrothermal growth was assessed by XRD rocking curve measurement. We performed XRC measurements in on-axis symmetric (ZnO (0002) plane) and off-axis skew-symmetric (ZnO ($10\bar{1}1$) plane) geometries to examine the out-of-plane and in-plane mosaic spreads of the ZnO epitaxial films. Figure 10 shows the results of the XRC analysis of the ZnO (0002) and ($10\bar{1}1$) planes for both samples. As expected, the XRC FWHM values ($\omega = 0.26°$ for (0002) plane; $\Delta\omega = 0.7°$ for ($10\bar{1}1$) plane) of the ZnO layer grown on Type B seeds were smaller than those of the film grown on Type A seeds ($\Delta\omega = 0.41°$ for (0002) plane; $\Delta\omega = 1.6°$ for ($10\bar{1}1$) plane). The crystal quality of the epitaxial ZnO layer grown on mica with Type B seeds was compared with those grown on the sapphire substrate under similar growth conditions [35,36].

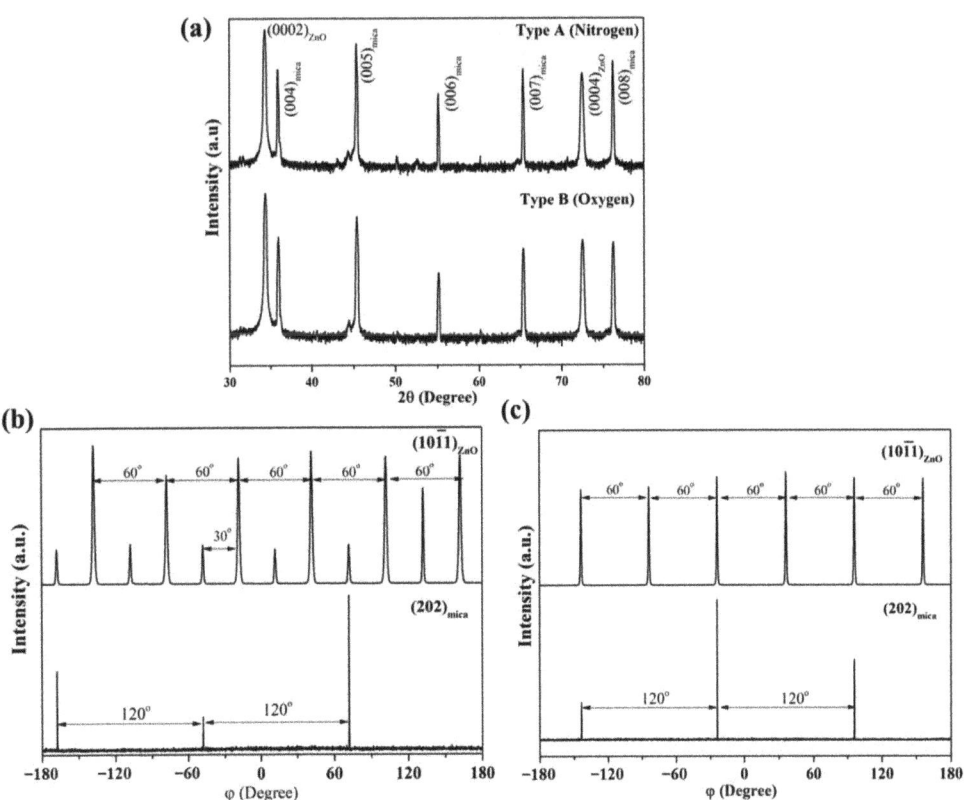

Figure 9. (a) XRD θ-2θ profiles of ZnO layers hydrothermally grown on Type A and Type B seeds. XRD φ-scan of the ZnO {10$\bar{1}$1} and mica {202} reflections for ZnO layers grown on (b) Type A and (c) Type B seeds, respectively.

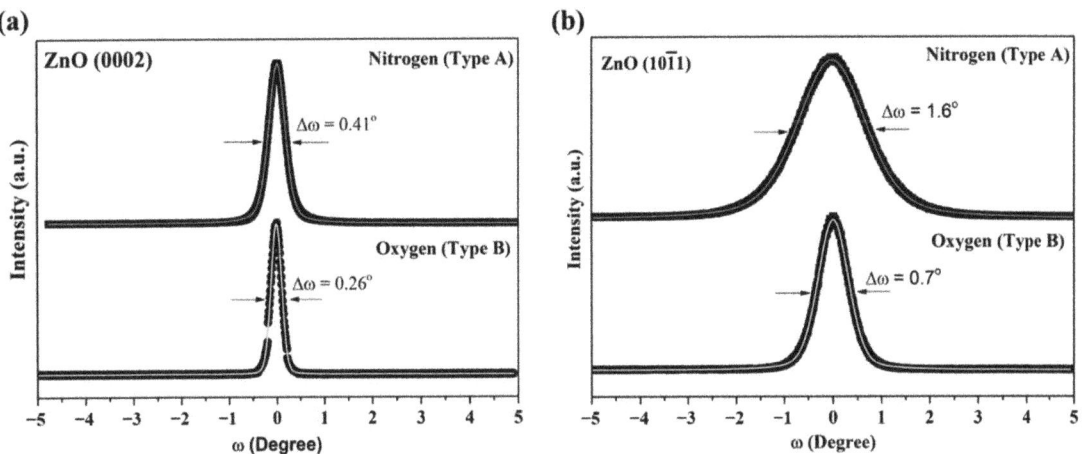

Figure 10. Rocking curve profiles of the (a) ZnO (0002) reflection and (b) ZnO (10$\bar{1}$1) reflection corresponding to ZnO layers hydrothermally grown on Type A and Type B seeds, respectively.

The characterization of optoelectrical and electrical properties is important for evaluating the potential application of vdW epitaxial ZnO layers in optoelectronic devices. We recorded the room-temperature photoluminescence (PL) and UV-Vis spectra of the vdW epitaxial ZnO films, corresponding to Figure 8c,d, hydrothermally grown on mica substrates with Type A and Type B seeds, respectively. In Figure 11a, the PL spectra of both samples exhibited a near-band edge (NBE) emission peak centered at 376 nm, and a broad deep-level emission (DLE) in the range of 450–700 nm, arising from defect-related emissions in the ZnO crystal. Based on the previous reports [44], this deep-level emission is attributed to the recombination of various intrinsic point defects, such as oxygen vacancies (V_o), zinc vacancies (V_{Zn}), and oxygen interstitials (O_i), which are usually abundant in the chemical solution-derived ZnO crystals. Figure 11b shows the UV-Vis spectra of the vdW epitaxial ZnO films grown on Type A and Type B seeds. An absorption edge near the UV region and a high transparency in the visible spectral range can be observed in both samples. The transmittance (T) of vdW epitaxial ZnO films in the visible region were observed to be in the range of 60% to 70% for Type A seeds and 75% to 85% for Type B seeds. The relatively low transmittance of the ZnO film grown on Type A seeds might be attributed to the light scattering occurring on the uneven surface of the film. The optical absorption coefficient (α) was calculated using the following equation:

$$\alpha = -\ln(T)/d \tag{1}$$

where d is the thickness of the film. Figure 11c shows the corresponding $(\alpha h\nu)^2$ versus $h\nu$ (Tauc) plots. The optical band gaps of both vdW epitaxial ZnO layers were estimated to be 3.25 eV through extrapolation of the linear region of the Tauc plots. Furthermore, the electrical properties of the vdW epitaxial ZnO layers grown on Type A and Type B seeds were analyzed by Hall-effect measurement, respectively. Both ZnO films showed n-type conductivity and the results of the electrical properties are summarized in Table 1. Based on the above-mentioned results, it is evident that the vdW epitaxial ZnO on the mica substrate is comparable, in terms of optical and electrical properties, with those grown on sapphire substrates through conventional solution-phase epitaxy techniques [44–46].

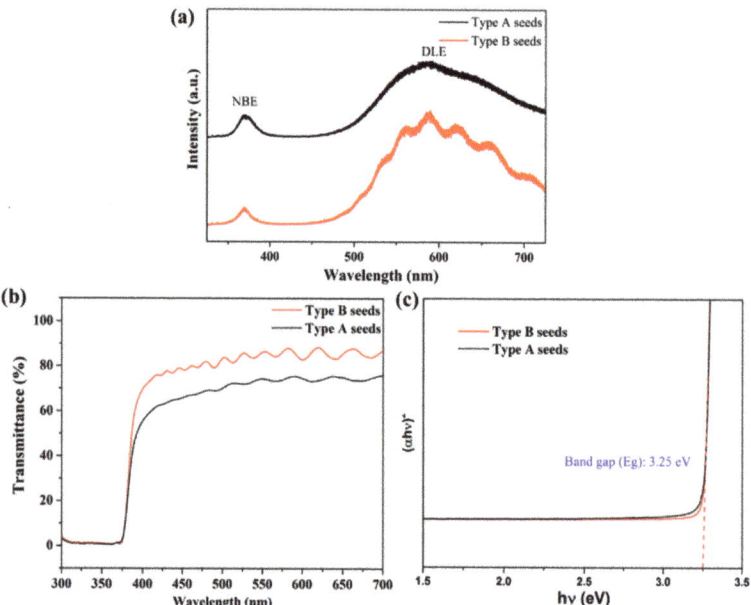

Figure 11. (**a**) Room-temperature PL spectra, (**b**) transmittance spectra, and (**c**) corresponding plots of $(\alpha h\nu)^2$ versus $h\nu$ of the ZnO layers hydrothermally grown on Type A and Type B seeds.

Table 1. Electrical properties of ZnO films hydrothermally grown on Type A and Type B seeds.

Type of Seeds	Electrical Resistivity (Ω-cm)	Carrier Concentration (cm^{-3})	Hall Mobility ($cm^2V^{-1}s^{-1}$)
Type A	0.84	1.5×10^{18}	2.92
Type B	1.05	6.0×10^{17}	11.55

4. Conclusions

In this work, we demonstrated the van der Waals epitaxial growth of ZnO films on mica substrates in a low-temperature aqueous solution. Proper annealing of the mica substrate was found to be critical for the solution-phase vdW epitaxy of ZnO micro-rods. A fully coalesced, continuous, ZnO epitaxial layer could be obtained through the addition of sodium citrate into the solution. The epitaxial relationship between the ZnO epitaxial film and mica substrate was $(0001)_{ZnO} || (001)_{mica}$ and $[11\bar{2}0]_{ZnO} || [10]_{mica}$, respectively. The analysis of HRTEM images revealed the near-strain relaxation of ZnO adjacent to the interface, and we, therefore, arrived at the outcome that the epitaxial paradigm of ZnO films on the mica substrate was the incommensurate vdW epitaxy. Furthermore, the seed-layer-assisted hydrothermal growth, implemented by combining the Mist-CVD process with low-temperature hydrothermal growth, improved significantly the crystal quality of the epitaxial ZnO layers on mica substrates. The in-plane alignment of Mist-CVD grown ZnO crystallites was affected by the growth ambience. The use of oxygen as a carrier gas promoted a high-density and even distribution of ZnO seeds on mica substrates, and it resulted in the growth of a ZnO epitaxial layer with a nearly flat and smooth surface morphology. Moreover, the optoelectronic and electrical properties of the vdW epitaxial ZnO films were comparable to those grown on sapphire substrates by conventional solution-based epitaxy techniques. Thus, these results led to the conclusion that the vdW epitaxial ZnO films grown on mica substrates had great potential for optoelectronic applications.

Author Contributions: Conceptualization, H.-G.C. and H.-S.W.; resources, H.-S.W. and S.-R.J.; writing—original draft, H.-G.C.; writing—review and editing, Y.-H.S.; methodology, T.-Y.Y. and S.-C.C.; investigation, T.-Y.Y. and S.-C.C.; formal analysis, H.-G.C., H.-S.W. and S.-R.J.; supervision, H.-G.C. All authors have read and agreed to the published version of the manuscript.

Funding: This research was funded by the Ministry of Science and Technology, Taiwan, R.O.C., grant numbers MOST 109-2221-E-214-018 and MOST 110-2221-E-214-004.

Institutional Review Board Statement: Not applicable.

Informed Consent Statement: Not applicable.

Data Availability Statement: Not applicable.

Acknowledgments: The authors gratefully acknowledge the use of EM000800 and XRD005100 of MOST 110-2731-M-006-001 belonging to the Core Facility Center, National Cheng Kung University, Taiwan, for HRTEM, FIB, XRD experiment, and thank the MANALAB at ISU for SEM and XRD.

Conflicts of Interest: The authors declare no conflict of interest.

References

1. Özgür, Ü.; Alivov, Y.I.; Liu, C.; Teke, A.; Reshchikov, M.A.; Doğan, S.; Avrutin, V.; Cho, S.-J.; Morkoç, H. A comprehensive review of ZnO materials and devices. *J. Appl. Phys.* **2005**, *98*, 041301. [CrossRef]
2. Kevin, M.; Tho, W.H.; Ho, G.W. Transferability of solution processed epitaxial Ga:ZnO films; tailored for gas sensor and transparent conducting oxide applications. *J. Mater. Chem.* **2012**, *22*, 16442–16447. [CrossRef]
3. Kang, Y.; Yu, F.; Zhang, L.; Wang, W.; Chen, L.; Li, Y. Review of ZnO-based nanomaterials in gas sensors. *Solid State Ion* **2021**, *360*, 115544. [CrossRef]
4. Lee, C.-T. Fabrication Methods and Luminescent Properties of ZnO Materials for Light-Emitting Diodes. *Materials* **2010**, *3*, 2218–2259. [CrossRef]
5. Janotti, A.; Van de Walle, C.G. Fundamentals of zinc oxide as a semiconductor. *Rep. Prog. Phys.* **2009**, *72*, 126501. [CrossRef]
6. Liu, K.; Sakurai, M.; Aono, M. ZnO-based ultraviolet photodetectors. *Sensors* **2010**, *10*, 8604–8634. [CrossRef]

7. Deka Boruah, B. Zinc oxide ultraviolet photodetectors: Rapid progress from conventional to self-powered photodetectors. *Nanoscale Adv.* **2019**, *1*, 2059–2085. [CrossRef]
8. Nandi, S.; Kumar, S.; Misra, A. Zinc oxide heterostructures: Advances in devices from self-powered photodetectors to self-charging supercapacitors. *Mater. Adv.* **2021**, *2*, 6768–6799. [CrossRef]
9. Xiang, Y.; Xie, S.; Lu, Z.; Wen, X.; Shi, J.; Washington, M.; Wang, G.-C.; Lu, T.-M. Domain boundaries in incommensurate epitaxial layers on weakly interacting substrates. *J. Appl. Phys.* **2021**, *130*, 065301. [CrossRef]
10. Koma, A. Van der Waals epitaxy—A new epitaxial growth method for a highly lattice-mismatched system. *Thin Solid Film.* **1992**, *216*, 72–76. [CrossRef]
11. Yang, T.; Zheng, B.; Wang, Z.; Xu, T.; Pan, C.; Zou, J.; Zhang, X.; Qi, Z.; Liu, H.; Feng, Y.; et al. Van der Waals epitaxial growth and optoelectronics of large-scale WSe_2/SnS_2 vertical bilayer p–n junctions. *Nat. Commun.* **2017**, *8*, 1906. [CrossRef] [PubMed]
12. Mattinen, M.; King, P.J.; Popov, G.; Hämäläinen, J.; Heikkilä, M.J.; Leskelä, M.; Ritala, M. Van der Waals epitaxy of continuous thin films of 2D materials using atomic layer deposition in low temperature and low vacuum conditions. *2D Mater.* **2019**, *7*, 011003. [CrossRef]
13. Dumcenco, D.; Ovchinnikov, D.; Marinov, K.; Lazic, P.; Gibertini, M.; Marzari, N.; Sanchez, O.L.; Kung, Y.C.; Krasnozhon, D.; Chen, M.W.; et al. Large-Area Epitaxial Mono layer MoS_2. *ACS Nano* **2015**, *9*, 4611–4620. [CrossRef] [PubMed]
14. Tantardini, C.; Kvashnin, A.G.; Gatti, C.; Yakobson, B.I.; Gonze, X. Computational modeling of 2D materials under high pressure and their chemical bonding: Silicene as possible field-effect transistor. *ACS Nano* **2021**, *15*, 6861–6871. [CrossRef]
15. Chu, Y.-H. Van der Waals oxide heteroepitaxy. *Npj Quantum Mater.* **2017**, *2*, 67. [CrossRef]
16. Li, C.-I.; Lin, J.-C.; Liu, H.-J.; Chu, M.-W.; Chen, H.-W.; Ma, C.-H.; Tsai, C.-Y.; Huang, H.-W.; Lin, H.-J.; Liu, H.-L.; et al. van der Waal Epitaxy of Flexible and Transparent VO_2 Film on Muscovite. *Chem. Mater.* **2016**, *28*, 3914–3919. [CrossRef]
17. Bitla, Y.; Chen, C.; Lee, H.C.; Do, T.H.; Ma, C.H.; Van Qui, L.; Huang, C.W.; Wu, W.W.; Chang, L.; Chiu, P.W.; et al. Oxide Heteroepitaxy for Flexible Optoelectronics. *ACS Appl. Mater. Interfaces* **2016**, *8*, 32401–32407. [CrossRef]
18. Liu, H.-J.; Wang, C.-K.; Su, D.; Amrillah, T.; Hsieh, Y.-H.; Wu, K.-H.; Chen, Y.-C.; Juang, J.-Y.; Eng, L.M.; Jen, S.-U.; et al. Flexible Heteroepitaxy of $CoFe_2O_4$/Muscovite Bimorph with Large Magnetostriction. *ACS Appl. Mater. Interfaces* **2017**, *9*, 7297–7304. [CrossRef]
19. Jiang, J.; Bitla, Y.; Huang, C.-W.; Do, T.H.; Liu, H.-J.; Hsieh, Y.-H.; Ma, C.-H.; Jang, C.-Y.; Lai, Y.-H.; Chiu, P.-W.; et al. Flexible ferroelectric element based on van der Waals heteroepitaxy. *Sci. Adv.* **2017**, *3*, e1700121. [CrossRef]
20. Amrillah, T.; Bitla, Y.; Shin, K.; Yang, T.; Hsieh, Y.-H.; Chiou, Y.-Y.; Liu, H.-J.; Do, T.H.; Su, D.; Chen, Y.-C.; et al. Flexible Multiferroic Bulk Heterojunction with Giant Magnetoelectric Coupling via van der Waals Epitaxy. *ACS Nano* **2017**, *11*, 6122–6130. [CrossRef]
21. Wu, P.-C.; Chen, P.-F.; Do, T.H.; Hsieh, Y.-H.; Ma, C.-H.; Ha, T.D.; Wu, K.-H.; Wang, Y.-J.; Li, H.-B.; Chen, Y.-C.; et al. Heteroepitaxy of Fe_3O_4/Muscovite: A New Perspective for Flexible Spintronics. *ACS Appl. Mater. Interfaces* **2016**, *8*, 33794–33801. [CrossRef]
22. Ma, C.H.; Lin, J.C.; Liu, H.J.; Do, T.H.; Zhu, Y.M.; Ha, T.D.; Zhan, Q.; Juang, J.Y.; He, Q.; Arenholz, E.; et al. Van der Waals epitaxy of functional MoO_2 film on mica for flexible electronics. *Appl. Phys. Lett.* **2016**, *108*, 5. [CrossRef]
23. Utama, M.I.B.; Belarre, F.J.; Magen, C.; Peng, B.; Arbiol, J.; Xiong, Q.H. Incommensurate van der Waals Epitaxy of Nanowire Arrays: A Case Study with ZnO on Muscovite Mica Substrates. *Nano Lett.* **2012**, *12*, 2146–2152. [CrossRef] [PubMed]
24. Li, B.; Ding, L.; Gui, P.; Liu, N.; Yue, Y.; Chen, Z.; Song, Z.; Wen, J.; Lei, H.; Zhu, Z.; et al. Pulsed Laser Deposition Assisted van der Waals Epitaxial Large Area Quasi-2D ZnO Single-Crystal Plates on Fluorophlogopite Mica. *Adv. Mater. Interfaces* **2019**, *6*, 1901156. [CrossRef]
25. Zhu, Y.; Zhou, Y.; Utama, M.I.B.; de la Mata, M.; Zhao, Y.Y.; Zhang, Q.; Peng, B.; Magen, C.; Arbiol, J.; Xiong, Q.H. Solution phase van der Waals epitaxy of ZnO wire arrays. *Nanoscale* **2013**, *5*, 7242–7249. [CrossRef] [PubMed]
26. Fujita, S.; Kaneko, K.; Ikenoue, T.; Kawaharamura, T.; Furuta, M. Ultrasonic-assisted mist chemical vapor deposition of II-oxide and related oxide compounds. *Phys. Status Solidi C* **2014**, *11*, 1225–1228. [CrossRef]
27. Hÿtch, M.J.; Snoeck, E.; Kilaas, R. Quantitative measurement of displacement and strain fields from HREM micrographs. *Ultramicroscopy* **1998**, *74*, 131–146. [CrossRef]
28. Feng, S.; Xu, Z. Strain Characterization in Two-Dimensional Crystals. *Materials* **2021**, *14*, 4460. [CrossRef]
29. Peters, J.J.P.; Beanland, R.; Alexe, M.; Cockburn, J.W.; Revin, D.G.; Zhang, S.Y.Y.; Sanchez, A.M. Artefacts in geometric phase analysis of compound materials. *Ultramicroscopy* **2015**, *157*, 91–97. [CrossRef]
30. Peters, J.J.P. Strain++; Version 1.7.1. 2022. Available online: https://jjppeters.github.io/Strainpp/ (accessed on 1 March 2022).
31. Utama, M.I.B.; de la Mata, M.; Magen, C.; Arbiol, J.; Xiong, Q.H. Twinning-, Polytypism-, and Polarity-Induced Morphological Modulation in Nonplanar Nanostructures with van der Waals Epitaxy. *Adv. Funct. Mater.* **2013**, *23*, 1636–1646. [CrossRef]
32. Utama, M.I.B.; Peng, Z.; Chen, R.; Peng, B.; Xu, X.; Dong, Y.; Wong, L.M.; Wang, S.; Sun, H.; Xiong, Q. Vertically Aligned Cadmium Chalcogenide Nanowire Arrays on Muscovite Mica: A Demonstration of Epitaxial Growth Strategy. *Nano Lett.* **2011**, *11*, 3051–3057. [CrossRef] [PubMed]
33. Shan, H.Y.; Li, J.; Li, S.A.; Zhang, Q.Y. Epitaxial ZnO films grown on ZnO-buffered c-plane sapphire substrates by hydrothermal method. *Appl. Surf. Sci.* **2010**, *256*, 6743–6747. [CrossRef]
34. Wang, Y.; Wang, S.; Zhou, S.; Xu, J.; Ye, G.; Gu, S.; Zhang, R.; Ren, Q. Effects of sapphire substrate annealing on ZnO epitaxial films grown by MOCVD. *Appl. Surf. Sci.* **2006**, *253*, 1745–1747. [CrossRef]

35. Chen, H.-G.; Wang, C.-W.; Tu, Z.-F. Hydrothermal epitaxial growth of ZnO films on sapphire substrates presenting epitaxial ZnAl$_2$O$_4$ buffer layers. *Mater. Chem. Phys.* **2014**, *144*, 199–205. [CrossRef]
36. Chen, H.-G.; Tu, Z.-F.; Wang, C.-W.; Yu, M.-Y. Buffer Layer Assisted Epitaxial Growth of ZnO Films on Sapphire Substrates in Low Temperature Aqueous Solution. *Sci. Adv. Mater.* **2014**, *6*, 1858–1868. [CrossRef]
37. Urgessa, Z.N.; Oluwafemi, O.S.; Botha, J.R. Hydrothermal synthesis of ZnO thin films and its electrical characterization. *Mater. Lett.* **2012**, *79*, 266–269. [CrossRef]
38. Kim, J.H.; Kim, E.M.; Andeen, D.; Thomson, D.; Denbaars, S.P.; Lange, F.F. Growth of heteroepitaxial ZnO thin films on gan-buffered Al$_2$O$_3$(0001) substrates by low-temperature hydrothermal synthesis at 90 degrees C. *Adv. Funct. Mater.* **2007**, *17*, 463–471. [CrossRef]
39. Kim, J.H.; Andeen, D.; Lange, F.F. Hydrothermal growth of periodic, single-crystal ZnO microrods and microtunnels. *Adv. Mater.* **2006**, *18*, 2453–2457. [CrossRef]
40. Das, S.; Dutta, K.; Pramanik, A. Morphology control of ZnO with citrate: A time and concentration dependent mechanistic insight. *Crystengcomm* **2013**, *15*, 6349–6358. [CrossRef]
41. Ravadgar, P.; Horng, R.H.; Ou, S.L. A visualization of threading dislocations formation and dynamics in mosaic growth of GaN-based light emitting diode epitaxial layers on (0001) sapphire. *Appl. Phys. Lett.* **2012**, *101*, 231911. [CrossRef]
42. Liu, C.; Chang, S.H.; Noh, T.W.; Abouzaid, M.; Ruterana, P.; Lee, H.H.; Kim, D.W.; Chung, J.S. Initial growth behavior and resulting microstructural properties of heteroepitaxial ZnO thin films on sapphire (0001) substrates. *Appl. Phys. Lett.* **2007**, *90*, 011906. [CrossRef]
43. Zeng, Z.Q.; Liu, Y.Z.; Yuan, H.T.; Mei, Z.X.; Du, X.L.; Jia, J.F.; Xue, Q.K.; Zhang, Z. Surface modification of MgAl$_2$O$_4$ (111) for growth of high-quality ZnO epitaxial films. *Appl. Phys. Lett.* **2007**, *90*, 081911. [CrossRef]
44. Le, H.Q.; Goh, G.K.L.; Liew, L.L. Nanorod assisted lateral epitaxial overgrowth of ZnO films in water at 90 degrees C. *Crystengcomm* **2014**, *16*, 69–75. [CrossRef]
45. Miyake, M.; Fukui, H.; Doi, T.; Hirato, T. Preparation of Low-Resistivity Ga-Doped ZnO Epitaxial Films from Aqueous Solution Using Flow Reactor. *J. Electrochem. Soc.* **2014**, *161*, D725–D729. [CrossRef]
46. Tezel, F.M.; Karİper, İ.A. Structural and optical properties of undoped and silver, lithium and cobalt-doped ZnO thin films. *Surf. Rev. Lett.* **2020**, *27*, 1950138. [CrossRef]

Article

Optical and Structural Characterization of Cd-Free Buffer Layers Fabricated by Chemical Bath Deposition

William Vallejo [1,*], Carlos Diaz-Uribe [1] and Cesar Quiñones [2]

1. Grupo de Investigación en Fotoquímica y Fotobiología, Facultad de Ciencias Básicas, Universidad del Atlántico, Puerto Colombia 081007, Colombia; carlosdiaz@mail.uniatlantico.edu.co
2. Institución Universitaria Politécnico Gran Colombiano, Bogotá 110231, Colombia; caquinones@poligran.edu.co
* Correspondence: williamvallejo@mail.uniatlantico.edu.co; Tel.: +57-5359-9484

Abstract: Chemical bath deposition (CBD) is a suitable, inexpensive, and versatile synthesis technique to fabricate different semiconductors under soft conditions. In this study, we deposited Zn(O;OH)S thin films by the CBD method to analyze the effect of the number of thin film layers on structural and optical properties of buffer layers. Thin films were characterized by X-ray diffraction (XRD) and UV-Vis transmittance measurements. Furthermore, we simulated a species distribution diagram for Zn(O;OH)S film generation during the deposition process. The optical results showed that the number of layers determined the optical transmittance of buffer layers, and that the transmittance reduced from 90% (with one layer) to 50% (with four layers) at the visible range of the electromagnetic spectrum. The structural characterization indicated that the coatings were polycrystalline (α-ZnS and β-Zn(OH)$_2$ to four layers). Our results suggest that Zn(O;OH)S thin films could be used as buffer layers to replace CdS thin films as an optical window in thin-film solar cells.

Keywords: solar cells; buffer layers; thin films; chemical bath deposition

Citation: Vallejo, W.; Diaz-Uribe, C.; Quiñones, C. Optical and Structural Characterization of Cd-Free Buffer Layers Fabricated by Chemical Bath Deposition. *Coatings* **2021**, *11*, 897. https://doi.org/10.3390/coatings11080897

Academic Editor: Alessandro Latini

Received: 17 June 2021
Accepted: 24 July 2021
Published: 27 July 2021

Publisher's Note: MDPI stays neutral with regard to jurisdictional claims in published maps and institutional affiliations.

Copyright: © 2021 by the authors. Licensee MDPI, Basel, Switzerland. This article is an open access article distributed under the terms and conditions of the Creative Commons Attribution (CC BY) license (https://creativecommons.org/licenses/by/4.0/).

1. Introduction

Nowadays, the main primary energy source for electricity generation remains fossil fuels (oil, coal, natural gas), which are a non-renewable resource presenting a negative environmental impact due to emissions of greenhouse gases and other polluting by-products [1–3]. This traditional energy source does not warrant a long-term supply of the growing demand for energy created by population and industry growth [4–6]. This trend and the anthropogenic effect of human activities on the atmosphere, aquatic quality, and biodiversity are challenges for the near future [7,8]. This situation has generated great global interest in the search for new energy sources, preferably renewable ones. The Renewable Energy Police Network for the 21st Century (REN21) reported that 79.9% of the world's total energy consumption was supplied by fossil fuels, 2.2% by nuclear energy, 6.9% by traditional biomass, and 11% by modern renewables (e.g., biomass/solar/geothermal heat (4.3%), hydropower (3.6%), wind/solar/biomass/ocean/geothermal power (2.1%), and biofuels for transport (1.0%). Among modern renewables, the global solar installed photovoltaic (PV) capacity grew more than 200 gigawatts (GW) in 2019 [9]. In the last 6 years, the increasing adoption rate of photovoltaic systems has also led to a price drop in excess of 80% [10]. Currently, crystalline silicon PV cells represent more than 85% of world PV cell market; however, the thin-film chalcogenide PV technology has shown a rapid growth compared to that of silicon due in part to its low cost of production [11]. The Cu(In,Ga)(Se,S)$_2$ (CIGS) is one of the most researched materials as absorbent layers within thin-film PV technologies. In general, most of the CIGS-based solar cells include a very thin CdS (<100 nm) as a buffer layer to reduce crystalline mismatch between the chalcopyrite absorber layer and the transparent ZnO front electrode [12,13]. In the last two decades, serious efforts have been made to replace the CdS buffer layer by other nontoxic material [14–16]. Actually, in their last report, Green et al. reported an efficiency value of 23.3% for CIGS cells free of Cd [17]. Different compounds have been reported as alternatives to fabricate Cd-free buffer layers (e.g., ZnS [18], Zn(O,S) [15], ZnSe [19], In_2S_3 [20], CdS [21]). Among the semiconductor options, Zn(O,OH)S coatings are widely used as thin films in the fabrication of luminescent materials, light-emitting diodes, electroluminescent devices, optical covers, reflectors, and dielectric filters [22–27]. Since Zn(O;OH)S films have n-type conductivity and a large direct Eg bandgap (3.6–3.9 eV), they are suitable for use as buffer layers for thin-film solar cells [28].

Various deposition methods have been reported for the fabrication of semiconductor thin films: (i) thermal evaporation [29,30], (ii) sputtering [31], (iii) atomic layer deposition [32], (iv) electrochemical [33], (v) chemical vapor deposition [34], and (vi) chemical bath deposition (CBD) [35–37]. Among the physical and chemical methods of thin film deposition, the CBD process is the most economic and technically suitable one (e.g., regarding lab equipment and temperature and pressure requirements) [38]. In the typical procedure, with a temperature below that of the boiling point of water and under atmospheric pressure, a source of metal and chalcogenide are mixed in a vessel. First, temperature and pH are adjusted. Then, the solid substrate is immersed inside the reaction vessel, and the thin deposition starts. CBD is a convenient method for buffer layer deposition. However, thin film semiconductors grown using the CBD process produce large amounts of waste solvent and chemicals that then require costly waste processing [39]. The possibility of increasing optical transmission and deleting a toxic element (Cd) has directed the research in the field to study different synthesis parameters to optimize the CBD process (e.g., the effect of temperature, chalcogenide and metal source, complexing agents, pH, stirring). The main reports studied the physical–chemistry properties in thin films with only one coating (one layer) [40]. In this paper, we studied the effect of the number of buffer layers on the optical and structural properties of Zn(O;OH)S coatings deposited by CBD.

2. Materials and Methods

2.1. Thin Film Deposition by the CBD Process

Although the CBD process has been used in thin-film semiconductor synthesis for several decades, most reports do not explain the mechanism of film formation. Thus, different explanations have emerged, and two models are discussed in the literature to explain this process: (i) the ion–ion mechanism, which is a process that occurs by the direct reaction of the ions present in the solution on the surface of the substrate; and (ii) growth via cluster–cluster collisions. Figure 1a shows the ion–ion mechanism. In the first stage, the diffusion processes of metal ions (e.g., Zn^{2+} or In^{3+}) and S^{2-} ions occur on the surface of the substrate (Figure 1(a1)). In the second stage, the first semiconductor nuclei are generated on the surface of the substrate (Figure 1(a2)). In the third stage, the nuclei grow by adsorbing more ions, while new semiconductor nuclei are generated (Figure 1(a3)). Finally, the crystals grow and adhere to each other to generate the film (Figure 1(a4)) [41,42].

Figure 1b shows the cluster–cluster mechanism. In the first stage, colloidal size particles are generated in metal sulfide solution (e.g., ZnS or In_2S_3) or a possible intermediate ($Zn(OH)_2$ or $In(OH)_3$); then, these particles diffuse onto the substrate (Figure 1(bi)). In the second stage, the first nuclei are generated on the surface of the substrate (Figure 1(bii)). In the third stage, the nuclei grow by absorbing more Zn^{2+} and S^{2-} ions, and the reaction continues until the possible intermediates transform into the respective sulfide through interchange reactions (Figure 1(biii)). Particle growth occurs both inside the solution and on the substrate surface. Finally, the particles adhere to each other on the surface of the substrate and form the film [41,42]. In the CBD process, both mechanisms may be present, generating the film and allowing the addition of colloidal aggregates for the subsequent growth of the film. The control of one of the two mechanisms is established by the extent of homogeneous and heterogeneous nucleation.

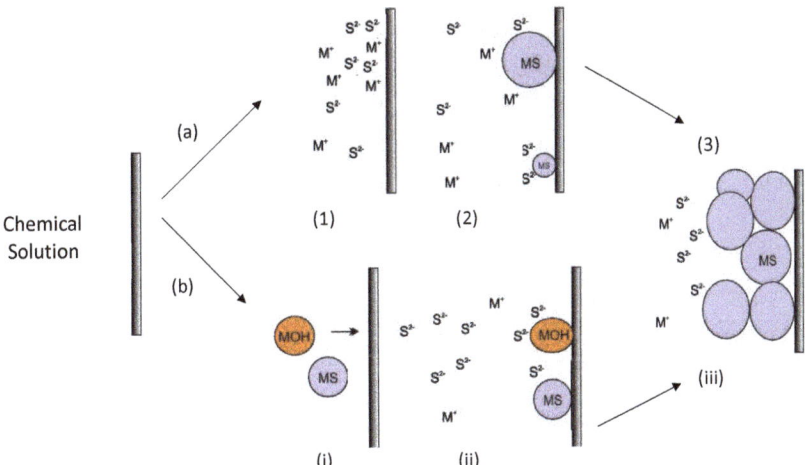

Figure 1. (a) Schematic diagram of the ion–ion mechanism: (1) Ion diffusion (M^+ represents Zn^{2+} or In^{3+}) from bulk solution to substrate surface. (2) Heterogeneous nucleation on the substrate surface and crystal growth on substrate surface. (3) Coalescence and growth of films. (b) Cluster–cluster mechanism: (i) metal sulfide and metal hydroxide particle formation into bulk solution, followed by solid particle diffusion to substrate surface. (ii) Diffusion of sulfide ions and reaction with solid particles located on the substrate surface and crystal growth by interchange reactions (metal hydroxide) and by sulfide addition (metal sulfide). (iii) Particle growth occurs both inside the solution and on the substrate surface. Bars represent solid substrates.

In the Zn(O;OH)S thin film deposition process, we used Thiourea as the S^{2+}-ions source (150 mM), Zn acetate as the Zn^{2+}-ions source (15 mM), and ammonia (350 mM) and sodium citrate (30 mM) as a complex agent with a temperature at 80 °C and pH at 10.5. In the CBD, soda-lime glass was used as substrate (SLG; 2 cm × 1.5 cm). After 45 min, the CBD was stoped, and the substrate was withdrawn and washed with distilled water. The thin layer was dried at ambient temperature. This first layer was immersed in a new chemical bath system, and the deposition process was repeated using this first layer as substrate to the second layer. This process was repeated to obtain different numbers of layers. Finally, the Zn(O;OH)S thin films were annealed in air at 400 °C for 30 min.

2.2. Thin Film Characterization

The thickness of the films was measured using a Veeco Dektak 150 profilometer (Plainview, NY, USA). The optical properties of the thin films were studied through transmittance measurements between 300 and 800 nm (Perkin Elmer Lambda 2S spectrophotometer, Waltham, MA, USA). The structural assay was carried out using a Shimadzu 6000 diffractometer (Tokyo, Japan) with a source of CuKα radiation (λ = 0.15418 nm) within the 2θ range of 20°–60°. Finally, the chemical surface assay was carried out using a Perkin-Elmer ESCA/SAM model 560 (Waltham, MA, USA).

3. Results and Discussion

3.1. Thin Film Depostion

Figure 2 shows the Zn(O;OH)S thin-film thickness as a function of the deposition time. In the typical growth trend during CBD, two regions are distinguished: (a) linear growth, a stage in which the film thickness increases linearly with time, and (b) the saturation zone, a stage in which the growth rate decreases significantly as a consequence of consumption of the reagents inside the solution [43,44]. In our case, the data on thin film thickness within a 60 min time were suitable with linear fitting (R^2 = 0.993); after this time, it is typical, in the CBD process, that linear kinetics change to due to the consumption of the reagents inside the solution. We performed the deposition of each layer within a 45-min deposition time (linear growth, Figure 2). Table 1 lists Zn(O;OH)S thin film thickness as a function of the number of layers. The results show a typical trend: as the number of layers increases, so does the thickness of the films. After the second layer, the coating thickness exceeds 100 nm. The typical thickness used to reduce the mechanical stress between the absorbent layer and the transparent conductor oxide (TCO) in thin-film solar cells is smaller than 100 nm. For three and four layers, the thickness values are 350 and 480 nm, respectively. The next sections will present the effect of the number of layers on the optical and structural thin films.

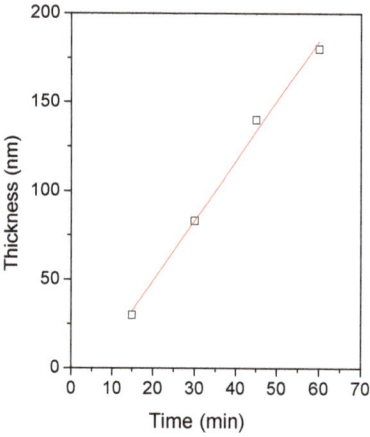

Figure 2. Zn(O;OH)S thin-film thickness as a function of CBD deposition time. White squares represent data and the red line represents the linear fitting. Fitting data: R^2 = 0.993. Fitting equation: $y = 3.38(x) - 18.5$.

Table 1. Optical and structural properties of the thin films, varying the number of layers deposited by CBD.

Number Layers	Thickness (nm)	Energy Band Gap (eV) [1]	Crystal Properties
1	90	3.81	Amorphous
2	190	3.77	Amorphous
3	350	3.70	Cubic/hexagonal phases
4	480	3.65	Cubic/hexagonal phases

[1] Value as a result of extrapolating the linear portion of the graph onto the x-axis in Figure 5b.

Although CBD is widely studied as a film deposition method, most reports are limited to exposing the different deposition parameters of the coatings; few reports show analytical studies about the chemical systems used during the deposition of the coatings. In a recent study, Gonzales et al. [45] reported the species distribution diagrams for CDB-ZnS film deposition. Due to chemical conditions (e.g., reagent concentration, pH, temperature), they did not report the formation of a ternary complex. It is known that the use of complexing agents reduces the homogeneous precipitation to obtain uniform and adherent coatings. The typical complexing agents in CBD include ammonia, hydrazine, ethanolamine, triethanolamine, tartaric acid, sodium citrate, and EDTA [46–49]. In the present study, we used sodium citrate to replace hydrazine in the CBD process. After CBD begins, the sulfide ion is generated as a free species in the solution due to hydrolysis of thiourea in a basic medium [50]:

$$SC(NH_2)_2 + OH^- \rightarrow NCNH_2 + SH^-_{(aq)} + H_2O. \tag{1}$$

Once the dissociated species of hydrogen sulfide (HS-) is formed in the medium, the S^{2-} ion is formed as follows [50]:

$$SH^- + OH^- \rightleftharpoons S^{2-}{}_{(aq)} + H_2O. \tag{2}$$

In the CBD solution, two solids can precipitate [50]:

$$Zn^{2+}{}_{(aq)} + S^{2-}{}_{(aq)} \rightleftharpoons ZnS_{(s)} \qquad K_{sp} = 10^{-24.7} \tag{3}$$

$$Zn^{2+}{}_{(aq)} + 2OH^-{}_{(aq)} \rightleftharpoons Zn(OH)_{2(s)} \qquad K_{sp} = 10^{-16} \tag{4}$$

where K_{sp} is the solubility product constant (the equilibrium constant for the chemical equilibrium of solid dissolving in aqueous solution) [51]. The presence of a complexing agent is necessary to prevent the excessive formation of ZnS/Zn(OH)$_2$. The complex formation reactions are as follows [52,53]:

$$Zn^{2+} + 3Cit \rightleftharpoons [Zn(Cit)_3]^{7-} \qquad \beta = 10^{5.5} \tag{5}$$

$$Zn^{2+} + 4NH_3 \rightleftharpoons [Zn(NH_3)_4]^{2+} \qquad \beta = 10^{9.46} \tag{6}$$

$$Zn^{2+} + 4OH^- \rightleftharpoons [Zn(OH)_4]^{2-} \qquad \beta = 10^{15} \tag{7}$$

$$[Zn(NH_3)_4]^{2+} + Cit \rightleftharpoons [Zn(NH_3)_3Cit]^- + NH_3 \qquad \beta = 10^{15} \tag{8}$$

$$[Zn(NH_3)_3Cit]^- + Cit \rightleftharpoons [Zn(NH_3)_2(Cit)_2]^{4-} + NH_3 \qquad \beta = 10^{10} \tag{9}$$

$$[Zn(NH_3)_2(Cit)_2]^{4-} + Cit \rightleftharpoons [Zn(NH_3)(Cit)_3]^{7-} + NH_3 \qquad \beta = 10^{7.46} \tag{10}$$

where Cit represents citrate anion ($[C_6H_5O_7]^{3-}$), β is the stability constant (the equilibrium constant for the formation of the complex between metallic-ion and complexing agent) [51]. Based on the chemical equilibrium Equations (1)–(10) and single complexing equilibrium, we simulated the species distribution diagram for Zn(O;OH)S synthesized by the CBD. We calculated the fraction molar (x_i) for each species at equilibrium, with the x_i value being determined by physical–chemical conditions of the mixture (e.g., pH, reagent concentration, temperature). The theory and details of the methodology used to simulate the distribution diagram can be found in previous reports [54,55]. In the equation, x_i represents the molar fraction of each species under specific chemical conditions, for which case we studied the pH effect on the x_i value [45]:

$$x_i = \frac{[s_i]}{\sum_i [s_i]} \tag{11}$$

where S_i is the concentration of the species i at a specific pH. Regarding the stability constants of ternary compounds (4–6), Figure 3a shows the distribution diagram of the species involved in the thin films deposited by CBD in the pH range 2–14. Figure 3a shows that the hydroxyl complexes are important from pH values higher than 12. Furthermore, the complexing agent used (sodium citrate) participates in the generation of ternary complexes. Figure 3b shows that when under the pH conditions used during CBD (pH = 10–11), the ternary complexes (ammonium-Zn-citrate) are present (near 70% of the species in the CBD solution). Conventionally, CBD processes use hydrazine as a reducing and/or complexing agent due to its chemical and physical properties [56]. However, this compound poses serious risks for the environment and health. The challenge is to combine inexpensive and green chemical routes for semiconductor fabrication and to direct synthetic routes to develop hydrazine-free chemical bath depositions [57]. The results show that the chemical reagents used in CBD are suitable for Zn(O;OH)S thin-films synthesis; furthermore, the simulation indicates that the citrate complex acts as a complexing agent during CBD.

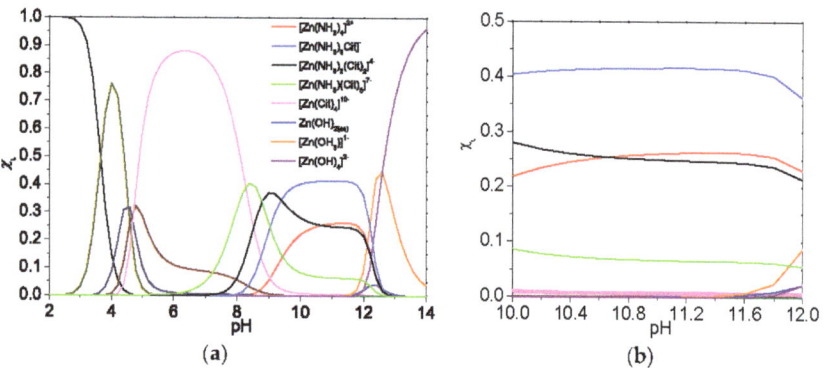

Figure 3. Species distribution diagram for Zn(O;OH)S synthesized by the CBD method under experimental conditions. (**a**) pH range 2–14. (**b**) zoom range pH 10–12. Inside figure, Cit represents citrate anion ($[C_6H_5O_7]^{3-}$). (x_i) represents fraction molar for each species at equilibrium.

3.2. Structural Characterization

The effect of the number of thin film layers on structural properties was studied by X-ray diffraction. After the annealing procedure, the coatings with one and two layers did not show signals in the diffraction pattern (patterns are not shown), suggesting that these coatings could be formed by small-sized grains (on a nanoscopic scale); besides, the coatings with three and four layers were polycrystalline. Figure 4 shows the X-ray diffraction patterns for Zn(O;OH)S thin films (three and four layers).

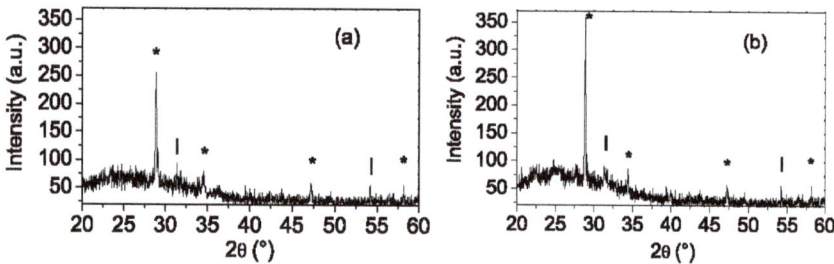

Figure 4. X-ray diffraction patterns of Zn(O;OH)S-thin film thickness as a function of the number of the layers. (**a**) Zn(O;OH)S coatings three layers. (**b**) Zn(O;OH)S coatings four layers. Inside figure: (*) The signal could be assigned to Wurtzite and/or Zincblende structure. (l) The signal could be assigned to Zn(OH)$_2$ structure.

Due to the chemical conditions of CBD, (i) the signals located at 2θ = 28.8°, 31.6°, 47.2°, and 58.2° were assigned to reflections of two different crystalline phases: (i) (002), (011), (110), and (0.21) planes of the hexagonal structure (Wurtzite; JCPDS 79-2204), and (ii) the signals located at 2θ = 28.8° and 47.2° were assigned to reflections of (111), (022) planes of the cubic structure (Zincblende; JCPDS 77-2100)) [58]. Furthermore, the signals located at 2θ = 34.4° and 58.22° were assigned to reflection β-Zn(OH)$_2$ (JCPDS 77-2100) [59], and the presence of this signal suggests the formation of Zn(OH)$_2$ or ZnO inside the buffer layer. The possible mechanism of generation and growth of the film occurs according to the formation of Zn(OH)$_2$ (Equation (4)). This process is not eliminated from the reaction medium despite the fact that a low concentration of the metal ion and the addition of the complexing agent were used during CBD [50]. The competition between the formation of ZnS and Zn(OH)$_2$ during the Zn-buffer layer deposition by CBD has been previously reported [45,60]. In a previous work [61], we identified that the Zn^{2+}, sulfur, and oxygen were present inside the Zn(O;OH)S-thin films deposited by CBD, suggesting that ZnS and ZnO could be generated during the CBD process. Sáez-Araoz et al. [62] reported that in the early Zn-buffer layer deposition by CBD, a very thin film of ZnS is generated, which is followed by a mixture of ZnS and ZnO. Finally, other authors report the generation of a mixture of chemical phases during Zn-based buffer layers synthesized by CBD [63–65].

3.3. Optical Characterization

Figure 5a shows the transmittance curves of the Zn(O;OH)S films deposited on SLG by the CBD process. The number of layers significantly affected the spectral transmittance. In general, for one (90 nm) and two (190 nm) layers, transmittances are greater than 80% in the visible region of the electromagnetic spectrum (zone of low absorbance λ > 310 nm). Such a high transmittance value indicates that these coatings are optically suitable for use as a buffer layer. Furthermore, Figure 5a shows a deterioration in the optical transmittance for three and four layers, reducing until 60–70% (three layers) and 50–60% (four layers). High optical transmittance in the visible range of the electromagnetic spectrum is a critical requirement for buffer layers inside solar cells. The band-gap energy value was determined for all samples using the Tauc model. For films with reduced thickness and that do not present sufficient interference patterns, the absorption coefficient of each layer can be determined using the following equation [66,67]:

$$\alpha(hv) = \frac{1}{d} Ln \frac{1}{T(hv)} \qquad (12)$$

where α is the absorption coefficient, d corresponds to the thickness of the buffer layer and, T corresponds to the spectral transmittance as a function of the wavelength. The absorption coefficient is related to the band gap according to the following equation [68]:

$$\alpha(hv) = A(hv - E_g)^2 \qquad (13)$$

where E_g corresponds to the band gap of the material, and A is a constant. The value of E_g can be obtained by extrapolating the linear part of the graph of $(\alpha hv)^2$ vs. (hv) to zero. This method was used to determine the E_g of the buffer layers. Figure 5b shows plots α^2 versus (hv) for the transmittance spectra of Figure 5b. The optical band gap of the films was determined by extrapolating the linear portion of the graph onto the x-axis [68]. Table 1 lists the optical properties of the thin films. During the chemical bath deposition of semiconductors, it is common to produce a mixture of different chemical compounds [41,42]. In the case of Zn-buffer layers, depending on experimental parameters, the mixture of compounds (e.g., sulfide, hydroxide, and oxide) could be generated [69,70]. Furthermore, the semiconductor thin films band gap is affecting by thickness, crystalline structure, and physical–chemical composition [71,72]. The ZnO has a wide and direct band gap (≈3.37 eV) [73]. In the case of ZnS thin films, this is a semiconductor with a wide optical band gap (≈3.6 eV) [74,75]. However, the thin films band gap is affected by the deposition method and the post-treatment process. Mursal et al. [76], reported the optical band gap of ZnO thin films varying between 3.82 and 3.69 eV deposited by the sol–gel spin-coating method after changed the sintering temperature. The difference in the experimental and theorical band gap is attributed to the intrinsic defects in ZnO (e.g., O vacancy (V_O), Zn vacancy (V_{Zn}), Zn interstitial (Zn_i), O interstitial (O_i) and anti-site Zn (Zn_O)) [77–79]. Table 1 shows that the coating's band gap changes between 3.65 and 3.81 eV. The band-gap energy of the Zn(O;OH)S coating synthesized by CBD has values between the band gap of ZnS and ZnO. This result could be related to the possibility of the generation of a mixture of chemical phases during buffer layers deposition (Section 3.2).

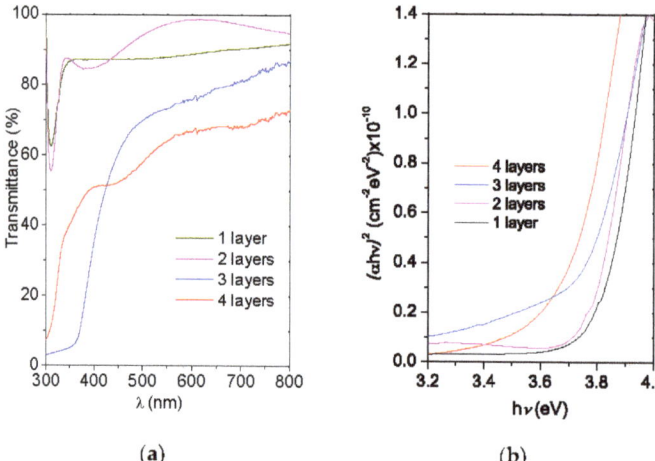

Figure 5. (a) Zn(O;OH)S thin-film transmittance as a function of the number of layers deposited by CBD. (b) Tauc plots and band-gap energy estimation for the Zn(O;OH)S thin films as a function of the number of layers deposited by CBD.

Furthermore, Table 1 shows that reducing the layer thickness affects the bulk properties and changes the optical properties, and that the thinner layers have a higher band-gap energy value, which is consistent with other reports. Das et al. reported band-gap variation with thickness for CdS thin films deposited by CBD and found that the band gap decreased from 3.2 eV (CdS thin films of 153 nm in thickness) until 2.54 eV (CdS thin films of 205 nm in thickness) [80]. Hossain et al. studied the In_2S_3 buffer layer band-gap variation from 2.0 to 2.9 eV to observe the effects thereof on electrical performance, and they reported that the band gap increased up to 2.9 eV with a layer of 50 nm in thickness, and this value decreased to 2.0 eV with a layer of 1 μm in thickness [81]. Such results verified that the number of layers in the coatings determines the physical–chemical properties of buffer layers (e.g., band gap, crystalline structure). Although all the coatings had energy values higher than 3.65 eV (requirement for the buffer layer in solar cells), only the coatings with one and two layers were suitable for use as buffer layers.

4. Conclusions

In this paper, we studied the structural and optical properties of Zn(O;OH)S coatings grown by CBD with different numbers of layers. We presented the simulation of the species distribution diagram for the Zn(O;OH)S synthesized by CBD. The thickness of the buffer layer films varied from 90 to 480 nm. The Zn(O;OH)S thin films with three and four layers represented a mixture of the cubic and hexagonal phases. Furthermore, the species diagram simulation indicates that the citrate complex could play an important role as a complexing agent during CBD. Finally, optical properties indicate that only the coatings with one and two layers were suitable for use as buffer layers, since these coatings had higher transmittance values. Furthermore, although all coatings had energy values higher than 3.65 eV, the coatings with three and four layers showed reduced transmittance in the visible region (50–70%).

Author Contributions: Conceptualization, W.V.; methodology, W.V., C.D.-U. and C.Q.; software, W.V.; validation, W.V., C.D.-U. and C.Q.; formal analysis, W.V., C.D.-U. and C.Q.; investigation, W.V.; resources, W.V.; data curation, W.V.; writing—original draft preparation, W.V., C.D.-U. and C.Q.; writing—review and editing, W.V., C.D.-U. and C.Q.; visualization, W.V.; supervision, W.V.; project administration, W.V.; funding acquisition, W.V. All authors have read and agreed to the published version of the manuscript.

Funding: This research was funded by Universidad del Atlántico RES. 002887.

Institutional Review Board Statement: Not applicable.

Informed Consent Statement: Not applicable.

Data Availability Statement: All data are contained within the article.

Acknowledgments: The authors thank Universidad del Atlántico.

Conflicts of Interest: The authors declare no conflict of interest.

References

1. Vohra, K.; Vodonos, A.; Schwartz, J.; Marais, E.A.; Sulprizio, M.P.; Mickley, L.J. Global mortality from outdoor fine particle pollution generated by fossil fuel combustion: Results from GEOS-Chem. *Environ. Res.* **2021**, *195*, 110754. [CrossRef]
2. Zhu, L.; Mickley, L.J.; Jacob, D.J.; Marais, E.A.; Sheng, J.; Hu, L.; Abad, G.G.; Chance, K. Long-term (2005–2014) trends in formaldehyde (HCHO) columns across North America as seen by the OMI satellite instrument: Evidence of changing emissions of volatile organic compounds. *Geophys. Res. Lett.* **2017**, *44*, 7079–7086. [CrossRef]
3. Perera, F. Pollution from fossil-fuel combustion is the leading environmental threat to global pediatric health and equity: Solutions exist. *Int. J. Environ. Res. Public Health* **2018**, *15*, 16. [CrossRef] [PubMed]
4. Dudley, B. *BP Statistical Review of World Energy 2020*; BP: London, UK, 2020.
5. Martins, F.; Felgueiras, C.; Smitkova, M.; Caetano, N. Analysis of fossil fuel energy consumption and environmental impacts in european countries. *Energies* **2019**, *12*, 964. [CrossRef]
6. Bebbington, J.; Schneider, T.; Stevenson, L.; Fox, A. Fossil fuel reserves and resources reporting and unburnable carbon: Investigating conflicting accounts. *Crit. Perspect. Account.* **2020**, *66*, 102083. [CrossRef]
7. Ceballos, G.; Ehrlich, P.R.; Raven, P.H. Vertebrates on the brink as indicators of biological annihilation and the sixth mass extinction. *Proc. Natl. Acad. Sci. USA* **2020**, *117*, 13596–13602. [CrossRef]
8. Bradshaw, C.J.A.; Ehrlich, P.R.; Beattie, A.; Ceballos, G.; Crist, E.; Diamond, J.; Dirzo, R.; Ehrlich, A.H.; Harte, J.; Harte, M.E.; et al. Underestimating the Challenges of Avoiding a Ghastly Future. *Front. Conserv. Sci.* **2021**, *1*, 615419. [CrossRef]
9. REN21. *Renewables 2020—Global Status Report*; 2020 Update; REN21: Paris, France, 2020; ISBN 9783948393007.
10. Castellanos, S.; Santibañez-Aguilar, J.E.; Shapiro, B.B.; Powell, D.M.; Peters, I.M.; Buonassisi, T.; Kammen, D.M.; Flores-Tlacuahuac, A. Sustainable silicon photovoltaics manufacturing in a global market: A techno-economic, tariff and transportation framework. *Appl. Energy* **2018**, *212*, 704–719. [CrossRef]
11. Wilson, G.M.; Al-Jassim, M.; Metzger, W.K.; Glunz, S.W.; Verlinden, P.; Xiong, G.; Mansfield, L.M.; Stanbery, B.J.; Zhu, K.; Yan, Y.; et al. Exceeding 200 ns lifetimes in polycrystalline CdTe solar cells. *Solar RRL* **2020**, *53*, 493001.
12. Lee, T.Y.; Lee, I.H.; Jung, S.H.; Chung, C.W. Characteristics of CdS thin films deposited on glass and Cu(In,Ga)Se$_2$ layer using chemical bath deposition. *Thin Solid Films* **2013**, *548*, 64–68. [CrossRef]
13. Merdes, S.; Malinen, V.; Ziem, F.; Lauermann, I.; Schüle, M.; Stober, F.; Hergert, F.; Papathanasiou, N.; Schlatmann, R. Zn(O,S) buffer prepared by atomic layer deposition for sequentially grown Cu(In,Ga)(Se,S)$_2$ solar cells and modules. *Sol. Energy Mater. Sol. Cells* **2014**, *126*, 120–124. [CrossRef]
14. Choubey, R.K.; Kumar, S.; Lan, C.W. Shallow chemical bath deposition of ZnS buffer layer for environmentally benign solar cell devices. *Adv. Nat. Sci. Nanosci. Nanotechnol.* **2014**, *5*, 025015. [CrossRef]
15. Choi, W.J.; Park, W.W.; Kim, Y.; Son, C.S.; Hwang, D. The effect of AlD-Zn(O,S) buffer layer on the performance of CIGSSe thin film solar cells. *Energies* **2020**, *13*, 412. [CrossRef]
16. Vallejo, W.; Quiñones, C.; Gordillo, G. A comparative study of thin films of Zn(O;OH)S and In(O;OH)S deposited on CuInS$_2$ by chemical bath deposition method. *J. Phys. Chem. Solids* **2012**, *73*, 573–578. [CrossRef]
17. Green, M.A.; Dunlop, E.D.; Hohl-Ebinger, J.; Yoshita, M.; Kopidakis, N.; Hao, X. Solar cell efficiency tables (version 56). *Prog. Photovolt. Res. Appl.* **2020**, *28*, 629–638. [CrossRef]
18. Hong, J.; Lim, D.; Eo, Y.J.; Choi, C. Chemical bath deposited ZnS buffer layer for Cu(In,Ga)Se$_2$ thin film solar cell. *Appl. Surf. Sci.* **2018**, *432*, 250–254. [CrossRef]
19. Chen, Y.; Mei, X.; Liu, X.; Wu, B.; Yang, J.; Yang, J.; Xu, W.; Hou, L.; Qin, D.; Wang, D. Solution-processed CdTe thin-film solar cells using ZnSe nanocrystal as a buffer layer. *Appl. Sci.* **2018**, *8*, 1195. [CrossRef]
20. Mughal, M.A.; Engelken, R.; Sharma, R. Progress in indium (III) sulfide (In$_2$S$_3$) buffer layer deposition techniques for CIS, CIGS, and CdTe-based thin film solar cells. *Sol. Energy* **2015**, *120*, 131–146. [CrossRef]
21. Salomé, P.M.P.; Keller, J.; Törndahl, T.; Teixeira, J.P.; Nicoara, N.; Andrade, R.R.; Stroppa, D.G.; González, J.C.; Edoff, M.; Leitão, J.P.; et al. CdS and Zn$_{1-x}$Sn$_x$O$_y$ buffer layers for CIGS solar cells. *Sol. Energy Mater. Sol. Cells* **2017**, *159*, 272–281. [CrossRef]
22. Sabitha, C.; Joe, I.H.; Kumar, K.D.A.; Valanarasu, S. Investigation of structural, optical and electrical properties of ZnS thin films prepared by nebulized spray pyrolysis for solar cell applications. *Opt. Quantum Electron.* **2018**, *50*, 1–18. [CrossRef]
23. Fathima, M.I.; Wilson, K.S.J. Antireflection coating application of zinc sulfide thin films by nebulizer spray pyrolysis technique. In *AIP Conference Proceedings*; AIP Publishing LLC: Melville, NY, USA, 2019; Volume 2115, p. 030327.

24. Włodarski, M.; Chodorow, U.; Jóźwiak, S.; Putkonen, M.; Durejko, T.; Sajavaara, T.; Norek, M. Structural and optical characterization of ZnS ultrathin films prepared by low-temperature ALD from diethylzinc and 1.5-pentanedithiol after various annealing treatments. *Materials* **2019**, *12*, 3212. [CrossRef] [PubMed]
25. Muchuweni, E.; Sathiaraj, T.S.; Nyakotyo, H. Synthesis and characterization of zinc oxide thin films for optoelectronic applications. *Heliyon* **2017**, *3*, e00285. [CrossRef]
26. Shaban, M.; Zayed, M.; Hamdy, H. Nanostructured ZnO thin films for self-cleaning applications. *RSC Adv.* **2017**, *7*, 617–631. [CrossRef]
27. Opasanont, B.; Van, K.T.; Kuba, A.G.; Choudhury, K.R.; Baxter, J.B. Adherent and conformal Zn(S,O,OH) thin films by rapid chemical bath deposition with hexamethylenetetramine additive. *ACS Appl. Mater. Interfaces* **2015**, *7*, 11516–11525. [CrossRef] [PubMed]
28. Johnston, D.A.; Carletto, M.H.; Reddy, K.T.R.; Forbes, I.; Miles, R.W. Chemical bath deposition of zinc sulfide based buffer layers using low toxicity materials. In Proceedings of the Thin Solid Films; Elsevier: Amsterdam, The Netherlands, 2002; Volume 403–404, pp. 102–106.
29. Memarian, N.; Rozati, S.M.; Concina, I.; Vomiero, A. Deposition of nanostructured Cds thin films by thermal evaporation method: Effect of substrate temperature. *Materials* **2017**, *10*, 773. [CrossRef]
30. Kim, S.Y.; Rana, T.R.; Kim, J.H.; Yun, J.H. Cu(In,Ga)Se$_2$ solar cells with In$_2$S$_3$ buffer layer deposited by thermal evaporation. *J. Korean Phys. Soc.* **2017**, *71*, 1012–1018. [CrossRef]
31. Miliucci, M.; Lucci, M.; Colantoni, I.; De Matteis, F.; Micciulla, F.; Clozza, A.; Macis, S.; Davoli, I. Characterization of CdS sputtering deposition on low temperature pulsed electron deposition Cu(In,Ga)Se$_2$ solar cells. *Thin Solid Film.* **2020**, *697*, 137833. [CrossRef]
32. Ramanathan, K.; Mann, J.; Glynn, S.; Christensen, S.; Pankow, J.; Li, J.; Scharf, J.; Mansfield, L.; Contreras, M.; Noufi, R. A comparative study of Zn(O,S) buffer layers and CIGS solar cells fabricated by CBD, ALD, and sputtering. In Proceedings of the 2012 38th IEEE Photovoltaic Specialists Conference, Austin, TX, USA, 4 October 2012; pp. 1677–1680.
33. Prabukanthan, P.; Harichandran, G. Electrochemical deposition of n-type ZnSe Thin Film Buffer Layer for Solar Cells. *J. Electrochem. Soc.* **2014**, *161*, D736–D741. [CrossRef]
34. Palve, A.M. Deposition of zinc sulfide thin films from Zinc(II) thiosemicarbazones as single molecular precursors using aerosol assisted chemical vapor deposition technique. *Front. Mater.* **2019**, *6*, 46. [CrossRef]
35. Li, J. Preparation and properties of CdS thin films deposited by chemical bath deposition. *Ceram. Int.* **2015**, *41*, S376–S380. [CrossRef]
36. Maria, K.H.; Sultana, P.; Asfia, M.B. Chemical bath deposition of aluminum doped zinc sulfide thin films using non-toxic complexing agent: Effect of aluminum doping on optical and electrical properties. *AIP Adv.* **2020**, *10*, 65315. [CrossRef]
37. Stumph, P.S.; Baranova, K.A.; Rogovoy, M.S.; Bunakov, V.V.; Maraeva, E.V.; Tulenin, S.S. Chemical bath deposition of In$_2$S$_3$ thin films as promising material and buffer layer for solar cells. In *AIP Conference Proceedings*; AIP Publishing LLC: Melville, NY, USA, 2019; Volume 2063, p. 040057.
38. Aguilera, M.L.A.; Márquez, J.M.F.; Trujillo, M.A.G.; Kuwahara, Y.M.; Morales, G.R.; Galán, O.V. Influence of CdS thin films growth related with the substrate properties and conditions used on CBD technique. *Energy Procedia* **2014**, *44*, 111–117. [CrossRef]
39. Chu, V.B.; Siopa, D.; Debot, A.; Adeleye, D.; Sood, M.; Lomuscio, A.; Melchiorre, M.; Guillot, J.; Valle, N.; El Adib, B.; et al. Waste- and Cd-free inkjet-printed Zn(O,S) buffer for Cu(In,Ga)(S,Se)$_2$ thin-film solar cells. *ACS Appl. Mater. Interfaces* **2021**, *13*, 13009–13021. [CrossRef]
40. Mugle, D.; Jadhav, G. Short review on chemical bath deposition of thin film and characterization. In *AIP Conference Proceedings*; AIP Publishing LLC: Melville, NY, USA, 2016; Volume 1728, p. 020597.
41. Kashchiev, D. *Nucleation Basic Theory with Applications*, 1st ed.; Elsevier: Amsterdam, The Netherlands, 2000; ISBN 9780750646826.
42. Hodes, G. Semiconductor and ceramic nanoparticle films deposited by chemical bath deposition. *Phys. Chem. Chem. Phys.* **2007**, *9*, 2181–2196. [CrossRef]
43. Froment, M.; Lincot, D. Phase formation processes in solution at the atomic level: Metal chalcogenide semiconductors. *Electrochim. Acta* **1995**, *40*, 1293–1303. [CrossRef]
44. Vallejo, W.; Clavijo, J.; Gordillo, G. CGS based solar cells with In$_2$S$_3$ buffer layer deposited by CBD and coevaporation. *Braz. J. Phys.* **2010**, *40*, 30–37. [CrossRef]
45. González-Chan, I.J.; Oliva, A.I. Physicochemical analysis and characterization of chemical bath deposited zns films at near ambient temperature. *J. Electrochem. Soc.* **2016**, *163*, D421–D427. [CrossRef]
46. Goudarzi, A.; Aval, G.M.; Sahraei, R.; Ahmadpoor, H. Ammonia-free chemical bath deposition of nanocrystalline ZnS thin film buffer layer for solar cells. *Thin Solid Films* **2008**, *516*, 4953–4957. [CrossRef]
47. Lădar, M.; Popovici, E.J.; Baldea, I.; Grecu, R.; Indrea, E. Studies on chemical bath deposited zinc sulphide thin films with special optical properties. *J. Alloy. Compd.* **2007**, *434–435*, 697–700. [CrossRef]
48. Karakawa, M.; Sugahara, T.; Hirose, Y.; Suganuma, K.; Aso, Y. Thin film of amorphous zinc hydroxide semiconductor for optical devices with an energy-efficient beneficial coating by metal organic decomposition process. *Sci. Rep.* **2018**, *8*, 10839. [CrossRef] [PubMed]
49. Rodríguez, C.A.; Flores, M.; Sandoval-Paz, M.; Delplancke, M.P.; Cabello-Guzmán, G.; Carrasco, C. Study of the early growth stages of chemically deposited ZnS thin films from a non-toxic solution. *Mater. Res. Express* **2018**, *5*, 076404. [CrossRef]

50. Oladeji, I.O.; Chow, L. A study of the effects of ammonium salts on chemical bath deposited zinc sulfide thin films. *Thin Solid Films* **1999**, *339*, 148–153. [CrossRef]
51. Wagh, A.S. Dissolution Characteristics of Metal Oxides and Kinetics of Ceramic Formation. In *Chemically Bonded Phosphate Ceramics*; Elsevier: Amsterdam, The Netherlands, 2016; pp. 61–73.
52. Meites, L. *Handbook of Analytical Chemistry*, 1st ed.; McGraw-Hill: New York, NY, USA, 1963.
53. Clavijo Díaz, A. *Fundamentos de Química Analítica: Equilibrio Iónico y Análisis Químico*; Universidad Nacional de Colombia: Bogotá, Colombia, 2002; Volume 1.
54. González-Panzo, I.J.; Martín-Várguez, P.E.; Oliva, A.I. Physicochemical conditions for ZnS films deposited by chemical bath. *J. Electrochem. Soc.* **2014**, *161*, D181–D189. [CrossRef]
55. Reinisch, M.; Perkins, C.L.; Steirer, K.X. Quantitative study on the chemical solution deposition of zinc oxysulfide. *ECS J. Solid State Sci. Technol.* **2016**, *5*, P58–P66. [CrossRef]
56. Kim, J. Comparison of ZnS film growth on glass and CIGS substrates via hydrazine-assisted chemical bath deposition for solar cell application. *Appl. Sci. Converg. Technol.* **2019**, *28*, 229–233. [CrossRef]
57. Guerrero, G.A.; Rodríguez, A.G.; Moreno-García, H. Hydrazine-free chemical bath deposition of WSe_2 thin films and bi-layers for photovoltaic applications. *Mater. Res. Express* **2019**, *6*, 105906. [CrossRef]
58. Zhao, H.; Liu, W.; Zhu, J.; Shen, X.; Xiong, L.; Li, Y.; Li, X.; Liu, J.; Wang, R.; Jin, C.; et al. Structural transition behavior of ZnS nanotetrapods under high pressure. *High Press. Res.* **2015**, *35*, 9–15. [CrossRef]
59. Mousavi-Kamazani, M.; Zinatloo-Ajabshir, S.; Ghodrati, M. One-step sonochemical synthesis of $Zn(OH)_2/ZnV_3O_8$ nanostructures as a potent material in electrochemical hydrogen storage. *J. Mater. Sci. Mater. Electron.* **2020**, *31*, 17332–17338. [CrossRef]
60. Hubert, C.; Naghavi, N.; Etcheberry, A.; Roussel, O.; Hariskos, D.; Powalla, M.; Kerrec, O.; Lincot, D. A better understanding of the growth mechanism of Zn(S,O,OH) chemical bath deposited buffer layers for high efficiency $Cu(In,Ga)(S,Se)_2$ solar cells. *Phys. Status Solidi* **2008**, *205*, 2335–2339. [CrossRef]
61. Vallejo, W.; Diaz-Uribe, C.; Hurtado, M. Caracterización fisicoquímica del sistema $Mo/CuInS_2/Zn(O,OH)S/ZnO$ por medio de espectroscopía fotoelectrónica XPS y AES. *Elementos* **2014**, *4*. [CrossRef]
62. Sáez-Araoz, R.; Abou-Ras, D.; Niesen, T.P.; Neisser, A.; Wilchelmi, K.; Lux-Steiner, M.C.; Ennaoui, A. In situ monitoring the growth of thin-film ZnS/Zn(S,O) bilayer on Cu-chalcopyrite for high performance thin film solar cells. *Thin Solid Film.* **2009**, *517*, 2300–2304. [CrossRef]
63. Buffière, M.; Harel, S.; Arzel, L.; Deudon, C.; Barreau, N.; Kessler, J. Fast chemical bath deposition of Zn(O,S) buffer layers for $Cu(In,Ga)Se_2$ solar cells. *Thin Solid Film.* **2011**, *519*, 7575–7578. [CrossRef]
64. Ennaoui, A.; Siebentritt, S.; Lux-Steiner, M.C.; Riedl, W.; Karg, F. High-efficiency Cd-free CIGSS thin-film solar cells with solution grown zinc compound buffer layers. *Sol. Energy Mater. Sol. Cells* **2001**, *67*, 31–40. [CrossRef]
65. Platzer-Björkman, C.; Törndahl, T.; Abou-Ras, D.; Malmström, J.; Kessler, J.; Stolt, L. Zn(O,S) buffer layers by atomic layer deposition in $Cu(In,Ga)Se_2$ based thin film solar cells: Band alignment and sulfur gradient. *J. Appl. Phys.* **2006**, *100*, 044506. [CrossRef]
66. Makuła, P.; Pacia, M.; Macyk, W. How to correctly determine the band gap energy of modified semiconductor photocatalysts based on UV-Vis spectra. *J. Phys. Chem. Lett.* **2018**, *9*, 6814–6817. [CrossRef] [PubMed]
67. Escobedo-Morales, A.; Ruiz-López, I.I.; Ruiz-Peralta, M.D.; Tepech-Carrillo, L.; Sánchez-Cantú, M.; Moreno-Orea, J.E. Automated method for the determination of the band gap energy of pure and mixed powder samples using diffuse reflectance spectroscopy. *Heliyon* **2019**, *5*, e01505. [CrossRef]
68. Viezbicke, B.D.; Patel, S.; Davis, B.E.; Birnie, D.P. Evaluation of the tauc method for optical absorption edge determination: ZnO thin films as a model system. *Phys. Status Solidi* **2015**, *252*, 1700–1710. [CrossRef]
69. Wang, L.P.; De Han, P.; Zhang, Z.X.; Zhang, C.L.; Xu, B.S. Effects of thickness on the structural, electronic, and optical properties of MgF_2 thin films: The first-principles study. *Comput. Mater. Sci.* **2013**, *77*, 281–285. [CrossRef]
70. Zhou, L.; Xue, Y.; Li, J. Study on ZnS thin films prepared by chemical bath deposition. *J. Environ. Sci.* **2009**, *21*, S76–S79. [CrossRef]
71. Chaves, A.; Azadani, J.G.; Alsalman, H.; da Costa, D.R.; Frisenda, R.; Chaves, A.J.; Song, S.H.; Kim, Y.D.; He, D.; Zhou, J.; et al. Bandgap engineering of two-dimensional semiconductor materials. *Npj 2D Mater. Appl.* **2020**, *4*, 29. [CrossRef]
72. Gong, Y.; Liu, Z.; Lupini, A.R.; Shi, G.; Lin, J.; Najmaei, S.; Lin, Z.; Elías, A.L.; Berkdemir, A.; You, G.; et al. band gap engineering and layer-by-layer mapping of selenium-doped molybdenum disulfide. *Nano Lett.* **2013**, *14*, 442–449. [CrossRef]
73. Kamarulzaman, N.; Kasim, M.F.; Rusdi, R. Band gap narrowing and widening of ZnO nanostructures and doped materials. *Nanoscale Res. Lett.* **2015**, *10*, 346. [CrossRef] [PubMed]
74. D'Amico, P.; Calzolari, A.; Ruini, A.; Catellani, A. New energy with ZnS: Novel applications for a standard transparent compound. *Sci. Rep.* **2017**, *7*, 16805. [CrossRef] [PubMed]
75. Jiang, P.; Jie, J.; Yu, Y.; Wang, Z.; Xie, C.; Zhang, X.; Wu, C.; Wang, L.; Zhu, Z.; Luo, L. Aluminium-doped n-type ZnS nanowires as high-performance UV and humidity sensors. *J. Mater. Chem.* **2012**, *22*, 6856–6861. [CrossRef]
76. Jalil, Z. Structural and optical properties of zinc oxide (ZnO) based thin films deposited by sol-gel spin coating method. *J. Phys. Conf. Ser.* **2018**, *1116*, 032020.
77. Xiu, F.; Xu, J.; Joshi, P.C.; Bridges, C.A.; Paranthaman, M.P. ZnO doping and defect engineering—A review. *Springer Ser. Mater. Sci.* **2016**, *218*, 105–140.

78. Lyons, J.L.; Varley, J.B.; Steiauf, D.; Janotti, A.; Walle, C.G. Van de first-principles characterization of native-defect-related optical transitions in ZnO. *J. Appl. Phys.* **2017**, *122*, 035704. [CrossRef]
79. Liang, Y.-C.; Hung, C.-S.; Zhao, W.-C. Thermal annealing induced controllable porosity and photoactive performance of 2D ZnO sheets. *Nanomaterials* **2020**, *10*, 1352. [CrossRef]
80. Das, N.S.; Ghosh, P.K.; Mitra, M.K.; Chattopadhyay, K.K. Effect of film thickness on the energy band gap of nanocrystalline CdS thin films analyzed by spectroscopic ellipsometry. *Phys. E Low-Dimens. Syst. Nanostruct.* **2010**, *42*, 2097–2102. [CrossRef]
81. Istiaque Hossain, M.; Chelvanathan, P.; Zaman, M.; Karim, M.R.; Alghoul, M.A.; Amin, N. Prospects of indium sulphide as an alternative to cadmium sulphide buffer layer in CIS based solar cells from numerical analysis. *Chalcogenide Lett.* **2011**, *8*, 315–324.

Article

Enhanced Methanol Oxidation Activity of PtRu/C$_{100-x}$MWCNTs$_x$ (x = 0–100 wt.%) by Controlling the Composition of C-MWCNTs Support

Dang Long Quan [1,2,3] and Phuoc Huu Le [4,*]

[1] Faculty of Applied Science, Ho Chi Minh City University of Technology (HCMUT), 268 Ly Thuong Kiet Street, District 10, Ho Chi Minh City 700000, Vietnam; dlquan.sdh20@hcmut.edu.vn
[2] Vietnam National University Ho Chi Minh City, Linh Trung Ward, Thu Duc District, Ho Chi Minh City 700000, Vietnam
[3] Department of Physics, College of Natural Sciences, Can Tho University, Can Tho City 900000, Vietnam
[4] Department of Physics and Biophysics, Faculty of Basic Sciences, Can Tho University of Medicine and Pharmacy, 179 Nguyen Van Cu Street, Can Tho City 94000, Vietnam
* Correspondence: lhuuphuoc@ctump.edu.vn; Tel.: +84-29-2373-9730

Abstract: PtRu nanoparticles decorated on carbon-based supports are of great interest for direct methanol fuel cells (DMFCs). In this study, PtRu alloy nanoparticles decorated on carbon Vulcan XC-72 (C), multi-walled carbon nanotubes (MWCNTs), and C-MWCNTs composite supports were synthesized by co-reduction method. As a result, PtRu nanoparticles obtained a small mean size (d_{mean} = 1.8–3.8 nm) with a size distribution of 1–7 nm. We found that PtRu/C$_{60}$MWCNTs$_{40}$ possesses not only high methanol oxidation activity, but also excellent carbonaceous species tolerance ability, suggesting that C-MWCNTs composite support is better than either C or MWCNTs support. Furthermore, detailed investigation on PtRu/C$_{100-x}$MWCNTs$_x$ (x = 10–50 wt.%) shows that the current density (J_f), catalyst tolerance ratio (J_f/J_r), and electron transfer resistance (R_{et}) are strongly affected by C-MWCNTs composition. The highest J_f is obtained for PtRu/C$_{70}$MWCNTs$_{30}$, which is considered as an optimal electrocatalyst. Meanwhile, both PtRu/C$_{70}$MWCNTs$_{30}$ and PtRu/C$_{60}$MWCNTs$_{40}$ exhibit a low R_{et} of 5.31–6.37 Ω·cm². It is found that C-MWCNTs composite support is better than either C or MWCNTs support in terms of simultaneously achieving the enhanced methanol oxidation activity and good carbonaceous species tolerance.

Keywords: composite support; electrocatalysts; methanol oxidation; multi-walled carbon nanotubes; PtRu nanoparticles

Citation: Quan, D.L.; Le, P.H. Enhanced Methanol Oxidation Activity of PtRu/C$_{100-x}$MWCNTs$_x$ (x = 0–100 wt.%) by Controlling the Composition of C-MWCNTs Support. *Coatings* **2021**, *11*, 571. https://doi.org/10.3390/coatings11050571

Academic Editor: Adriano Sacco

Received: 1 April 2021
Accepted: 11 May 2021
Published: 14 May 2021

Publisher's Note: MDPI stays neutral with regard to jurisdictional claims in published maps and institutional affiliations.

Copyright: © 2021 by the authors. Licensee MDPI, Basel, Switzerland. This article is an open access article distributed under the terms and conditions of the Creative Commons Attribution (CC BY) license (https://creativecommons.org/licenses/by/4.0/).

1. Introduction

Currently, green energy research is more urgent than ever due to environmental pollution and the gradual depletion of fossil energy. As a green and clean electric power source, fuel cells are able to directly transform chemical energy into electrical energy, and used as power generation in portable, stationary, and transportation applications [1–3]. Among various types of fuel cells, direct methanol fuel cells (DMFCs) have been considered as a promising energy source owing to their low operation temperature (below 100 °C), operation safety, superior specific energy, and durability [3,4]. The commercialization of DMFCs requires further reducing the cost and increasing the performance of the electrode catalyst. However, precious and expensive metal catalyst (i.e., Pt or Pt alloy) has been generally used as the catalyst for methanol oxidation reaction (MOR) in DMFCs [5–10]. Moreover, carbon monoxide (CO) gas released in MOR poisons Pt catalyst and limits the performance of DMFCs. In order to solve these problems, various Pt-based alloys have been developed to reduce Pt usage and enhance the catalyst activity, including PtRu [11,12], PtCo [13], PtMo [14], PtRuNi [15], PtRuMo [16], etc. Among the catalyst systems, PtRu is well known, owing to its superior performance in preventing the poisoning of the Pt surface by CO gas. It is because Ru forms an oxygenated species at lower potentials than that of Pt, and thus Ru promotes the oxidation of CO gas produced during MOR [17].

Membrane electrode assembly (MEA), which is the main component of DMFCs, has a significant effect on fuel cell performance. An effective anode catalyst is one of the prerequisites for an ideal MEA [18]. In DMFCs, supporting materials (or substrates) for catalysts play a certain important role that can significantly affect the catalyst activity. Carbon black (CB) is commonly used for supporting catalyst nanoparticles in DMFCs because of its large specific surface area and high electrical conductivity [19], but it poses several drawbacks—poor corrosion resistance and limitation of mass transfer due to its dense structure [20]. Carbon nanotubes (CNTs) with high chemical stability and high electrical and thermal conductivities are considered as an excellent support material [21]. Indeed, the catalysts/CNTs possess 1.3–1.6 times greater methanol oxidation activity than that of catalysts/Vulcan carbon [22–24]. However, the electrical and thermal conductivities at CNT–CNT inter-tube junctions are at least an order of magnitude lower than those of individual CNTs [25,26]. In recent years, the development of carbon-based nanomaterial supports with different structures–morphologies to enhance DMFCs performance has attracted much attention. Graphene, carbon xerogels, carbon nanofiber, mesoporous carbon, and functionalized or doped carbon supports exhibited the enhancements in methanol oxidation efficiency as compared to the traditional carbon supports [22–24,27–33]. Interestingly, PtRu nanoclusters decorated on three-dimensional porous composite support of graphene sheets (GS) and CNTs presented ~3.2 times higher current intensity than the catalysts on carbon substrate, which was attributed to the decreased aggregation of metallic nanoparticles in the PtRu/GS-CNT [24]. In addition, Yang et al. reported that Pt nanoparticles decorated on a composite support of 10 wt.% MWCNTs and CB resulted in an enhanced power density of 1.5 and 2 times greater than those of the Pt catalyst on MWCNTs- and CB-supports, respectively [34]. Moreover, MWCNTs support showed better durability than CB support [34]. These results suggest that, besides developing catalyst materials, studies on composite carbon-based supporting nanomaterials are a new promising approach toward further enhancement of DMFCs performance.

In this study, PtRu alloy nanoparticles synthesized by a co-reduction method were decorated on $C_{100-x}MWCNTs_x$ composites with various mixing weight percentages (i.e., x = 0–100 wt.%). The structure and morphology of the PtRu/$C_{100-x}MWCNTs_x$ (x = 0–100 wt.%) samples were studied by X-ray diffraction (XRD) and transmission electron microscope (TEM) analyses. The effect of MWCNTs content on the electrocatalytic activity of the nanomaterials was investigated by cyclic voltammetry (CV) and electrochemical impedance spectroscopy (EIS) analyses. This study provides the optimal composition of C-MWCNTs support for the enhanced catalytic activity of PtRu/C-MWCNTs, and gains insight into the support role to the overall performance of an electrocatalyst.

2. Materials and Methods

PtRu/$C_{100-x}MWCNTs_x$ (x = 0, 10, 20, 30, 40, 50, and 100 wt.%) were synthesized using the following procedure. First, solutions containing $C_{100-x}MWCNTs_x$ (x = 0, 10, 20, 30, 40, 50, and 100 wt.%) were prepared by mixing accurate amounts of the commercial C (carbon Vulcan XC-72, Fuelcellstore, College Station, TX, USA) and MWCNTs (>95%, OD: 10–20 nm, US Research Nanomaterials, Inc., Houston, TX, USA) with 10 mL deionized (DI) water in an ultrasonic bath for 15 min. Noticeably, since the pristine MWCNTs lacks bonding sites, namely –COOH, =O, and –OH groups, the deposition of metal nanoparticles on the surface of MWCNTs is difficult. Thus, it requires functionalizing MWCNTs to keep metal nanoparticles on its surface by several developed methods [35–37]. The common method is the treatment of MWCNTs with HNO_3 and H_2SO_4 acids under suitable time and temperature to activate MWCNTs. To our knowledge, MWCNTs are activated effectively under refluxing condition in 65% HNO_3 and 98% H_2SO_4 acids (1:1) at 50 °C for 5 h. Next, 30 mL ethylene glycol and 15 mL sulfuric acid (H_2SO_4 98%) solutions were added to the $C_{100-x}MWCNTs_x$ solutions, and the mixture solutions were then stirred at 170 °C for 30 min.

The PtRu precursor solutions were prepared by mixing 3.5 mL $H_2PtCl_6·6H_2O$ 0.02 M and 3.5 mL $RuCl_3·xH_2O$ 0.02 M (corresponding to 13 mg Pt and 7 mg Ru, the atomic ratio of Pt:Ru = 1:1) in an ultrasonic bath for 15 min. Next, the mixture of PtRu precursor solution was slowly dropped into the $C_{100-x}MWCNTs_x$ solutions, following by a sprinkle of 0.2 M $NaBH_4$ solution. The pH solution was adjusted to 10 by using 10 M NaOH. The mixture was stirred at room temperature for 8 h. Finally, the $PtRu/C_{100-x}MWCNTs_x$ (x = 0, 10, 20, 30, 40, 50, and 100 wt.%) products were collected by filtration, washed thoroughly with DI water, and dried overnight at 90 °C. The $H_2PtCl_6·6H_2O$, $RuCl_3·xH_2O$, H_2SO_4 98%, HNO_3 65%, $NaBH_4$, CH_3OH, and NaOH were purchased from Merck KGaA of Darmstadt, Germany. The mass of each support type, Pt and Ru metals, and their proportions in each sample are listed in Table 1.

Table 1. Detailed preparation conditions of the catalytic samples investigated.

Sample	C-MWCNTs Support					Pt and Ru Metal		
	Carbon Vucal XC-72 Mass (mg)	MWCNTs Mass (mg)	Mass Ratio of C and C-MWCNTs (wt.%)	Mass Ratio of MWCNTs and C-MWCNTs (wt.%)	Mass Ratio of C-MWCNTs and Sample (wt.%)	Pt Mass (mg)	Ru Mass (mg)	Mass Ratio of Pt + Ru and Sample (wt.%)
PtRu/C	80	0	100	0	80	13	7	20
PtRu/MWCNTs	0	80	0	100	80	13	7	20
$PtRu/C_{90}MWCNTs_{10}$	72	8	90	10	80	13	7	20
$PtRu/C_{80}MWCNTs_{20}$	64	16	80	20	80	13	7	20
$PtRu/C_{70}MWCNTs_{30}$	56	24	70	30	80	13	7	20
$PtRu/C_{60}MWCNTs_{40}$	48	32	60	40	80	13	7	20
$PtRu/C_{50}MWCNTs_{50}$	40	40	50	50	80	13	7	20

The crystalline orientations of the nanomaterials were studied via XRD analysis using Bruker D8 and Cu Kα (1.5406 Å) radiation. Structural characterization at atomic scale was performed by using a TEM (JEOL JEM1010, Hanoi, Vietnam). A three-electrode test cell configuration using an Ag/AgCl reference electrode was used for electrochemical analyses. The electrolyte was a mixture solution of 0.5 M H_2SO_4 98% and 1.0 M CH_3OH. The working electrode was made using 4 mg of catalytic powder (Pt and Ru masses were 0.52 mg and 0.28 mg, respectively) mixed with 1 mL of 2-propanol (Merck, Darmstadt, Germany) in an ultrasonic bath. Afterward, the catalytic powder was swept onto 1 cm^2 carbon paper using Nafion 117 binder solution (Aldrich, Darmstadt, Germany). The carbon paper was then scanned into catalytic powder, which was assembled in a sealed plastic frame with a blank area of 1 cm^2. This active area was completely immersed in the electrolyte during CV measurements. CV curves were recorded using an Autolab 302N system (Ho Chi Minh City, Vietnam) within a potential range of −0.2–1.2 V vs. Ag/AgCl (3M KCl) at a scan rate of 50 mV·s^{-1}. Electrochemical impedance spectroscopy (EIS) measurements were performed using the same system with a potential amplitude of ±10 mV in a frequency range of 0.1–100 kHz.

3. Results and Discussion

3.1. Effect of Catalyst Supports on Structure–Composition and Methanol Oxidation Performance of $PtRu/C_{100-x}MWCNTs_x$ (x = 0, 40, 100%)

Figure 1 shows the XRD patterns of PtRu/C, PtRu/MWCNTs, and $PtRu/C_{60}MWCNTs_{40}$. Clearly, the XRD patterns of all three types of nanomaterials are similar, whose peaks can be indexed to the (111), (200), (220), and (311) planes of a face-centered cubic (f.c.c) lattice structure of platinum. In addition, the patterns of the hexagonal closest packed (h.c.p) structure of ruthenium should be included inside these diffraction peaks, namely Ru (002) at 42.2°, Ru (101) at 44.1°, Ru (110) at 69.5°, and Ru (112) at 84.8°. In addition, a broad diffraction peak at approximately 26° is attributed to hexagonal graphite structure [C (002)], suggesting that these supports could have good electrical conductivity [38]. The present XRD results confirm for the PtRu alloy on C-based supports, and they are similar to those results reported in [39–41]. It was found in a PtRu alloy that if Ru content in PtRu alloy is lower than 60 wt.%, the alloys will stay in the f.c.c structure of platinum; inversely, if Ru content in PtRu alloy is higher than 60 wt.%, the PtRu alloy will exhibit h.c.p structure of ruthenium [40,41]. Since the XRD patterns matched better with the f.c.c structure of platinum, the Ru content in the PtRu alloys in this study should be lower than 60 wt.%. Furthermore, no XRD peak shift was observed, thus the composition of PtRu should be stable among the prepared samples.

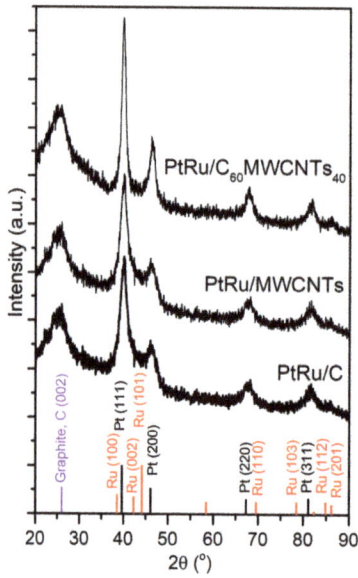

Figure 1. XRD patterns of PtRu/C, PtRu/MWCNTs, and PtRu/C$_{60}$MWCNTs$_{40}$.

Figure 2 shows the typical TEM images and the particle size distribution histograms of PtRu/C, PtRu/MWCNTs, and PtRu/C$_{60}$MWCNTs$_{40}$. Obviously, PtRu nanoparticles were well distributed and decorated on the C$_{100-x}$MWCNTs$_x$ (x = 0, 40, 100%) supports. All the samples had narrow size distributions and small mean values (d_{mean}) of 2.4 ± 0.2, 3.8 ± 0.1, and 2.2 ± 0.1 nm for PtRu/C, PtRu/MWCNTs, and PtRu/C$_{60}$MWCNTs$_{40}$, respectively (Figure 2). It is known that particle size of catalysts can affect the methanol oxidation activity of PtRu alloy catalysts. It is plausible that when the present PtRu sizes with d_{mean} between 2.2 ± 0.1 and 3.8 ± 0.1 nm are close to the reported optimal PtRu size of ~3 nm, high methanol oxidation activity is achieved [42]. Importantly, the uniformities in size and distribution of PtRu nanoparticles among the samples investigated were sufficient to allow for studying the compositional effects of C$_{100-x}$MWCNTs$_x$ supports on the electrocatalytic activity of PtRu/C$_{100-x}$MWCNTs$_x$ (x = 0–100 wt.%), as described in a later section. Interestingly, by considering both d_{mean} values and the upper tails of the size distributions, the C$_{60}$MWCNTs$_{40}$ composite support with high porosity likely hinders the growth of PtRu nanoparticles. This finding agreed well with that in [11], in which the mesoporous carbon support was found to restrict the crystal growth of PtRu nanoparticles.

Cyclic voltammograms (CV) of PtRu/C, PtRu/MWCNTs, and PtRu/C$_{60}$MWCNTs$_{40}$ are shown in Figure 3a,b. These samples exhibited different current density (J); the peak current density values of forward and reverse scans (J_f and J_r) and J_f/J_r ratio were determined and are listed in Table 2. The J_f (J_r) values are 21.6 (6.0) mA/mg$_{PtRu}$, 67.0 (36.0) mA/mg$_{PtRu}$, and 65.4 (19.4) mA/mg$_{PtRu}$ for PtRu/C, PtRu/MWCNTs, and PtRu/C$_{60}$MWCNTs$_{40}$, respectively. Obviously, PtRu/C had lowest J_f value, while PtRu/MWCNTs and PtRu/C-MWCNTs achieved J_f values more than three times higher. In contrast, the PtRu/C obtained the highest J_f/J_r ratio of 3.6, and PtRu/C$_{60}$MWCNTs$_{40}$ also reached a high J_f/J_r value of

3.4, whereas PtRu/MWCNTs exhibited a low J_f/J_r ratio of 1.9. The J_f/J_r ratio is used to describe the catalyst tolerance to carbonaceous species accumulation [43]. The larger J_f/J_r value indicates a better CO resistant catalyst. In MOR, CO is a critical intermediate that reduces both fuel cell potential and energy conversion efficiency. A forward scan is attributable to methanol oxidation, forming Pt-adsorbed carbonaceous intermediates (e.g., carbon monoxide). The adsorbed carbon monoxide causes a suppression of the electrocatalyst activity. The reverse oxidation peak is attributed to the additional oxidation of the adsorbed carbonaceous species to carbon dioxide (CO_2) [44].

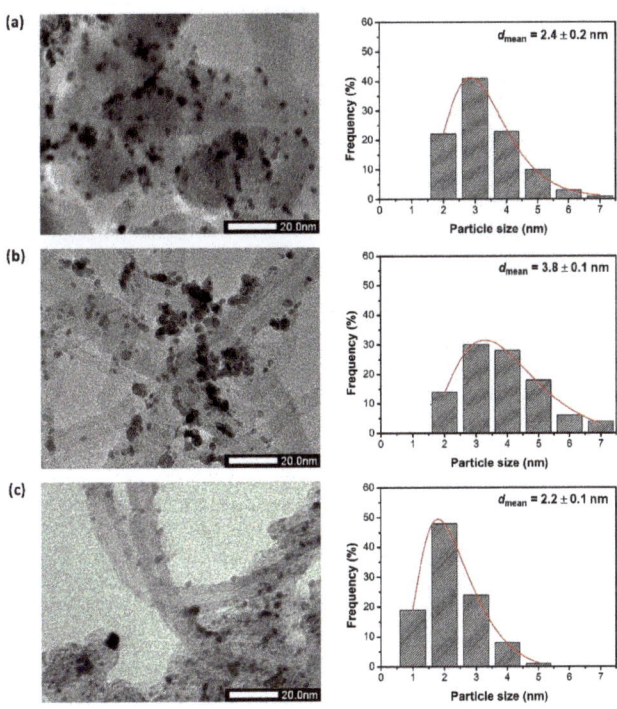

Figure 2. TEM images and particle size distributions of (**a**) PtRu/C, (**b**) PtRu/MWCNTs, and (**c**) PtRu/C_{60}MWCNTs$_{40}$.

Table 2. Electrochemical parameters of PtRu/C, PtRu/MWCNTs, and PtRu/C_{60}MWCNTs$_{40}$.

Sample	J_f		J_r		J_f/J_r	R_{et}
	(mA·cm^{-2})	(mA/mg$_{PtRu}$)	(mA·cm^{-2})	(mA/mg$_{PtRu}$)	-	($\Omega \cdot cm^2$)
PtRu/C	17.3	21.6	4.8	6.0	3.6	7.39
PtRu/MWCNTs	53.6	67.0	28.8	36.0	1.9	11.19
PtRu/C_{60}MWCNTs$_{40}$	52.3	65.4	15.5	19.4	3.4	6.37

Based on these results, the best carbonaceous species tolerance ability (J_f/J_r = 3.6) was achieved in the PtRu/C. For the support effect, the J_f value of PtRu/MWCNTs is higher than that of PtRu/C (Figure 3a and Table 2), which could be attributed to the higher porosity and better catalyst dispersion–distribution of MWCNTs support than those of C support [21]. However, the carbonaceous species tolerance ability of PtRu/MWCNTs (J_f/J_r = 1.9) is very limited. Nevertheless, PtRu/C_{60}MWCNTs$_{40}$ can simultaneously achieve a high J_f of 52.3 mA·cm^{-2} (or 65.4 mA/mg$_{PtRu}$) and a high J_f/J_r ratio of 3.4. This means that PtRu/C_{60}MWCNTs$_{40}$ possesses not only high MOR activity, but also excellent carbonaceous species tolerance ability, suggesting that the composite of C-MWCNTs support is better than either C or MWCNTs support. Thus, it is necessary to further optimize the composite composition of $C_{100-x}MWCNTs_x$ with x = 10–50 wt.% toward achieving the highest methanol oxidation activity and excellent carbonaceous species tolerance ability, as described in the next section.

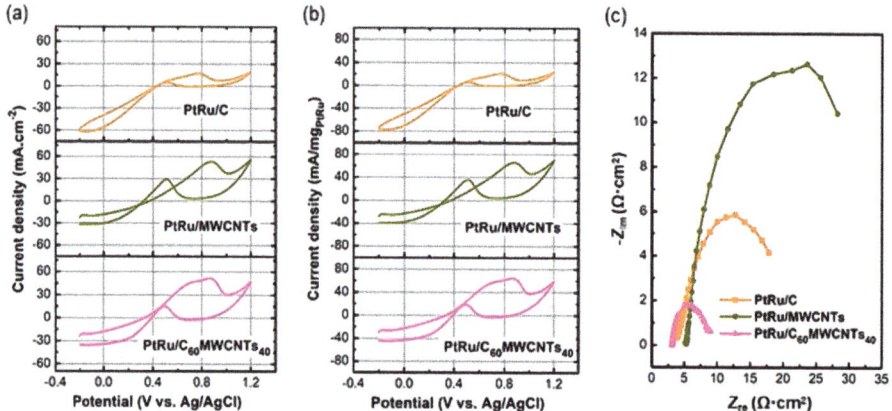

Figure 3. (**a**,**b**) Cyclic voltammograms (CV) of PtRu/C, PtRu/MWCNTs, and PtRu/C_{60}MWCNTs$_{40}$ with current density units of mA·cm^{-2} and mA/mg$_{PtRu}$. (**c**) EIS Nyquist plots of the samples in a frequency range from 0.1 Hz to 100 kHz. The electrolyte was a mixture solution of 0.5 M H$_2$SO$_4$ and 1.0 M CH$_3$OH.

Figure 3c shows the Nyquist plot curves of PtRu/C, PtRu/MWCNTs, and PtRu/C_{60}MWCNTs$_{40}$ at a potential of 0.8 V (vs. Ag/AgCl). Although the Nyquist curves of the samples do not show full semicircles, it is still clear that the total complex impedance Z of the samples has a decreasing order of PtRu/MWCNTs < PtRu/C < PtRu/C_{60}MWCNTs$_{40}$. For quantitative analysis of the EIS data, we employed the Randles equivalent circuit model to fit the plots with one semicircle [45], whose model was also used to fit the EIS data of Fe$_2$O$_3$ films [46], Pt–MnO$_2$ nanoparticles decorated on reduced graphene oxide sheets [47], and carbon-supported Ru-Pt nanoparticles [48]. The circuit included the electron transfer resistance (R_{et}), solution resistance (R_s), and double layer capacity (C_{dl}). The circular fits with the Randles circuit yield R_{et} values of 7.39, 11.19, and 6.37 Ω·cm^2 for PtRu/C, PtRu/MWCNTs, and PtRu/C_{60}MWCNTs$_{40}$, respectively. The lowest R_{et} was obtained for PtRu/C_{60}MWCNTs$_{40}$, indicating composite C-MWCNTs support offered excellent electron transfer that was attributed to the highest current density at 0.8 V. The highest electrocatalytic activity and the smallest R_{et} of PtRu/C_{60}MWCNTs$_{40}$ indicates that the composite of C-MWCNTs is a superior support over either C or MWCNTs.

These results can be explained by considering the structure–morphology of the three types of supports. Indeed, MWCNTs are chemically inert, so PtRu nanoparticles will be unable to stick well on the tube walls if MWCNTs are not treated. In this study, after being treated with 98% H$_2$SO$_4$ + 65% HNO$_3$ at 50 °C for 5 h, functional groups =O, –OH, –COOH are formed on the surfaces of MWCNTs, and consequently PtRu nanoparticles resides on these functional groups (Figure 4). Although commercial carbon Vulcan XC-72 has high porosity, it clumps easily and reduces the surface area if this C support is used in a large quantity. The blending of carbon Vulcan XC-72 with the activated MWCNTs allows dispersed C sheets/ bulks into the gaps between MWCNTs (Figure 4). Therefore, this composite support can exhibit high porosity and large surface area. Consequently, the deposition of PtRu nanoparticles on C-MWCNTs will be easier and distributed more evenly, which is attributed to the enhanced methanol oxidation catalytic efficiency (Figure 4). Furthermore, C-MWCNTs support can combine the advantages but limit the disadvantages of each component (i.e., C, MWCNTs).

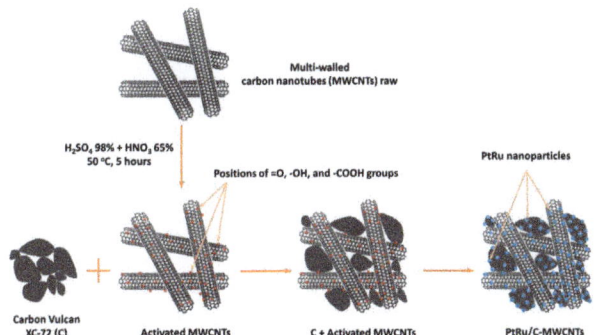

Figure 4. Schematic diagram of PtRu nanoparticles deposited on C-MWCNTs composite support, where Vulcan XC-72 carbon dispersed into the gaps between activated MWCNTs.

3.2. Effect of MWCNTs Percentage (x = 10–50 wt.%) in PtRu/C_{100-x}MWCNTs$_x$ Samples on the Methanol Oxidation Activity

To determine the most suitable C-MWCNTs supports, a series of PtRu/C_{100-x}MWCNTs$_x$ (x = 10–50 wt.%) were prepared and characterized. The PtRu/C_{100-x}MWCNTs$_x$ samples had the same amount of PtRu nanoparticles with 20 wt.% and Pt/Ru atomic ratio of 1; meanwhile, the C_{100-x}MWCNTs$_x$ supports had various weight percentages of MWCNTs (x) from 10 to 50%. Figure 5 shows TEM images and particle size distributions of the PtRu/C_{100-x}MWCNTs$_x$ samples. It can be seen clearly that Vulcan XC-72 carbon substrate was dispersed into MWCNTs. In addition, PtRu nanoparticles were deposited and evenly distributed on both carbon Vulcan XC-72 and MWCNTs for all the samples. Moreover, particle size distribution of all the samples was in the range 1–6 nm with the peak of the distribution at 2 nm. Intriguingly, when the MWCNTs content was small (x = 10 wt.%) or large (x = 50 wt.%), the particle sizes were less uniform with wider size distribution as compared to those of the other three samples (x = 20–40 wt.%).

Figure 6a,b present the CV curves of PtRu/C_{100-x}MWCNTs$_x$ (x = 10–50 wt.%) in the electrolyte solution. Clearly, the MWCNTs content strongly affected the current density of PtRu/C_{100-x}MWCNTs$_x$ samples. The value of J_f, J_r, and J_f/J_r ratio are summarized in Table 3. The J_f and J_r of PtRu/C_{100-x}MWCNTs$_x$ increased when x increased from 10% to 30%, and then decreased with further increasing the MWCNTs weight percentage (x = 40–50 wt.%). The PtRu/C_{70}MWCNTs$_{30}$ obtained the highest J_f value, 115.8 mA/mg$_{PtRu}$ (92.6 mA·cm^{-2}), or achieving the highest MOR activity. Meanwhile, PtRu/C_{90}MWCNTs$_{10}$ obtained the highest J_f/J_r value of 4.4, and PtRu/C_{80}MWCNTs$_{20}$ exhibited a high J_f/J_r of 4.0, whose values were higher than that of PtRu/C (J_f/J_r = 3.6, Tables 2 and 3), indicating that C_{100-x}MWCNTs$_x$ (x = 10–20 wt.%) achieved the enhanced carbonaceous species tolerance. Moreover, Figure 6c shows the Nyquist plot of PtRu/C_{100-x}MWCNTs$_x$ (x = 10–50 wt.%) at a potential of 0.8 V (vs. Ag/AgCl). The semicircle of PtRu/C_{70}MWCNTs$_{30}$ and PtRu/C_{60}MWCNTs$_{40}$ are smaller as compared with that of the other samples. By fitting the semicircles with the Randles equivalent circuit model, PtRu/C_{70}MWCNTs$_{30}$ has the smallest R_{et}, 5.31 Ω·cm^2 (Table 3). Owing to reaching the highest electrocatalytic activity and the smallest R_{et}, PtRu/C_{70}MWCNTs$_{30}$ is proposed as the best sample.

Table 4 summarizes the methanol oxidation activity results of PtRu nanoparticles decorated on various carbon-related supports in the literature and in this study, in which current density (J_f) and current density ratio (J_f/J_{fC}) are used to evaluate the activity. Here, J_f/J_{fC} is the ratio between J_f of the present sample and J_{fC} of the conventional carbon-supported PtRu nanoparticle sample in each reference and in this study; thus J_f/J_{fC} indicates the current density enhancement of a particular sample relative to the conventional PtRu/C. By observing J_f/J_{fC} > 1 in Table 4, CNTs or MWCNTs, graphene-related, or composite support generally exhibited the enhancements in methanol oxidation activity over the conventional carbon-supported sample. In other words, PtRu catalytic materials on the modified supports or composite supports present higher catalytic results than samples using traditional carbon support (e.g., 82.7 mA·cm^{-2} for PtRu/N-CNTs vs. 27.5 mA·cm^{-2} for PtRu/CB [23]; and 136.7 mA·mg^{-1} for PtRu/CTNs-GS vs. 42.7 mA·mg^{-1} for PtRu/C [24]). In our study, the current intensity of catalyst with PtRu nanoparticles deposited on C$_{70}$MWCNTs$_{30}$ composite support is 5.35 times higher than that of Vulcan XC-72 carbon support. Therefore, C$_{70}$MWCNTs$_{30}$ composite support is recommended to use in the electrodes of DMFC.

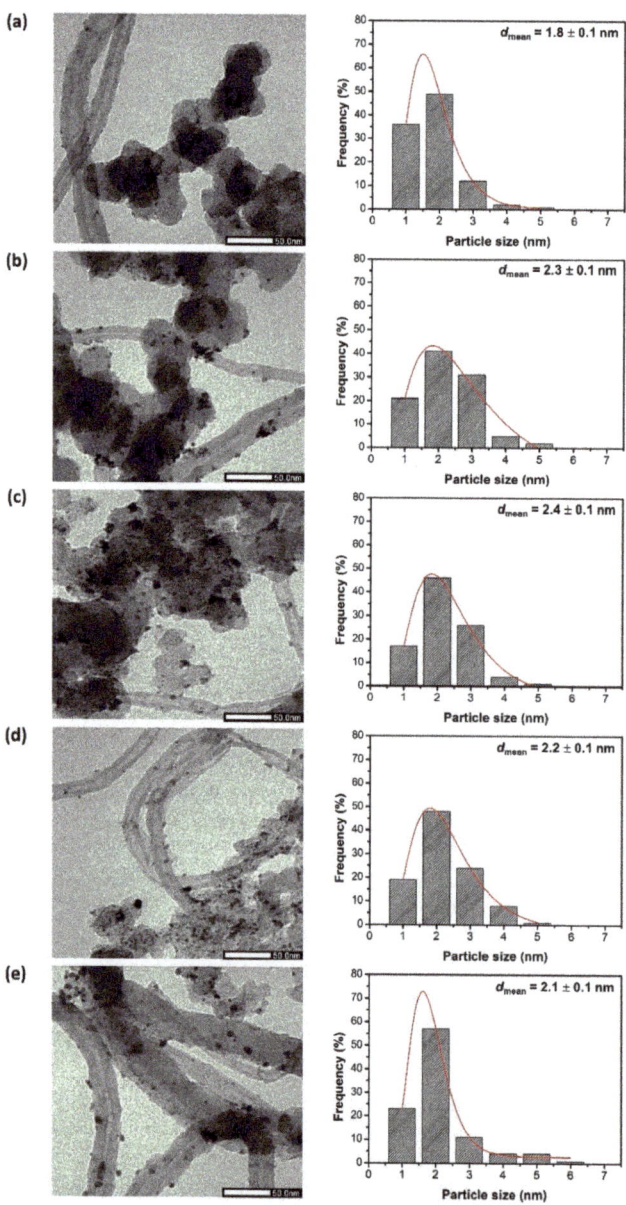

Figure 5. TEM images and particle size distributions of (**a**) PtRu/C$_{90}$MWCNTs$_{10}$, (**b**) PtRu/C$_{80}$MWCNTs$_{20}$, (**c**) PtRu/C$_{70}$MWCNTs$_{30}$, (**d**) PtRu/C$_{60}$MWCNTs$_{40}$, and (**e**) PtRu/C$_{50}$MWCNTs$_{50}$.

Table 3. Electrochemical parameters of PtRu/C$_{90}$MWCNTs$_{10}$, PtRu/C$_{80}$MWCNTs$_{20}$, PtRu/C$_{70}$MWCNTs$_{30}$, PtRu/C$_{60}$MWCNTs$_{40}$, and PtRu/C$_{50}$MWCNTs$_{50}$.

Sample	J_f		J_r		J_f/J_r	R_{et}
	(mA/cm^2)	(mA/mg$_{PtRu}$)	(mA/cm^2)	(mA/mg$_{PtRu}$)	-	($\Omega \cdot$cm^2)
PtRu/C$_{90}$MWCNTs$_{10}$	29.8	37.3	6.7	8.4	4.4	29.2
PtRu/C$_{80}$MWCNTs$_{20}$	52.8	66.0	13.5	16.9	4.0	6.16
PtRu/C$_{70}$MWCNTs$_{30}$	92.6	115.8	42.9	53.6	2.2	5.31
PtRu/C$_{60}$MWCNTs$_{40}$	52.3	65.4	15.5	19.4	3.4	6.37
PtRu/C$_{50}$MWCNTs$_{50}$	38.6	48.3	9.7	12.1	3.9	8.87

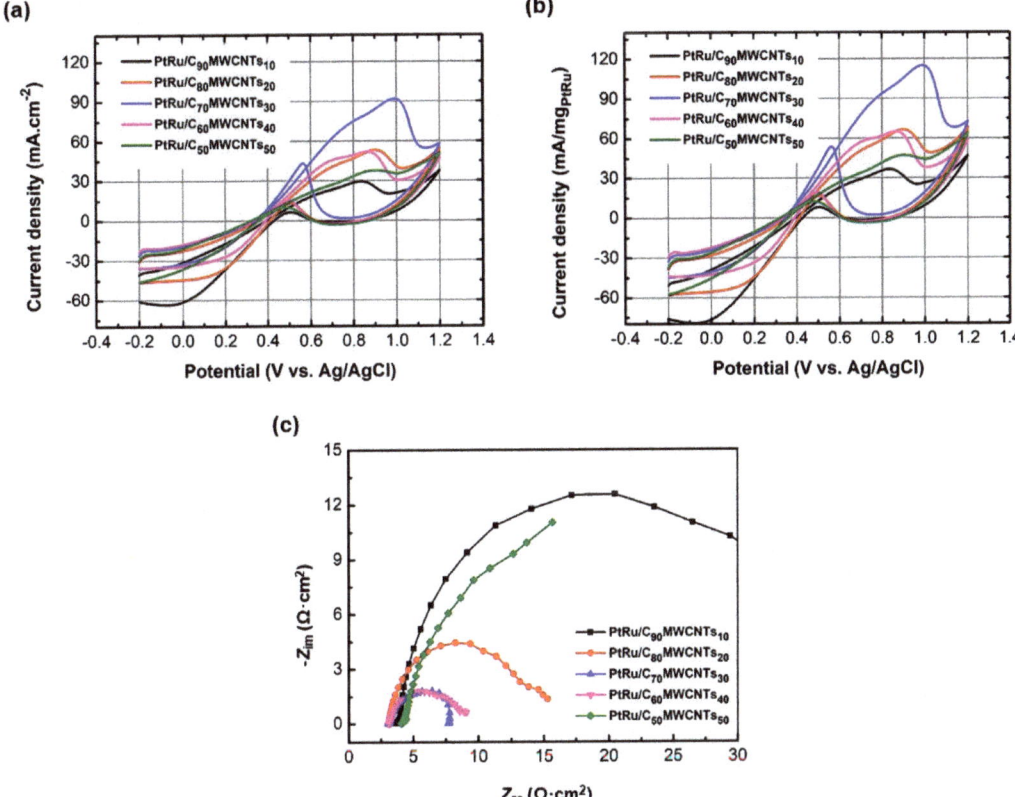

Figure 6. (**a**,**b**) Cyclic voltammograms of PtRu/C$_{90}$MWCNTs$_{10}$, PtRu/C$_{80}$MWCNTs$_{20}$, PtRu/C$_{70}$MWCNTs$_{30}$, PtRu/C$_{60}$MWCNTs$_{40}$, and PtRu/C$_{50}$MWCNTs$_{50}$ with current density units of mA·cm^{-2} and mA/mg$_{PtRu}$. (**c**) Nyquist plots of the EIS for these samples in a frequency range from 0.1 Hz to 100 kHz. The electrolyte was a mixture solution of 0.5 M H$_2$SO$_4$ and 1.0 M CH$_3$OH.

Table 4. Comparison of CV results of the selected studies of PtRu nanoparticles decorated on novel carbon-related substrates in the literature and in this study. Current density ratio (J_f/J_{fC}) is the ratio between J_f of a PtRu/modified support and the J_f of a PtRu/conventional carbon support in each reference and in this study.

Sample	Support Material	Measurement Condition	Evaluate Performance		Reference
			Current Density (J_f)	Current Density Ratio (J_f/J_{fC})	
PtRu/E-Tek PtRu/CX	Vulcan XC-72R Carbon xerogels	2 M CH_3OH + 0.5 M H_2SO_4, 0.02 V·s^{-1}	0.29 mA·cm^{-2} 0.36 mA·cm^{-2}	– 1.44	[27]
PtRu/C PtRu 70%/CNF	Carbon Carbon nanofiber	1 M CH_3OH + 0.5 M H_2SO_4, 2 mV·s^{-1}	340 mA·cm^{-2} 390 mA·cm^{-2}	– 1.15	[28]
PtRu/XC-72 PtRu/CMK-8-II	Vulcan XC-72R Mesoporous Carbon	1 M CH_3OH + 0.5 M H_2SO_4, 50 mV·s^{-1}	27 mA·cm^{-2} 60 mA·cm^{-2}	– 2.22	[29]
PtRu/C PtRu/CNTs	Vulcan XC-72R Carbon nanotube	1 M CH_3OH + 0.5 M H_2SO_4, 20 mV·s^{-1}	22.5 mA·cm^{-2} 33.5 mA·cm^{-2}	– 1.49	[22]
PtRu/CB PtRu/CNT PtRu/N-CNTs	Carbon Carbon nanotube Carbon nanotube doping N	1 M CH_3OH + 0.5 M H_2SO_4, 50 mV·s^{-1}	27.5 mA·cm^{-2} 44.1 mA·cm^{-2} 82.7 mA·cm^{-2}	– 1.60 3.00	[23]
PtRu/C PtRu/CNTs PtRu/GS PtRu/CTNs-GS	Carbon Carbon nanotubes Graphene sheet Carbon nanotubes + Graphene sheet	1 M CH_3OH + 0.5 M H_2SO_4, 20 mV·s^{-1}	42.7 mA·mg^{-1} 56.0 mA·mg^{-1} 78.7 mA·mg^{-1} 136.7 mA·mg^{-1}	– 1.31 1.84 3.20	[24]
PtRu/C PtRu/RGO	Carbon Reduced graphene oxide (RGO)	0.5 M H_2SO_4 + 1 M CH_3OH	430 mA·mg^{-1} 570 mA·mg^{-1}	– 1.33	[31]
PtRu/C PtRu/FGSs	Carbon Functionalized graphene sheets	1 M CH_3OH + 0.5 M H_2SO_4, 50 mV·s^{-1}	8.21 mA·cm^{-2} 14.05 mA·cm^{-2}	– 1.71	[32]
PtRu/C PtRu/MWCNTs PtRu/C_{70}MWCNTs30	Carbon Vulcan XC-72 Multi-walled carbon nanotubes C + MWCNTs	1 M CH_3OH + 0.5 M H_2SO_4, 50 mV·s^{-1}	17.3 mA·cm^{-2} 53.6 mA·cm^{-2} 92.6 mA·cm^{-2}	– 3.10 5.35	Our results

4. Conclusions

PtRu alloy nanoparticles decorated on C, MWCNTs, and C-MWCNTs supports for high-performance methanol oxidation were synthesized by co-reduction method. The synthesized PtRu/C_{100-x}MWCNTs$_x$ (x = 0–100 wt.%) exhibited the crystal structure closed to the f.c.c lattice structure of platinum with (111), (200), (220), and (311) preferred orientations. In addition, PtRu nanoparticles obtained a narrow size distribution and a small mean size (d_{mean} = 1.8–3.8 nm). For the support effects, PtRu/C-MWCNTs offered higher J_f (or higher electrocatalytic activity) and lower R_{et} than those of PtRu/C and PtRu/MWCNTs. In addition, PtRu/C-MWCNTs has high J_f/J_r values, which means good carbonaceous species tolerance ability. To optimize the C-MWCNTs composition, PtRu/C_{100-x}MWCNTs$_x$ (x = 10–50 wt.%) were synthesized and characterized. The highest J_f, 115.8 mA/mg$_{PtRu}$, was obtained for PtRu/C_{70}MWCNTs$_{30}$, which was considered an optimal nanomaterial system. Meanwhile, both PtRu/C_{70}MWCNTs$_{30}$ and PtRu/C_{60}MWCNTs$_{40}$ exhibited low resistances with R_{et} of 5.31–6.37 Ω·cm^2. The results of this study demonstrate that C-MWCNTs composite support is better than either C or MWCNTs support; significant enhancements in methanol oxidation activity and carbonaceous species tolerance ability can be achieved by controlling the MWCNTs content in C-MWCNTs support.

Author Contributions: D.L.Q. performed the experiments, analyzed the data, and wrote the first draft; P.H.L. revised and edited the paper, and supervised the project. Both authors have read and agreed to the published version of the manuscript.

Funding: This research is funded by Ho Chi Minh City University of Technology (HCMUT), VNU-HCM, under grant number BK-SDH-2021-2080906, and Vietnam National Foundation for Science and Technology Development (NAFOSTED) under Grant No. 103.02-2019.374.

Institutional Review Board Statement: Not applicable.

Informed Consent Statement: Not applicable.

Data Availability Statement: Not applicable.

Acknowledgments: We acknowledge the support of time and facilities from Ho Chi Minh City University of Technology (HCMUT), VNU-HCM, Can Tho University, and Can Tho University of Medicine and Pharmacy for this study.

Conflicts of Interest: The authors declare no conflict of interest.

References

1. Bokach, D.; de la Fuente, J.L.G.; Tsypkin, M.; Ochal, P.; Endsjø, I.C.; Tunold, R.; Sunde, S.; Seland, F. High-temperature electrochemical characterization of Ru core Pt shell fuel cell catalyst. *Fuel Cells* **2011**, *11*, 735–744. [CrossRef]
2. Zhao, S.; Yin, H.; Du, L.; Yin, G.; Tang, Z.; Liu, S. Three dimensional N-doped graphene/PtRu nanoparticle hybrids as high performance anode for direct methanol fuel cells. *J. Mater. Chem. A* **2014**, *2*, 3719–3724. [CrossRef]
3. Almeida, T.S.; Garbim, C.; Silva, R.G.; De Andrade, A.R. Addition of iron oxide to Pt-based catalyst to enhance the catalytic activity of ethanol electrooxidation. *J. Electroanal. Chem.* **2017**, *796*, 49–56. [CrossRef]
4. Lu, X.; Hu, J.; Foord, J.S.; Wang, Q. Electrochemical deposition of Pt–Ru on diamond electrodes for the electrooxidation of methanol. *J. Electroanal. Chem.* **2011**, *654*, 38–43. [CrossRef]
5. Moura, A.S.; Fajín, J.L.C.; Mandado, M.; Cordeiro, M.N.D.S. Ruthenium-platinum catalysts and direct methanol fuel cells (DMFC): A review of theoretical and experimental breakthroughs. *Catalysts* **2017**, *7*, 47. [CrossRef]
6. Liu, M.; Zhang, R.; Chen, W. Graphene-supported nanoelectrocatalysts for fuel cells: Synthesis, properties, and applications. *Chem. Rev.* **2014**, *114*, 5117–5160. [CrossRef] [PubMed]
7. Muthuswamy, N.; de la Fuente, J.L.G.; Tran, D.T.; Walmsley, J.; Tsypkin, M.; Raaen, S.; Sunde, S.; Rønning, M.; Chen, D. Ru@Pt core–shell nanoparticles for methanol fuel cell catalyst: Control and effects of shell composition. *Int. J. Hydrogen Energy* **2013**, *38*, 16631–16641. [CrossRef]
8. Arun, A.; Gowdhamamoorthi, M.; Ponmani, K.; Kiruthika, S.; Muthukumaran, B. Electrochemical characterization of Pt-Ru-Ni/C anode electrocatalyst for methanol electrooxidation in membraneless fuel cells. *RSC Adv.* **2015**, *5*, 49643–49650. [CrossRef]
9. Ávila-García, I.; Ramírez, C.; Hallen López, J.M.; Arce Estrada, E.M. Electrocatalytic activity of nanosized Pt alloys in the methanol oxidation reaction. *J. Alloy. Compd.* **2010**, *495*, 462–465. [CrossRef]

10. Salgado, J.R.C.; Paganin, V.A.; Gonzalez, E.R.; Montemora, M.F.; Tacchini, I.; Anson, A.; Salvador, M.A.; Ferreira, P.; Figueiredo, F.M.L.; Ferreira, M.G.S. Characterization and performance evaluation of Pt–Ru electrocatalysts supported on different carbon materials for direct methanol fuel cells. *Int. J. Hydrog. Energy* **2013**, *38*, 910–920. [CrossRef]
11. Liu, Z.; Lee, J.Y.; Chen, W.; Han, M.; Gan, L.M. Physical and electrochemical characterizations of microwave-assisted polyol preparation of carbon-supported PtRu nanoparticles. *Langmuir* **2004**, *20*, 181–187. [CrossRef]
12. Neto, A.O.; Dias, R.R.; Tusi, M.M.; Linardi, M.; Spinacé, E.V. Electro-oxidation of methanol and ethanol using PtRu/C, PtSn/C and PtSnRu/C electrocatalysts prepared by an alcohol-reduction process. *J. Power Sources* **2017**, *166*, 87–91. [CrossRef]
13. Baronia, R.; Goel, J.; Tiwari, S.; Singh, P.; Singh, D.; Singh, S.P.; Singhal, S.K. Efficient electro-oxidation of methanol using PtCo nanocatalysts supported reduced graphene oxide matrix as anode for DMFC. *Int. J. Hydrog. Energy* **2017**, *42*, 10238–10247. [CrossRef]
14. Maya-Cornejo, J.; Garcia-Bernabé, A.; Compañ, V. Bimetallic Pt-M electrocatalysts supported on single-wall carbon nanotubes for hydrogen and methanol electrooxidation in fuel cells applications. *Int. J. Hydrog. Energy* **2018**, *43*, 872–884. [CrossRef]
15. Wang, Z.N.; Huo, S.H.; Zhou, P.X. Rich-grain-boundary PtRuNi with network structure as efficient catalysts for methanol oxidation reaction. *Int. J. Electrochem. Sci.* **2019**, *14*, 10576–10581. [CrossRef]
16. Wang, Z.B.; Zuo, P.J.; Yin, G.P. Investigations of compositions and performance of PtRuMo/C ternary catalysts for methanol electrooxidation. *Fuel Cells* **2009**, *9*, 106–113. [CrossRef]
17. Ramli, Z.A.C.; Kamarudin, S.K. Platinum-based catalysts on various carbon supports and conducting polymers for direct methanol fuel cell applications: A review. *Nanoscale Res. Lett.* **2018**, *13*, 1–25. [CrossRef]
18. Rambabu, G.; Bhat, D.S.; Figueiredo, F.M. Carbon nanocomposite membrane electrolytes for direct methanol fuel cells—A concise review. *Nanomaterials* **2019**, *9*, 1292. [CrossRef] [PubMed]
19. Yaldagard, M.; Jahanshahi, M.; Seghatoleslami, N. Carbonaceous nanostructured support materials for low temperature fuel cell electrocatalysts—A review. *World J. Nano Sci. Eng.* **2013**, *3*, 121–153. [CrossRef]
20. Maass, S.; Finsterwalder, F.; Frank, G.; Hartmann, R.; Merten, C. Carbon support oxidation in PEM fuel cell cathodes. *J. Power Sources* **2008**, *176*, 444–451. [CrossRef]
21. Kannan, R.; Pillai, V.K. Applications of carbon nanotubes in polymer electrolyte membrane fuel cells. *J. Indian Inst. Sci.* **2009**, *89*, 425–436.
22. Cui, Z.; Liu, C.; Liao, J.; Xing, W. Highly active PtRu catalysts supported on carbon nanotubes prepared by modified impregnation method for methanol electro-oxidation. *Electrochim. Acta* **2008**, *53*, 7807–7811. [CrossRef]
23. Fard, H.F.; Khodaverdi, M.; Pourfayaz, F.; Ahmadi, M.H. Application of N-doped carbon nanotube-supported Pt–Ru as electrocatalyst layer in passive direct methanol fuel cell. *Int. J. Hydrog. Energy* **2020**, *45*, 25307–25316. [CrossRef]
24. Wang, Y.S.; Yang, S.Y.; Li, S.M.; Tien, H.W.; Hsiao, S.T.; Liao, W.H.; Liu, C.H.; Chang, K.H.; Ma, C.C.M.; Hu, C.C. Three-dimensionally porous graphene–carbon nanotube composite-supported PtRu catalysts with an ultrahigh electrocatalytic activity for methanol oxidation. *Electrochim. Acta* **2013**, *87*, 261–269. [CrossRef]
25. Hu, L.; Hecht, D.S.; Grüner, G. Percolation in transparent and conducting carbon nanotube networks. *Nano Lett.* **2004**, *4*, 2513–2517. [CrossRef]
26. Do, J.W.; Chang, N.N.; Estrada, D.; Lian, F.; Cha, H.; Duan, X.J.; Haasch, R.T.; Pop, E.; Girolami, G.S.; Lyding, J.W. Solution-mediated selective nanosoldering of carbon nanotube junctions for improved device performance. *ACS Nano* **2015**, *9*, 4806–4813. [CrossRef]
27. Alegre, C.; Calvillo, L.; Moliner, R.; González-Expósito, J.A.; Guillén-Villafuerte, O.; Huerta, M.M.; Pastor, E.; Lázaro, M.J. Pt and PtRu electrocatalysts supported on carbon xerogels for direct methanol fuel cells. *J. Power Sources* **2011**, *196*, 4226–4235. [CrossRef]
28. Kang, S.; Lim, S.; Peck, D.H.; Kim, S.K.; Jung, D.H.; Hong, S.H.; Jung, H.G.; Shul, Y. Stability and durability of PtRu catalysts supported on carbon nanofibers for direct methanol fuel cells. *Int. J. Hydrog. Energy* **2012**, *37*, 4685–4693. [CrossRef]
29. Maiyalagan, T.; Alaje, T.O.; Scott, K.K. Highly stable Pt–Ru nanoparticles supported on three-dimensional cubic ordered mesoporous carbon (Pt–Ru/CMK-8) as promising electrocatalysts for methanol oxidation. *J. Phys. Chem. C* **2012**, *116*, 2630–2638. [CrossRef]
30. Lee, S.H.; Kakati, N.; Jee, S.H.; Maiti, J.; Yoon, Y.S. Hydrothermal synthesis of PtRu nanoparticles supported on graphene sheets for methanol oxidation in direct methanol fuel cell. *Mater. Lett.* **2011**, *65*, 3281–3284. [CrossRef]
31. Reddy, G.V.; ORaghavendra, P.; Ankamwar, B.; Chandana, P.S.; Kumar, S.S.; Sarma, L.S. Ultrafine Pt–Ru bimetallic nanoparticles anchored on reduced graphene oxide sheets as highly active electrocatalysts for methanol oxidation. *Mater. Chem. Front.* **2017**, *1*, 757–766. [CrossRef]
32. Zhao, J.; Zhang, L.; Xue, H.; Wang, Z.; Hu, H. Methanol electrocatalytic oxidation on highly dispersed platinum-ruthenium/graphene catalysts prepared in supercritical carbon dioxide–methanol solution. *RSC Adv.* **2012**, *2*, 9651–9659. [CrossRef]
33. Cheng, Y. Highly effective and CO-tolerant PtRu electrocatalysts supported on poly(ethyleneimine) functionalized carbon nanotubes for direct methanol fuel cells. *Electrochim. Acta* **2013**, *99*, 124–132. [CrossRef]
34. Yang, Z.; Berber, M.R.; Nakashima, N. Design of polymer-coated multi-walled carbon nanotube/carbon black-based fuel cell catalysts with high durability and performance under non-humidified condition. *Electrochim. Acta* **2015**, *170*, 1–8. [CrossRef]
35. Hsu, N.Y.; Chien, C.C.; Jeng, K.T. Characterization and enhancement of carbon nanotube-supported PtRu electrocatalyst for direct methanol fuel cell applications. *Appl. Catal. B* **2008**, *84*, 196–203. [CrossRef]

36. Zhang, S.; Shao, Y.; Yin, G.; Lin, Y. Carbon nanotubes decorated with Pt nanoparticles via electrostatic self-assembly: A highly active oxygen reduction electrocatalyst. *J. Mater. Chem.* **2010**, *20*, 2826–2830. [CrossRef]
37. Zhao, Y.; Fan, L.; Ren, J.; Hong, B. Electrodeposition of Pt–Ru and Pt–Ru–Ni nanoclusters on multi-walled carbon nanotubes for direct methanol fuel cell. *Int. J. Hydrogen Energy* **2014**, *39*, 4544–4557. [CrossRef]
38. Giordano, N.; Passalacqua, E.; Pino, L.; Arico, A.S.; Antonucci, V.; Vivaldi, M.; Kinoshita, K. Analysis of platinum particle size and oxygen reduction in phosphoric acid. *Electrochim. Acta* **1991**, *36*, 1979–1984. [CrossRef]
39. Antolini, E. Formation of carbon-supported PtM alloys for low temperature fuel cells: A review. *Mater. Chem. Phys.* **2003**, *78*, 563–573. [CrossRef]
40. Antolini, E.; Giorgi, L.; Cardellini, F.; Passalacqua, E. Physical and morphological characteristics and electrochemical behaviour in PEM fuel cells of PtRu/C catalysts. *J. Solid State Chem.* **2001**, *5*, 131–140. [CrossRef]
41. Arico, A.S.; Antonucci, P.L.; Modica, E.; Baglio, V.; Kim, H.; Antonucci, V. Effect of Pt–Ru alloy composition on high-temperature methanol electro-oxidation. *Electrochim. Acta* **2002**, *47*, 3723–3732. [CrossRef]
42. Takasu, Y.; Itaya, H.; Iwazaki, T.; Miyoshi, R.; Ohnuma, T.; Sugimoto, W.; Murakami, Y. Size effects of ultrafine Pt–Ru particles on the electrocatalytic oxidation of methano. *Chem. Commun.* **2001**, *4*, 341–342. [CrossRef]
43. Liu, Z.; Ling, X.Y.; Su, X.; Lee, J.Y. Carbon-supported Pt and PtRu nanoparticles as catalysts for a direct methanol fuel cell. *J. Phys. Chem. B* **2004**, *108*, 8234–8240. [CrossRef]
44. Manoharan, R.; Goodenough, J.B. Methanol oxidation in acid on ordered NiTi. *J. Mater. Chem.* **1992**, *2*, 875–887. [CrossRef]
45. Randles, J.E.B. Kinetics of rapid electrode reactions. *Discuss. Faraday Soc.* **1947**, *1*, 11–19. [CrossRef]
46. Wickman, B.; Fanta, A.B.; Burrows, A.; Hellman, A.; Wagner, J.B.; Iandolo, B. Iron oxide films prepared by rapid thermal processing for solar energy conversion. *Sci. Rep.* **2017**, *7*, 1–9. [CrossRef] [PubMed]
47. Vilian, A.T.E.; Rajkumar, M.; Chen, S.M.; Hu, C.C.; Boopathi, K.M.; Chu, C.W. High electrocatalytic performance of platinum and manganese dioxide nanoparticle decorated reduced graphene oxide sheets for methanol electro-oxidation. *RSC Adv.* **2014**, *4*, 41387–41397. [CrossRef]
48. Dang, Q.L.; Nguyen, T.M.; Truong, N.V.; Le, P.H.; Long, N.V. Investigation of carbon supported Ru–Pt nanoparticles for high–performance electrocatalytic oxidation of methanol. *Int. J. Electrochem. Sci.* **2017**, *12*, 10187–10198. [CrossRef]

Article

Fabrication of Carboxylated Carbon Nanotube Buckypaper Composite Films for Bovine Serum Albumin Detection

Kuo-Jung Lee [1,*], Ming-Husan Lee [1], Yung-Hui Shih [1,*], Chao-Ping Wang [2,3,*], Hsun-Yu Lin [1] and Sheng-Rui Jian [1]

[1] Department of Materials Science and Engineering, I-SHOU University, Kaohsiung 84001, Taiwan; e124857720yms@gmail.com (M.-H.L.); carbonfish028@gmail.com (H.-Y.L.); srjian@isu.edu.tw (S.-R.J.)
[2] Division of Cardiology, E-Da Hospital, Kaohsiung 84001, Taiwan
[3] School of Medicine for International Students, College of Medicine, I-SHOU University, Kaohsiung 84001, Taiwan
* Correspondence: krlee@isu.edu.tw (K.-J.L.); yhshi@isu.edu.tw (Y.-H.S.); ed100232@edah.org.tw (C.-P.W.); Tel.: +886-7-6577-711 (ext. 3126) (K.-J.L.); +886-7-6577-711 (ext. 3115) (Y.-H.S.); +886-7-6151-100 (ext. 5018) (C.-P.W.)

Abstract: The salient point of this study is to fabricate carbon nanotube (CNT) buckypaper composite films prepared through the methods of pumping filtration and spin coating. Firstly, carboxylated CNTs were used to make the original buckypaper specimen and further modify the buckypaper surface by incorporating different surface modifiers. Then, all of original (unmodified) and modified buckypaper composite films had different concentrations of bovine serum albumin (BSA) added, and differential pulse voltammetry (DPV) electrochemical measurement was used to measure the characteristics of the various buckypaper composite films, after adding different concentrations of BSA. The experimental results show that the contact angles for four modified specimens are smaller than that of the original unmodified S–BP specimen (62°). These results indicate that the four modifiers used in this study can improve the hydrophilic properties of the original, unmodified S–BP specimen, and benefit the subsequent bonding of a modified specimen with aqueous BSA. In addition to the improvement of the hydrophilic properties of the modified specimen, which affects the bonding with BSA, the bonding type produced by the modifier also plays an essential role in the bonding between specimen and BSA. Therefore, the S–BP–EDC/NHS and S–BP–TA specimens have better linear dependence between log (BSA concentration) and oxidation current data.

Keywords: buckypaper film; carbon nanotubes (CNTs); bovine serum albumin (BSA); modifiers; electrochemical; hydrophilic

1. Introduction

For biosensors, the efficiency of electronic interactions between the reacting target molecules and the substrate plays a key role in determining the accuracy and efficiency of measurement results. CNTs have the promising advantages of large surface area, high mobilities, and large electrical conductivities to electronic and electrochemical biosensors [1–3]. In addition, many biological molecules, such as proteins or DNA, can be conjugated with CNTs easily. This phenomenon also provides an advantageous solution for the application of CNTs in biosensors. In recent years, there has been a dramatic proliferation of research concerned with the application of CNTs in the biosensors of glucose [4–7], cholesterol [8–12], and DNA [13–18], that achieves excellent properties, including sensitivity, selectivity, and reproducibility.

However, due to the strong inter–tube van der Waals attraction, the aggregation of CNTs acts as an obstacle, in most applications, that deteriorates the mechanical and electrical properties of CNTs. Moreover, the size of CNTs is too small, which makes them difficult to control and disperse in processing, so their application is limited [19,20].

The buckypapers fabricated with CNTs have a macroscopic paper–like structure, which can be easily used and operated [21]. It is an important breakthrough to scale up CNTs to the macroscopic scale by using buckypapers to overcome the difficulties in manipulating CNTs [20–22]. In addition to their excellent CNT–like properties, buckypapers also cover a large area and are easy to control, so the current research uses for buckypapers are fairly extensive and are getting considerable attention. These include biosensors [23–25], electrical and thermal conductivity [26,27], gas filters [28], field emissions [29,30], etc. At present, the research on the application of buckypaper in biosensors has just begun, and there is not much relevant literature. It has been found that buckypaper made up of carboxyl–functionalized CNTs with incorporated gold nanoparticles could be highly sensitive glucose [24,25] and hydrogen peroxide [31] detectors. Buckypapers fabricated using functionalized CNTs are also used as working electrodes to detect dopamine [25].

According to early literature [32], the electrical conductivity of buckypapers made of carboxylated (purified) CNTs was 2–4 times higher than that made of as–grown CNTs. By means of carboxylation (functionalization), CNTs can not only be dispersed in solvents, but can attach to molecules physically or chemically without significantly changing their inherent unique properties [24].

This study provides a convenient, rapid, and accurate method to detect BSA, which is expected to be applied to the detection of other biomolecules in the future. In this study, buckypaper composite films were fabricated with carboxylated CNTs through pumping filtration and spin coating. Different modifiers were added by spin coating to modify the surfaces of buckypaper composite films. By combining the excellent conductive characteristics of buckypapers with its modified surface, it can attract and immobilize the BSA. This study will adopt the DPV electrochemical measurement to detect and quantitatively measure various BSA concentrations.

2. Materials and Methods

2.1. Fabrication of Buckypaper Film

First, 80 mg carboxylated multi–wall CNTs (Golden Innovation Business Co., AC tube-100LH, New Taipei City, Taiwan) with the addition of 1 wt.% EDC/NHS were added in 100 mL of 95% alcohol by the process of grinding, stirring, and sonication, respectively, to prepare the CNTs' suspension solution. Then the suspension solution was poured into the pumping filtration system (Figure 1a) containing Teflon filter paper (Rocker, 201PTFE–47–022–50, Kaohsiung, Taiwan), and the suction filtration process was carried out for about 20 min through the pressure difference formed by the vacuum pump. The obtained buckypaper film, with diameter about 4 cm and thickness of 0.37 ± 0.03 mm (Figure 1b), was dried at 100 °C for 60 min before being cut into square strips for various tests. In this study, the original buckypaper film prepared from CNTs was designated "S–BP" specimen.

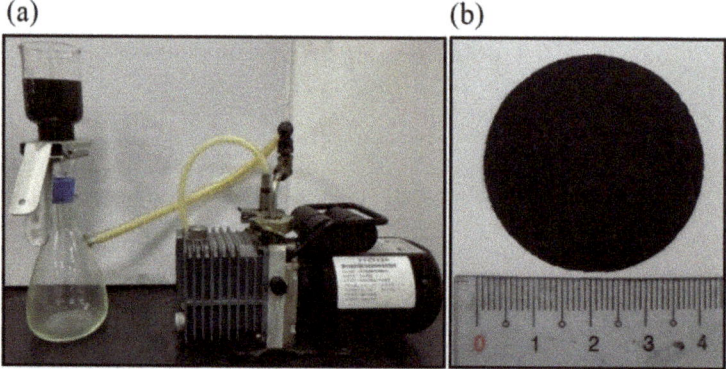

Figure 1. (**a**) Pumping filtration system; (**b**) Appearance of the fabricated buckypaper film.

2.2. Modification of Buckypaper Film

Four different modifiers of (1–(3 dimethylaminopropyl) ethylcarbodiimide (EDC)+ N–hydroxysuccinimide (NHS), Tannic acid (TA), Poly (Sodium 4–Styrenesulfonate (PSS), and Sodium Dodecyl Sulfate, (SDS)) were used to modify the surface of buckypaper film by spin coating, respectively. The preparation methods of different modifier solutions were mainly based on references [33–36]. During the spin–coating process, 100 µL of the modifier solution was dropped onto the surface of specimen, and rotated at about 1000 rpm for 30 s to make the modifier evenly coated on the surface of specimen. In the following study, the S–BP specimens were modified with four different modifiers. According to the different modifiers added, the modified buckypaper composite films were designated S–BP–EDC/NHS, S–BP–PSS, S–BP–SDS, and S–BP–TA specimens, respectively.

2.3. Addition of Bovine Serum Albumin (BSA)

Due to its cheap price and direct solubility in water, bovine serum albumin (BSA) is frequently used in research as a standard to quantify the concentration of other proteins or biomolecules. Before adding BSA to the buckypaper composite films, the aqueous solutions of 0.02 M NaH_2PO_4 (Sigma–Aldrich, VE–V900060, Beijing, China) and 0.1 M Na_2HPO_4 (Sigma–Aldrich, T–3828–01, Bengaluru, India) were mixed to prepare the 0.1 M phosphate buffered saline (PBS). Then, PBS solution and BSA powder (Sigma–Aldrich, A7906, St. Louis, MO, USA) were mixed to prepare BSA solutions with different concentrations (10, 100, 1000, 10,000 and 100,000 ng/mL). Next, in order to render the BSA to bond with the surfaces of specimens evenly, the various specimens and 10 mL of different BSA solutions were put into the serum bottle, respectively, then placed in a horizontal shaker (Dragon LAB, model SK–O180–E, Woodbury, CT, USA), and shaken at 100 rpm for 15 min. These specimens were then removed from the BSA solution and rinsed with PBS solution to wash away excess BSA that remained unbound on the specimens. Finally, these specimens were dried at 30 °C, and the subsequent measurement and microstructure observation were carried out.

2.4. Characterization

2.4.1. Contact Angle Measurement

Since the addition of BSA in this experiment is carried out in the form of an aqueous solution, the hydrophilic/hydrophobic properties of the specimen's surface may affect the adsorbed amount of BSA. Therefore, it is necessary to further measure the hydrophilic/hydrophobic properties of the specimen's surface by measuring the contact angle. The contact angles were measured using a contact–angle analyzer (Xinchuangda Technology Co., Ltd., Model 100, Taipei, Taiwan). A drop of deionized water (10 µL) was dropped on the surface of the specimen, and a camera was used to capture the contact angle of water droplet and specimen.

2.4.2. Electrochemical Measurement

In this experiment, an electrochemical workstation (CHI Instruments, 6114E, Austin, Texas, TX, USA) was used to carry out the measurement via Differential Pulse Voltammetry (DPV). The electrochemical system that has three electrodes, including the counter electrode of platinum, the reference electrode of Ag/AgCl (CHI Instruments, RE–1BP, Tokyo, Japan), and the working electrode of the specimen, was used in the experiment. During the measurement, a square specimen of 2×2 cm^2 with thickness of 0.37 ± 0.03 mm was used, and the electrolyte was the 5 mM $K_3Fe(CN)_6$ PBS solution.

2.4.3. Functional Group Detection

In this experiment, Fourier transform infrared spectroscopy (FTIR) (PerkinElmer, model Nicolet 460, Akron, OH, USA) was applied to the functional group detection (scanning range: 400–4000 cm^{-1}), which could detect the changes of functional groups on surfaces of specimens before and after modification. Because the specimens were black and opaque, the mode of Attenuated Total Reflection (ATR) was adopted for measurement.

2.4.4. Observation of Microstructure

Before observation, a gold plating machine (Hitachi, E–1010, Lbaraki, Japan) was used to coat the surface of specimen with a platinum film to increase the electrical conductivity of the specimen. The scanning electron microscope (Hitachi, FE-SEM4700, Lbaraki, Japan) with a cold–field emission emitter was used to observe microstructures of various specimens before and after the addition of different concentrations of BSA, respectively. Meanwhile, energy–dispersive X–ray spectroscopy (EDX, Horiba, Osaka, Japan) was also used to analyze the element composition of the modified specimen surface.

3. Results and Discussion

3.1. Contact Angle Measurement

The hydrophilic/hydrophobic properties of the specimen can be carefully researched by the contact angle measurement. The smaller the contact angle, the better the hydrophilicity of the specimen. In this study, if the specimen has fine hydrophilic properties, it will be beneficial to combine it with aqueous BSA. As shown in Figure 2, it can be found that the contact angle of the original, unmodified S–BP specimen (62°) is larger than that of all modified specimens. Among the modified specimens, the contact angle of the S–BP–EDC/NHS specimen (53°) is the largest, followed by the S–BP–PSS specimen (38°), the S–BP–TA specimen (22°), and lastly, the S–BP–SDS specimen (17°). It is important to note that all of the contact angles for these four modified specimens are smaller than that of the S–BP specimen (62°). Namely, it indicates that the four modifiers used in this study can improve the hydrophilic properties of the original S–BP specimen and benefit the subsequent bonding of the modified specimen with aqueous BSA.

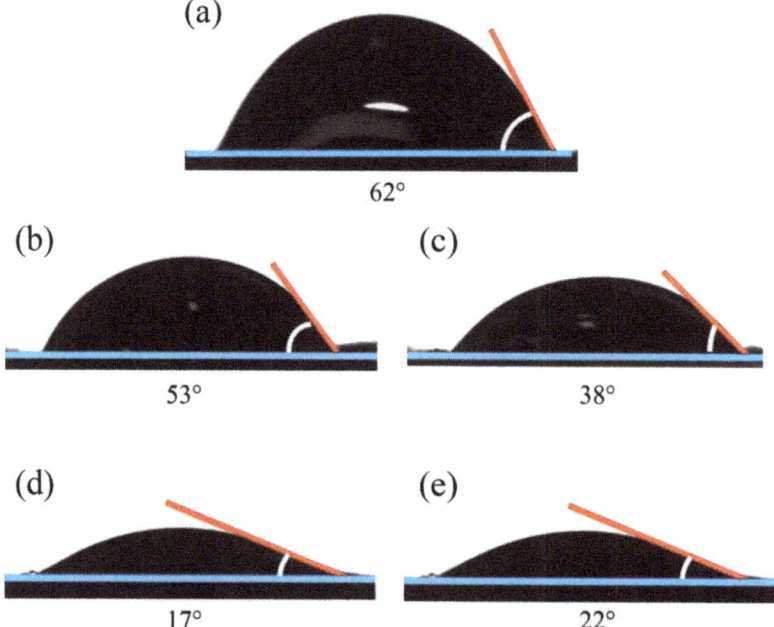

Figure 2. Contact angle measurement of various specimens. (**a**) S–BP; (**b**) S–BP–EDC/NHS; (**c**) S–BP–PSS; (**d**) S–BP–SDS; (**e**) S–BP–TA.

3.2. Differential Pulse Voltammetry (DPV) Measurement of Buckypaper Composite Films

This part mainly uses the DPV electrochemical method to measure the reaction feedback (oxidation current) of various specimens, adding different concentrations of BSA (0 ng/mL, 10 ng/mL, 100 ng/mL, 1000 ng/mL, 10,000 ng/mL, 100,000 ng/mL). The change rate of the oxidation current of the specimen before and after adding different concentrations of BSA was further calculated. The correlation between the logarithm BSA concentration (log (BSA) ng/mL) and the change of oxidation current was analyzed by linear regression analysis.

Figure 3a shows the DPV measurement results of the original, unmodified S–BP specimen after adding different concentrations of BSA. It can be found that there is no obvious difference in the oxidation current of the S–BP specimen when adding lower concentrations of BSA (10 ng/mL, 100 ng/mL). However, after adding a high concentration of BSA (1000 ng/mL, 10,000 ng/mL, 100,000 ng/mL), the oxidation current of the S–BP specimen begins to decrease gradually. When adding BSA with a low concentration, it has difficulty bonding stably to the surface of the S–BP specimen, leading to an inconspicuous change in the oxidation current. The main cause is conceivably due to the fact that the S–BP specimen only exhibits weak hydrophilic properties (contact angle: 62°). When the concentration of BSA is higher, the bonding chance of BSA to the specimen increases, so the oxidation current reflected by the electrochemical system also changes significantly.

From the diagram of linear regression analysis (Figure 4), it is evident that the oxidation current of all specimens roughly decreases with the increase of logarithm concentration of BSA (log (BSA) ng/mL). It can be found that there is no obvious difference in the oxidation current of various specimens when adding lower concentrations of BSA (10 ng/mL, 100 ng/mL). However, after adding a high concentration of BSA (1000 ng/mL, 10,000 ng/mL, 100,000 ng/mL), the oxidation currents of the various specimens apparently start to become different. It also can be found that the regression line slope of the unmodified S–BP specimen (−0.022) is "less tilted" than that of the modified specimens (S–BP–EDC/NHS specimen: −0.0318, S–BP–PSS: −0.0438, S–BP–SDS specimen: −0.034, and S–BP–TA specimen: −0.0388). This phenomenon implies that, with the increase of BSA concentration, the change in oxidation current of the unmodified S–BP specimen was less obvious than that of the modified specimens.

Figure 3b shows the DPV measurement results of the S–BP–EDC/NHS specimens after adding different concentrations of BSA. It can be found that the oxidation current of the S–BP–EDC/NHS specimen, compared with that of the S–BP specimen, trends prominently downwards with the log (BSA concentration). It is conceivable that an amine–reactive O–acylisourea intermediate develops [37] when the carboxyl groups on the surface of the S–BP specimens are modified by EDC/NHS. This O–acylisourea intermediate will form covalent amide bonds with the amine group on BSA. The covalent bond has strong bonding strength to conjugate with BSA more stably. Therefore, the measurement results also demonstrate that the S–BP–EDC/NHS specimens have a more obvious oxidation current feedback when adding different concentrations of BSA.

According to the linear regression analysis of the S–BP–EDC/NHS specimens (Figure 4b), it can be found that the added log (BSA concentration) presents a good inverse linear relationship with the feedback oxidation current, which represents that the S–BP–EDC/NHS specimens have a good linear dependence between log (BSA concentration) and oxidation current data.

The DPV measurements of the S–BP–PSS and S–BP–SDS specimens were quite similar (Figure 3c–d). After the concentration of BSA was higher than 100 ng/mL, the oxidation current of these specimens began to change significantly, and the change extent was also more obvious than that of the unmodified S–BP specimen. It is evident that these modified specimens, owning better hydrophilicity, improve the bonding effect between these specimens and aqueous BSA. Moreover, the oxidation currents measured by these specimens also differed with changes in BSA concentration. In the linear regression analysis of these specimens (Figure 4c–d), it is clear that the oxidation currents of these specimens also roughly decrease with the increase of the logarithm concentration of BSA.

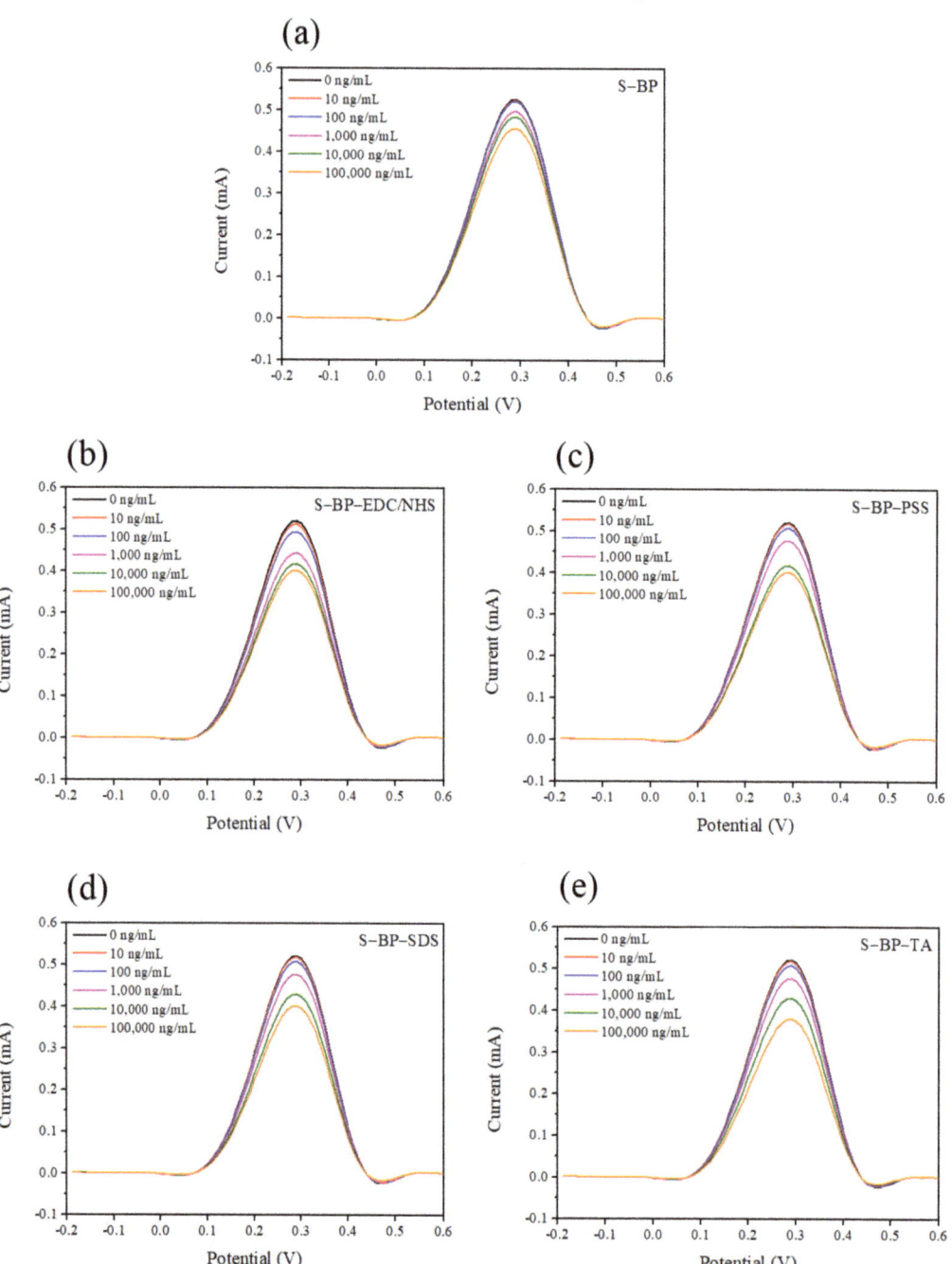

Figure 3. Differential Pulse Voltammetry (DPV) measurement of various specimens. (**a**) S–BP; (**b**) S–BP–EDC/NHS; (**c**) S–BP–PSS; (**d**) S–BP–SDS; (**e**) S–BP–TA.

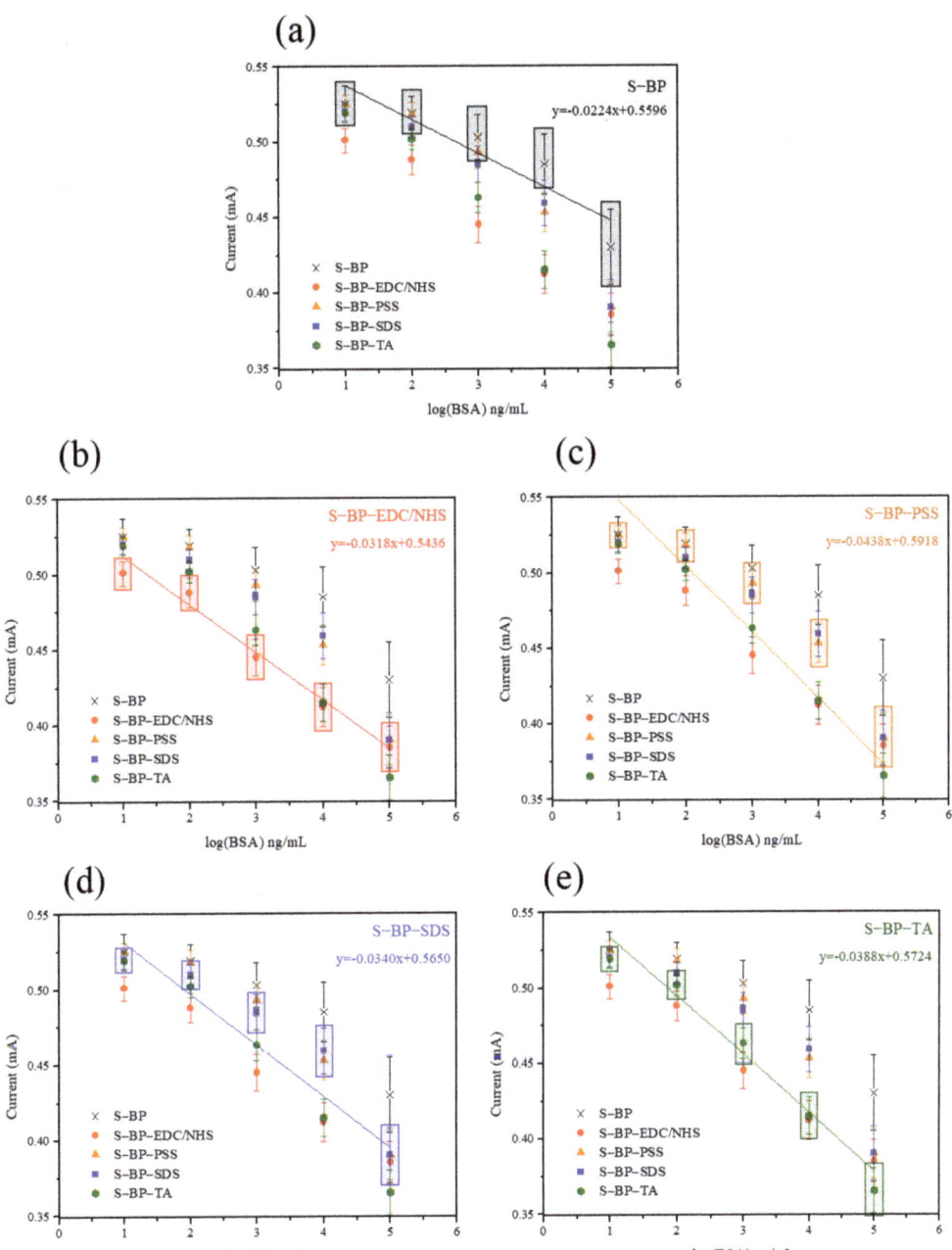

Figure 4. Linear regression analysis diagram of various specimens. (**a**) S–BP; (**b**) S–BP–EDC/NHS; (**c**) S–BP–PSS; (**d**) S–BP–SDS; (**e**) S–BP–TA.

When comparing these four modified specimens, it can be observed that the S–BP–PSS and S–BP–SD specimens have similar measurement data, while the S–BP–EDC/NHS and S–BP–TA specimens are still slightly different from the other specimens. The possible reason is that the EDC/NHS and TA modifier can not only improve the hydrophilic properties of the specimens, but also has more O–acylisourea intermediate and conjugated structure (benzene), which is also beneficial to further bonding with BSA with covalent amide bond or conjugated structure. Therefore, the measured data of the S–BP–EDC/NHS and S–BP–TA specimens in the linear regression analysis showed a better linear dependence.

By comprehensively analyzing the measurement results of each modified specimen, it can be found that the hydrophilic/hydrophobic properties of the modifier affect the bonding effect between the specimen and BSA. However, if the modifier can produce an amide bond or conjugate structure on the specimen, it will have a more significant and positive influence on the bonding effect between the specimen and BSA.

3.3. Functional Group Detection

In order to explore whether the various modifiers were successfully attached on the surface of S–BP specimens to achieve the effect of modification, the functional groups of various S–BP specimens before and after modification were analyzed by FTIR.

Figure 5 shows the FTIR analyses of various specimens. The C=C peak (1600~1650 cm^{-1}) and hydroxyl functional groups (–OH) peak (2900~3100 cm^{-1}) denote the CNTs and carboxylated CNTs respectively, which can be found in the unmodified S–BP specimens.

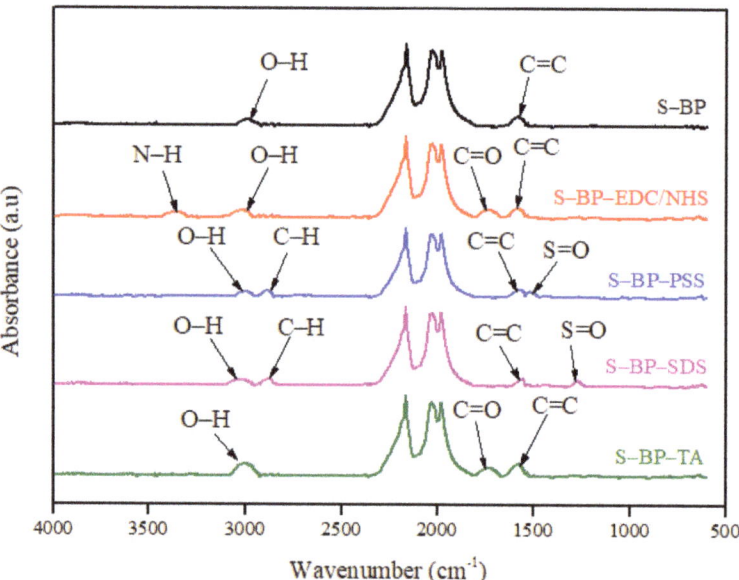

Figure 5. Fourier transform infrared spectroscopy (FTIR) analyses of various specimens.

The S–BP–EDC/NHS specimen has the C=O peak (1650~1700 cm^{-1}) and the N–H peak (3200~3400 cm^{-1}), which is consistent with the chemical formula of EDC/NHS ($C_8H_{17}N_3/C_4H_5NO_3$). It indicates that EDC/NHS is successfully attached to the surface of the specimen.

In the S–BP–PSS specimen, there are the S=O peaks (1380~1435 cm^{-1}) of sulfonates and the C–H peaks (2840~3000 cm^{-1}) of alkenes. Both peaks are consistent with the chemical formula of PSS ($C_8H_7NaO_3S)_n$. It is believed that the PSS is also successfully attached to the specimen.

The S–BP–SDS specimen has the S=O peak (1250~1315 cm^{-1}) and the C–H peak (2900~3000 cm^{-1}) that denote the elements of sulfate and alkane respectively. Both peaks are consistent with the SDS chemical formula of $NaC_{12}H_{25}SO_4$, which indicates that SDS can be successfully attached to the specimen.

In the S–BP–TA specimen, in addition to the C=O peak (1650~1700 cm^{-1}) representing the carboxyl group (–COOH), the intensity of the hydroxyl (–OH) peak (2900~3000 cm^{-1}) is also higher than that of the unmodified S–BP specimen. This means that the S–BP–TA specimen not only contains the hydroxyl group of the S–BP specimen itself, but also adsorbs the hydroxyl group in the modifier TA, and these two peaks are also consistent with the chemical formula of TA ($C_{76}H_{52}O_{46}$). It is speculated that TA is also successfully attached to the specimen.

3.4. Observation of Microstructure

Figure 6 shows the microstructure of various specimens before adding BSA. It can be found that in the original unmodified S–BP specimen (Figure 6a) prepared from carboxylated CNTs, the CNTs have some cutting surfaces due to the carboxylation treatment (as indicated by the arrow). The CNTs of the specimen, without agglomeration, intertwined and stacked with each other, as well as distributed evenly. This implies that using pumping filtration to make buckypaper films has a good effect. By observing the microstructure of various specimens modified by different modifier (Figure 6b–e), it is found that all modified specimens showed similar surface morphologies, and the differences cannot be distinguished clearly by SEM.

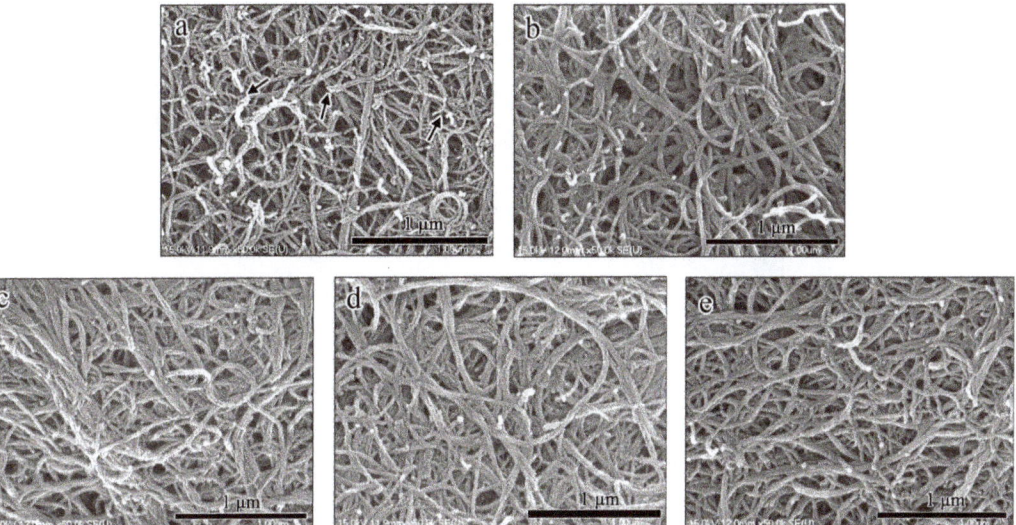

Figure 6. Microstructure of various specimens before adding BSA. (**a**) S–BP; (**b**) S–BP–EDC/NHS; (**c**) S–BP–PSS; (**d**) S–BP–SDS; (**e**) S–BP–TA. "Arrows" indicate some cutting surfaces of CNTs resulted from the carboxylation treatment.

In order to investigate whether the modifier is successfully attached to the specimen, this study not only uses FTIR to analyze the functional groups of each modified specimen, but also employs EDX to analyze the element compositions of the modified specimen surfaces. As shown in Figure 7, it can be found that in the unmodified S–BP specimen (Figure 7a), the EDX signal is mostly dominated by the element of carbon (C), and contains a weak signal of the element of oxygen (O). This weak signal is caused by the oxygen-containing functional groups on carboxylated CNTs.

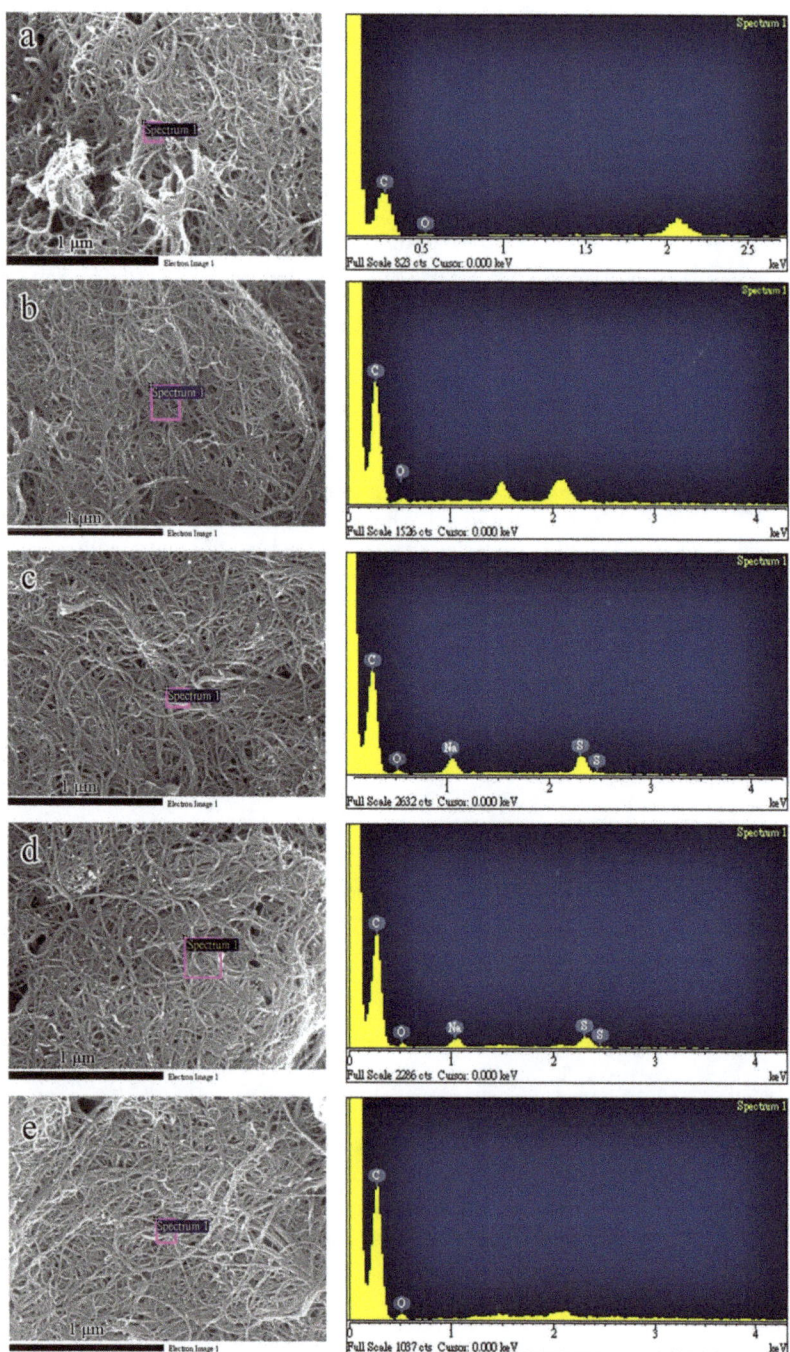

Figure 7. EDX analyses of various specimens. (**a**) S–BP; (**b**) S–BP–EDC/NHS; (**c**) S–BP–PSS; (**d**) S–BP–SDS; (**e**) S–BP–TA.

In the S–BP–EDC/NHS specimen (Figure 7b), it can be found that the signal of the element of oxygen (O) is slightly stronger than that of the unmodified S–BP specimen, but the element of nitrogen (N) in the chemical formula of EDC/NHS ($C_8H_{17}N_3/C_4H_5NO_3$) cannot be detected. It is believed to be due to the fact that the content of the nitrogen element (N) is too scant to be detected. Nevertheless, from the EDX signal of the oxygen element (O) and the analytic results of the functional groups in Figure 5, it can be inferred that EDC/NHS is successfully attached to the surface of the specimen. In the S–BP–PSS (Figure 7c) and S–BP–SDS specimens (Figure 7d), the detectable EDX signals of the sulfur element (S) and the sodium element (Na), corresponding to the chemical formula ($C_8H_7NaO_3S)_n$ of PSS and the chemical formula of SDS ($NaC_{12}H_{25}SO_4$), indicate that both PSS and SDS are attached to the specimen surface. In the EDX element analysis of the S–BP–TA specimen (Figure 7e), it can be found that the signal of the oxygen element (O) is stronger than that of the unmodified S–BP specimen, which also accords with the chemical formula of TA ($C_{76}H_{52}O_{46}$). Therefore, it is reasonable to infer that TA is attached to the surface of the specimen. Based on the analytic results of FTIR and EDX, it can be proven that the various modifiers can successfully attach to the surfaces of the specimens by spin coating.

Figure 8 shows the microstructure of various specimens after adding two different concentrations (100 ng/mL and 100,000 ng/mL) of BSA. It is clear that there is no obvious BSA particle adhesion on the surface of the S–BP specimen with 100 ng/mL added BSA (Figure 8a). The main cause is due to the low concentration of added BSA and the weak bonding strength between BSA and the surface of the S–BP specimen. When the added BSA concentration was 100,000 ng/mL, more BSA particles could be obviously observed on the surface of the S–BP specimen (as indicated by the arrow) (Figure 8b), so the corresponding electrochemical measurement results also showed more obvious changes.

However, the microstructures of the modified specimens after adding BSA are different from those of the S–BP specimens (Figure 8c–j). It can be found that even after adding a low concentration (100 ng/mL) of BSA to the S–BP–EDC/NHS specimen, some BSA particles can still be observed on the surface of the S–BP–EDC/NHS specimen (Figure 8c). This shows that the bonding effect between the S–BP–EDC/NHS specimen and BSA is better than that of the S–BP specimen. When the added concentration of BSA is 100,000 ng/mL (Figure 8d), it reveals that the surface of the S–BP–EDC/NHS specimen is covered with a large range of BSA particles, which also explains the reason for the most obvious change in electrochemical properties of the S–BP–EDC/NHS specimen after adding BSA.

In the microstructure of the S–BP–PSS, S–BP–SDS and S–BP–TA specimens with 100 ng/mL added BSA, it can be observed that BSA with a membrane–like structure is bonded to the surface of the specimens (Figure 8e,g,i). According to the contact angle measurements (Figure 2c–e), the modifiers of PSS, SDS, and TA can reduce the surface tension, improve the hydrophilic property of these three specimens, and avail the combination of the specimen and aqueous BSA to achieve a membrane–like structure [38]. When 100,000 ng/mL BSA is added (Figure 8f,h,j), a large area of membrane–like BSA can also be observed on the surface of these three specimens. These phenomena also provide the main reason why the electrochemical measurement performance of these three modified specimens is different from that of the unmodified S–BP specimen.

From the microstructure observation of all modified specimens with 100,000 ng/mL added BSA, it can be found that the BSA attached on the surfaces of the S–BP–EDC/NHS and S–BP–TA specimens seems to show a larger area and denser morphologies. Furthermore, the corresponding data of these two specimens in the linear regression analysis also exhibit a better linear dependence. For the modified specimens, this phenomenon once again shows that the hydrophilic property affects the bonding effect with BSA, and the bonding type (such as amide bond or conjugate structure, etc.) produced by the modifier on the specimens also performs an important role in the bonding effect between the specimen and BSA.

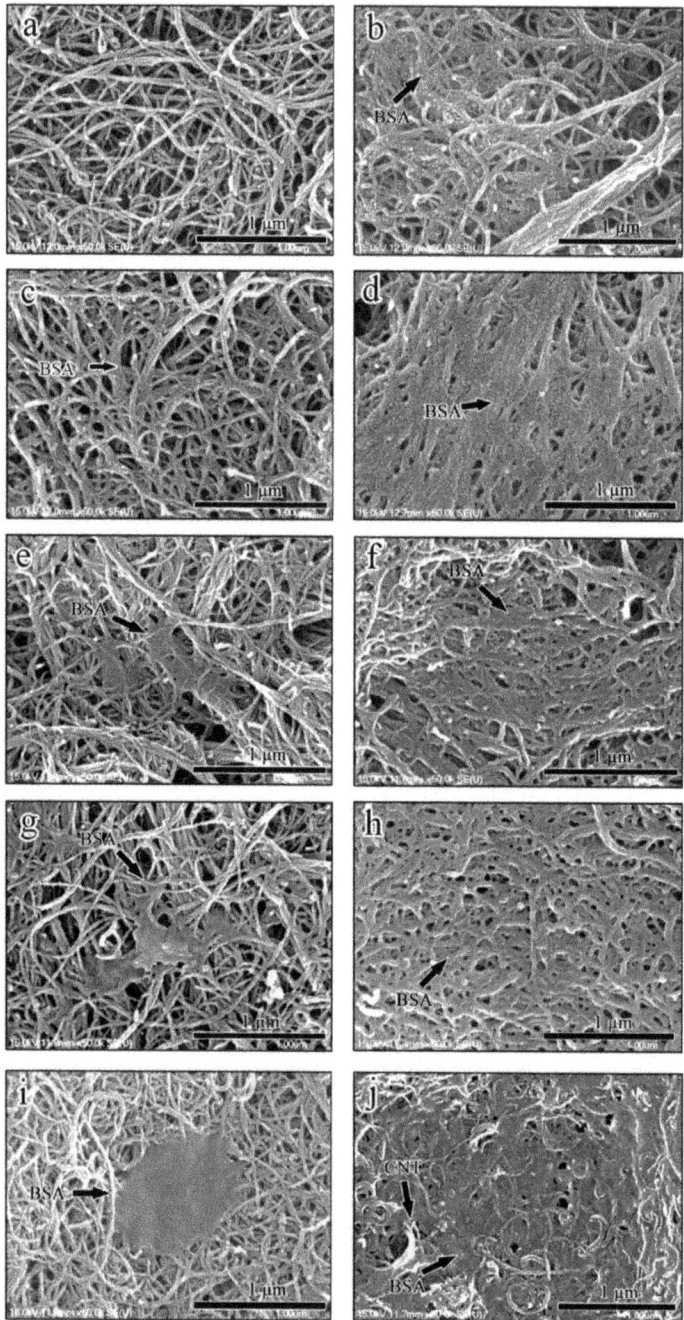

Figure 8. Microstructure of various specimens after adding two different concentrations (100 ng/mL and 100,000 ng/mL) of BSA. (**a**) S–BP, 100 ng/mL; (**b**) S–BP, 100,000 ng/mL; (**c**) S–BP–EDC/NHS, 100 ng/mL; (**d**) S–BP–EDC/NHS, 100,000 ng/mL; (**e**) S–BP–PSS, 100 ng/mL; (**f**) S–BP–PSS, 100,000 ng/mL; (**g**) S–BP–SDS, 100 ng/mL; (**h**) S–BP–SDS, 100,000 ng/mL; (**i**) S–BP–TA, 100 ng/mL; (**j**) S–BP–TA, 100,000 ng/mL. "Arrows" indicate the positions of part of the BSA covering on the specimen surface.

4. Conclusions

In this study, carboxylated CNTs were used to make buckypaper composite films and further modify them by incorporating different surface modifiers. After adding different concentrations of BSA, DPV electrochemical measurement was applied to measure the characteristics of the specimens. It is observed that the hydrophilic properties of all modified specimens are better than that of the original unmodified S–BP specimen. Succinctly, the changes brought about by using modifiers can improve the bonding effects between modified specimens and aqueous BSA. Therefore, the variation of electrochemical properties for the modified specimens after adding BSA is more apparent than that of the original S–BP specimen. In the linear regression analyses of all specimens, the oxidation current of a specimen roughly decreases with the increase of logarithm concentration of BSA (log (BSA) ng/mL). The regression line slope of the unmodified S–BP specimen is less tilted than that of the modified specimens. This phenomenon implies that, with the increase of BSA concentration, the change in oxidation current of the unmodified S–BP specimen was less obvious than that of the modified specimens. It is noteworthy that the bonding type produced by the modifier has a more prominent influence than the hydrophilic property on the bonding effect between the specimen and BSA.

Author Contributions: Conceptualization, K.-J.L. and M.-H.L.; Data curation, M.-H.L.; Funding acquisition, K.-J.L., C.-P.W. and S.-R.J.; Resources, C.-P.W.; Writing—original draft, K.-J.L.; Writing—review & editing, Y.-H.S. and H.-Y.L. All authors have read and agreed to the published version of the manuscript.

Funding: This work was funded by the Ministry of Science and Technology of Taiwan under the contract numbers MOST 110–2221–E–214–003 and MOST 109–2221–E–214–020 and I-SHOU University & E-DA Hospital of Taiwan under the contract numbers ISU 106–IUC–04.

Institutional Review Board Statement: Not applicable.

Informed Consent Statement: Not applicable.

Data Availability Statement: Data is contained within the article.

Acknowledgments: The authors gratefully acknowledge the use of SEM belonging to the Precious Instrument Center of I-SHOU University.

Conflicts of Interest: The authors declare no conflict of interest.

References

1. Zhou, S.; Shi, H.; Feng, X.; Xue, K.; Song, W. Design of templated nanoporous carbon electrode materials with substantial high specific surface area for simultaneous determination ofbiomolecules. *Biosens. Bioelectron.* **2013**, *42*, 163–169. [CrossRef] [PubMed]
2. Kara, P.; de la Escosura–Muñiz, A.; Maltez–da Costa, M.; Guix, M.; Ozsoz, M.; Merkoçi, A. Aptamers based electrochemical biosensor for protein detection using carbon nanotubes platforms. *Biosens. Bioelectron.* **2010**, *26*, 1715–1718. [CrossRef] [PubMed]
3. Valentini, F.; Orlanducci, S.; Terranova, M.L.; Amine, A.; Palleschi, G. Carbon nanotubes as electrode materials for the assembling of new electrochemical biosensors. *Sens. Actuators B Chem.* **2004**, *100*, 117–125. [CrossRef]
4. Li, F.; Wang, Z.; Shan, C.; Song, J.; Han, D.; Niu, L. Preparation of gold nanoparticles/functionalized multiwalled carbon nanotube nanocomposites and its glucose biosensing application. *Biosens. Bioelectron.* **2009**, *24*, 1765–1770. [CrossRef]
5. Guan, W.J.; Li, Y.; Chen, Y.Q.; Zhang, X.B.; Hu, G.Q. Glucose biosensor based on multi–wall carbon nanotubes and screen printed carbon electrodes. *Biosens. Bioelectron.* **2005**, *21*, 508–512. [CrossRef]
6. Huang, J.; Yang, Y.; Shi, H.; Song, Z.; Zhao, Z.; Anzai, J.I.; Osa, T.; Chen, Q. Multi–walled carbon nanotubes–based glucose biosensor prepared by a layer–by–layer technique. *Mater. Sci. Eng. C* **2006**, *26*, 113–117. [CrossRef]
7. Fatoni, A.; Numnuam, A.; Kanatharana, P.; Limbut, W.; Thammakhet, C.; Thavarungkul, P. A highly stable oxygen-independent glucose biosensor based on a chitosan–albumin cryogel incorporated with carbon nanotubes and ferrocene. *Sens. Actuators B Chem.* **2013**, *185*, 725–734. [CrossRef]
8. Li, G.; Liao, J.M.; Hu, G.Q.; Ma, N.Z.; Wu, P.J. Study of carbon nanotube modified biosensor for monitoring total cholesterol in blood. *Biosens. Bioelectron.* **2005**, *20*, 2140–2144. [CrossRef]
9. Tsai, Y.C.; Chen, S.Y.; Lee, C.A. Amperometric cholesterol biosensors based on carbon nanotube–chitosan–platinum–cholesterol oxidase nanobiocomposite. *Sens. Actuators B Chem.* **2008**, *135*, 96–101. [CrossRef]
10. Shi, Q.; Peng, T.; Cheng, J. A cholesterol biosensor based on cholesterol oxidase immobilized in a sol–gel on a platinum–decorated carbon nanotubes modified electrode. *Chin. J. Anal. Chem.* **2005**, *33*, 329–332.

11. Wisitsoraat, A.; Sritongkham, P.; Karuwan, C.; Phokharatkul, D.; Maturos, T.; Tuantranont, A. Fast cholesterol detection using flow injection microfluidic device with functionalized carbon nanotubes based electrochemical sensor. *Biosens. Bioelectron.* **2010**, *26*, 1514–1520. [CrossRef]
12. Tong, Y.; Li, H.; Guan, H.; Zhao, J.; Majeed, S.; Anjum, S.; Liang, F.; Xu, G. Electrochemical cholesterol sensor based on carbon nanotube@molecularly imprinted polymer modified ceramic carbon electrode. *Biosens. Bioelectron.* **2013**, *47*, 553–558. [CrossRef]
13. Dong, X.Y.; Mi, X.N.; Zhang, L.; Liang, T.M.; Xu, J.J.; Chen, H.Y. DNAzyme–functionalized Pt nanoparticles/carbon nanotubes for amplified sandwich electrochemical DNA analysis. *Biosens. Bioelectron.* **2012**, *38*, 337–341. [CrossRef]
14. Yang, K.; Zhang, C.Y. Simple detection of nucleic acids with a single–walled carbon–nanotube–based electrochemical biosensor. *Biosens. Bioelectron.* **2011**, *28*, 257–262. [CrossRef]
15. Brahman, P.K.; Dar, R.A.; Pitre, K.S. DNA–functionalized electrochemical biosensor for detection of vitamin B1 using electrochemically treated multiwalled carbon nanotube paste electrode by voltammetric methods. *Sens. Actuators B Chem.* **2013**, *177*, 807–812. [CrossRef]
16. Li, F.; Peng, J.; Wang, J.; Tang, H.; Tan, L.; Xie, Q.; Yao, S. Carbon nanotube–based label–free electrochemical biosensor for sensitive detection of miRNA–24. *Biosens. Bioelectron.* **2014**, *54*, 158–164. [CrossRef]
17. Wu, L.; Xiong, E.; Zhang, X.; Zhang, X.; Chen, J. Nanomaterials as signal amplification elements in DNA–based electrochemical sensing. *Nano Today* **2014**, *9*, 197–211. [CrossRef]
18. Kim, J.; Elsnab, J.; Gehrke, C.; Li, J.; Gale, B.K. Microfluidic integrated multi–walled carbon nanotube (MWCNT) sensor for electrochemical nucleic acid concentration measurement. *Sens. Actuators B Chem.* **2013**, *185*, 370–376. [CrossRef]
19. Rinzler, A.G.; Liu, J.; Dai, H.; Nikolaev, P.; Huffman, C.B.; Rodríguez–Macías, F.J.; Boul, P.J.; Lu, A.H.; Heymann, D.; Colbert, D.T.; et al. Large–scale purification of single–wall carbon nanotubes: Process, product, and characterization. *Appl. Phys. A* **1998**, *67*, 29–37. [CrossRef]
20. Li, Y.; Kröger, M. A theoretical evaluation of the effects of carbon nanotube entanglement and bundling on the structural and mechanical properties of buckypaper. *Carbon* **2012**, *50*, 1793–1806. [CrossRef]
21. Zhang, J.; Jiang, D.; Peng, H.X. A pressurized filtration technique for fabricating carbon nanotube buckypaper: Structure, mechanical and conductive properties. *Microporous Mesoporous Mater.* **2014**, *184*, 127–133. [CrossRef]
22. Shankar, K.R. *Preparation and Characterization of Magnetically Aligned Carbon Nanotube Buckypaper and Composite*; Florida State University: Tallahassee, FL, USA, 2003.
23. Desmet, C.; Marquette, C.A.; Blum, L.J.; Doumèche, B. Paper electrodes for bioelectrochemistry: Biosensors and biofuel cells. *Biosens. Bioelectron.* **2016**, *76*, 145–163. [CrossRef] [PubMed]
24. Papa, H.; Gaillard, M.; Gonzalez, L.; Chatterjee, J. Fabrication of Functionalized Carbon Nanotube Buckypaper Electrodes for Application in Glucose Biosensors. *Biosensors* **2014**, *4*, 449–460. [CrossRef] [PubMed]
25. Chupp, J.; Papa, H.; Gonzales, L.; Turgeon, O.; Chatterjee, J. Studies on Electrochemical Properties of Functionalized Carbon Nanotube Bucky Paper Electrodes for Biosensor Applications. *Res. Rev. J. Mater. Sci.* **2015**, *3*, 17–24. [CrossRef]
26. Kulesza, S.; Szroeder, P.; Patyk, J.K.; Szatkowski, J.; Kozanecki, M. High–temperature electrical transport properties of buckypapers composed of doped single–walled carbon nanotubes. *Carbon* **2006**, *44*, 2178–2183. [CrossRef]
27. Su, F.; Miao, M. Transition of electrical conductivity in carbon nanotube/silver particle composite buckypapers. *Particuology* **2014**, *17*, 15–21. [CrossRef]
28. Smajda, R.; Kukovecz, Á.; Kónya, Z.; Kiricsi, I. Structure and gas permeability of multi–wall carbon nanotube buckypapers. *Carbon* **2007**, *45*, 1176–1184. [CrossRef]
29. Chen, Y.; Miao, H.Y.; Zhang, M.; Liang, R.; Zhang, C.; Wang, B. Analysis of a laser post–process on a buckypaper field emitter for high and uniform electron emission. *Nanotechnology* **2009**, *20*, 325302. [CrossRef]
30. Knapp, W.; Schleussner, D. Carbon Buckypaper field emission investigations. *Vacuum* **2002**, *69*, 333–338. [CrossRef]
31. Chatterjee, S.; Chen, A. Functionalization of carbon buckypaper for the sensitive determination of hydrogen peroxide in human urine. *Biosens. Bioelectron.* **2012**, *35*, 302–307. [CrossRef]
32. Ansón–Casaos, A.; González–Domínguez, J.M.; Terrado, E.; Martínez, M.T. Surfactant–free assembling of functionalized single–walled carbon nanotube buckypapers. *Carbon* **2010**, *48*, 1480–1488. [CrossRef]
33. Lin, D.; Xing, B. Tannic Acid Adsorption and Its Role for Stabilizing Carbon Nanotube Suspensions. *Environ. Sci. Technol.* **2008**, *42*, 5917–5923. [CrossRef]
34. Pacios, M.; Yilmaz, N.; Martín–Fernández, I.; Villa, R.; Godignon, P.; Del Valle, M.; Bartrolí, J.; Esplandiu, M.J. A simple approach for DNA detection on carbon nanotube microelectrode arrays. *Sens. Actuators B Chem.* **2012**, *162*, 120–127. [CrossRef]
35. Wang, L.; Chen, W.; Xu, D.; Shim, B.S.; Zhu, Y.; Sun, F.; Liu, L.; Peng, C.; Jin, Z.; Xu, C.; et al. Simple, Rapid, Sensitive, and Versatile SWNT—Paper Sensor for Environmental Toxin Detection Competitive with ELISA. *Nano Lett.* **2009**, *9*, 4147–4152. [CrossRef]
36. Liu, G.; Craig, V.S.J. Improved Cleaning of Hydrophilic Protein–Coated Surfaces using the Combination of Nanobubbles and SDS. *ACS Appl. Mater. Interfaces* **2009**, *1*, 481–487. [CrossRef]
37. Gao, Y.; Kyratzis, I. Covalent Immobilization of Proteins on Carbon Nanotubes Using the Cross–Linker 1–Ethyl–3–(3–dimethylaminopropyl)carbodiimide—A Critical Assessment. *Bioconjugate Chem.* **2008**, *19*, 1945–1950. [CrossRef]
38. Lee, C.T.; Smith, K.A.; Hatton, T.A. Photocontrol of Protein Folding: The Interaction of Photosensitive Surfactants with Bovine Serum Albumin. *Biochemistry* **2005**, *44*, 524–536. [CrossRef]

MDPI
St. Alban-Anlage 66
4052 Basel
Switzerland
www.mdpi.com

Coatings Editorial Office
E-mail: coatings@mdpi.com
www.mdpi.com/journal/coatings

Disclaimer/Publisher's Note: The statements, opinions and data contained in all publications are solely those of the individual author(s) and contributor(s) and not of MDPI and/or the editor(s). MDPI and/or the editor(s) disclaim responsibility for any injury to people or property resulting from any ideas, methods, instructions or products referred to in the content.

www.ingramcontent.com/pod-product-compliance
Lightning Source LLC
LaVergne TN
LVHW070441100526
838202LV00014B/1643